CALF AND HEIFER REARING

Proceedings of Previous Easter Schools in Agricultural Science, published by Butterworths, London

SOIL ZOOLOGY Edited by D.K. McL, Kevan (1955)
THE GROWTH OF LEAVES Edited by F.L. Milthorpe (1956)
CONTROL OF PLANT ENVIRONMENT Edited by J.P. Hudson (1957)
NUTRITION OF THE LEGUMES Edited by E.G. Hallsworth (1958)
THE MEASUREMENT OF GRASSLAND AND PRODUCTIVITY Edited by J.D. Ivins (1959)
DIGESTIVE PHYSIOLOGY AND NUTRITION OF THE RUMINANT Edited by D. Lewis (1960)
NUTRITION OF PIGS AND POULTRY Edited by J.T. Morgan and D. Lewis (1961)
ANTIBIOTICS IN AGRICULTURE Edited by M. Woodbine (1962)
THE GROWTH OF THE POTATO Edited by J.D. Ivins and F.L. Milthorpe (1963)
EXPERIMENTAL PEDOLOGY Edited by E.G. Hallsworth and D.V. Crawford (1964)
THE GROWTH OF CEREALS AND GRASSES Edited by F.L. Milthorpe and J.D. Ivins (1965)
REPRODUCTION IN THE FEMALE ANIMAL Edited by G.E. Lamming and E.C. Amoroso (1967)
GROWTH AND DEVELOPMENT OF MAMMALS Edited by G.A. Lodge and G.E. Lamming (1968)
ROOT GROWTH Edited by W.J. Whittington (1968)
PROTEINS AS HUMAN FOOD Edited by R.A. Lawrie (1970)
LACTATION Edited by I.R. Falconer (1971)
PIG PRODUCTION Edited by D.J.A. Cole (1972)
SEED ECOLOGY Edited by W. Heydecker (1973)
HEAT LOSS FROM ANIMALS AND MAN: ASSESSMENT AND CONTROL Edited by J.L. Montieth and L.E. Mount (1974)
MEAT Edited by D.J.A. Cole and R.A. Lawrie (1975)
PRINCIPLES OF CATTLE PRODUCTION Edited by Henry Swan and W.H. Broster (1976)
LIGHT AND PLANT DEVELOPMENT Edited by H. Smith (1976)
PLANT PROTEINS Edited by G. Norton (1977)
ANTIBIOTICS AND ANTIBIOSIS AGRICULTURE Edited by M. Woodbine (1977)
CONTROL OF OVULATION Edited by D.B. Crighton, N.B. Haynes, G.R. Foxcroft and G.E. Lamming (1978)
POLYSACCHARIDES IN FOOD Edited by J.M.V. Blanshard and J.R. Mitchell (1979)
SEED PRODUCTION Edited by P.D. Hebblethwaite (1980)
PROTEIN DEPOSITION IN ANIMALS Edited by P.J. Buttery and D.B. Lindsay (1981)
PHYSIOLOGICAL PROCESSES LIMITING PLANT PRODUCTIVITY Edited by C. Johnson (1981)
ENVIRONMENTAL ASPECTS OF HOUSING FOR ANIMAL PRODUCTION Edited by J.A. Clark (1981)
EFFECTS OF GASEOUS AIR POLLUTION IN AGRICULTURE AND HORTICULTURE Edited by M.H. Unsworth and D.P. Ormrod (1982)
CHEMICAL MANIPULATION OF CROP GROWTH AND DEVELOPMENT Edited by J.S. McLaren (1982)
CONTROL OF PIG REPRODUCTION Edited by D.J.A. Cole and G.R. Foxcroft (1982)
SHEEP PRODUCTION Edited by W. Haresign (1983)
UPGRADING WASTE FOR FEEDS AND FOOD Edited by D.A. Ledward, A.J. Taylor and R.A. Lawrie (1983)
FATS IN ANIMAL NUTRITION Edited by J. Wiseman (1984)
IMMUNOLOGICAL ASPECTS OF REPRODUCTION IN MAMMALS Edited by D.B. Crighton (1984)
ETHYLENE AND PLANT DEVELOPMENT Edited by J.A. Roberts and G.A. Tucker (1985)
THE PEA CROP Edited by P.D. Hebblethwaite, M.C. Heath and T.C.K. Dawkins (1985)
PLANT TISSUE CULTURE AND ITS AGRICULTURAL APPLICATIONS Edited by Lindsay A. Withers and P.G. Alderson (1986)
CONTROL AND MANIPULATION OF ANIMAL GROWTH Edited by P.J. Buttery, N.B. Haynes and D.B. Lindsay (1986)
COMPUTER APPLICATIONS IN AGRICULTURAL ENVIRONMENTS Edited by J.A. Clark, K. Gregson and R.A. Saffell (1986)
MANIPULATION OF FLOWERING Edited by J.G. Atherton (1987)
NUTRITION AND LACTATION IN THE DAIRY COW Edited by P.C. Garnsworthy (1988)
MANIPULATION OF FRUITING Edited by C.J. Wright (1989)
APPLICATIONS OF REMOTE SENSING IN AGRICULTURE Edited by M.D. Steven and J.A. Clark (1990)
GENETIC ENGINEERING OF CROP PLANTS Edited by G.W. Lycett and D. Grierson (1990)
FEEDSTUFF EVALUATION Edited by J. Wiseman and D.J.A. Cole (1990)
THE CONTROL OF FAT AND LEAN DEPOSITION Edited by K.N. Boorman, P.J. Buttery and D.B. Lindsay (1992)

Proceedings of Previous Easter Schools in Agricultural Science, published by Nottingham University Press

THE GLASSY STATE IN FOODS Edited by J.M.V. Blanshard and P.J. Lillford (1993)
RESOURCE CAPTURE BY CROPS Edited by J.L. Montieth, R.K. Scott and M.H. Unsworth (1994)
PRINCIPLES OF PIG SCIENCE Edited by D.J.A. Cole, J. Wiseman and M.A. Varley (1994)
ISSUES IN AGRICULTURAL BIOETHICS Edited by T.B. Mepham, G.A. Tucker and J. Wiseman (1995)
BIOPOLYMER MIXTURES Edited by S.E. Harding, S.E. Hill and J.R. Mitchell (1995)
MECHANISMS AND APPLICATIONS OF GENE SILENCING Edited by D. Grierson, G.W. Lycett and G.A. Tucker (1996)
PROGRESS IN PIG SCIENCE Edited by J. Wiseman, M.A. Varley and J.P. Chadwick (1998)
PERSPECTIVES IN PIG SCIENCE Edited by J. Wiseman, M.A. Varley and B. Kemp (2003)

Calf and Heifer Rearing

P.C. Garnsworthy
University of Nottingham

Nottingham University Press
Manor Farm, Church Lane
Thrumpton, Nottingham, NG11 0AX, UK

www.nup.com

NOTTINGHAM

First published 2005

British Library Cataloguing in Publication Data
A catalogue record for this book is available from the British Library

ISBN 1-904761-22-4

Disclaimer

Every reasonable effort has been made to ensure that the material in this book is true, correct, complete and appropriate at the time of writing. Nevertheless the publishers, the editors and the authors do not accept responsibility for any omission or error, or for any injury, damage, loss or financial consequences arising from the use of the book.

Typeset by Nottingham University Press, Nottingham
Printed and bound by Biddles Ltd, King's Lynn, England

PREFACE

This book contains the proceedings of the 60th University of Nottingham Easter School in Agricultural Science. As with previous Easter Schools, experts were gathered from all over the world to give in-depth consideration to a subject of current relevance to international agriculture. Calf and heifer rearing was chosen as the subject because this area often has low priority on modern dairy units, where emphasis is usually placed on efficiency of milk production from the adult cows. However, young heifers will be the milk producers of the future, so it is important that they should be given due consideration.

The modern dairy cow has resulted from intensive selection for milk yield over the past 20 to 30 years. Increased milk yield has been accompanied by increased body weight, but there has been little change in calf birth weight, and the target age at first calving remains 24 months. This means that heifers have to grow faster during the rearing period, but growth rate is known to affect mammary development and subsequent milk yield, so an optimum rearing strategy needs to be established. In addition, nutrient requirements were determined over thirty years ago, so we need to establish if they are still applicable to the modern dairy heifer.

The main objective of this Easter School was to examine strategies for rearing the modern dairy replacement heifer from birth to calving. The approach taken was to start from first principles, to use these principles for scientific study of rearing systems, and to use findings from these studies to develop practical recommendations. Although the main focus is on rearing dairy heifers, the principles of calf rearing also apply to rearing animals for beef production.

Using this approach, the first chapters of this book are concerned with principles of growth, development of the gut, and current perspectives on nutrient requirements of calves before weaning. The next chapters examine how these nutrient requirements can be met using liquid and solid feeds, the importance of colostrum for acquisition of passive immunity, and changes in gut secretions that occur during weaning. The next chapters consider diseases and welfare of calves and heifers, and how early growth rate affects general health, mammary development and milk yield of heifers. The final chapters cover grazing systems and fertility of heifers, and general guidelines derived from all previous chapters.

This book should appeal to all those interested in calf and heifer rearing, whether they be research scientists, veterinarians, advisors, commercial producers or students. Each chapter contains much information that has direct application to commercial practice, but the conclusions are supported by sound scientific evidence.

Phil Garnsworthy
University of Nottingham
January 2005

ACKNOWLEDGEMENT

The 60th University of Nottingham Easter School in Agricultural Sciences was sponsored by:

SCA is a brand of Provimi Ltd

SCA Mill, Dalton Airfield Industrial Estate,
Dalton, Thirsk, North Yorkshire, Y07 3HE UK
Telephone: +44 (0) 1845 578125 Fax: +44 (0) 1845 578100
www.provimiltd.com

CONTENTS

1

MODERN CALVES AND HEIFERS: CHALLENGES FOR REARING SYSTEMS

P.C. GARNSWORTHY
University of Nottingham School of Biosciences, Sutton Bonington Campus, Loughborough, Leics. LE12 5RD, UK.

Introduction

The objective of this chapter is to consider how calves and heifers might have changed over the past fifty years, how rearing systems might have changed, and what are the challenges imposed by these changes on modern systems of calf and heifer rearing.

In many ways, the modern calf destined to be a dairy replacement is no different from calves reared twenty, fifty, or one hundred years ago. The calf has always been a young infant that requires special care and attention if it is to survive from birth to weaning and beyond. Under natural conditions, the calf would be suckled by its mother for up to 14 months (Reinhardt, 2002), during which time she would offer it protection and nutrition. Natural suckling is practiced in beef breeding herds throughout the world, and in some dual-purpose (milk and beef) herds, particularly in the tropics. Natural suckling is not practical, however, for large herds of dairy cows milked through parlours, so some form of artificial rearing is necessary. In addition, the cost of milk replacer is only 0.6 to 0.8 of the sale value of liquid milk on a dry-matter basis, thus providing a strong economic case for artificial rearing. The case is strengthened further when the calf is weaned onto solid feed that costs less than 0.1 of the value of whole milk.

Requirements of calves

From a nutritional point of view, the initial requirements of a calf are for colostrum to provide concentrated nutrients and immunity to diseases, followed by a suitable diet that can meet the nutrient requirements for optimum growth. In the pre-ruminant stage, the diet must have nutritional components and properties similar to those of milk, bearing in mind that the calf's digestive system is underdeveloped. Over the first 6 to 8 weeks of life, the diet must encourage the development of a fully functioning rumen to allow for the transition from liquid feed to solid feed at weaning. Post-weaning, the diet must be designed to achieve the appropriate targets for growth rate, and during grazing periods there is a need to minimise parasite burdens. These aspects are all thoroughly discussed in subsequent chapters.

From a housing point of view, the calf requires a clean dry bed, a well-ventilated environment, a minimum air capacity, and freedom from draughts. Group-housing and individual-housing both have advantages and disadvantages; group-housing allows more social interactions between calves and more exercise, but individual housing greatly reduces the spread of diseases between calves. The best disease control is found when calves are housed outside in individual kennels or hutches; such calves have an infinite supply of fresh air and there is no build up of disease-causing organisms if the kennels are moved to fresh ground between occupants. In the EU, calves in kennels must be provided with a run because tethering of calves is prohibited, except for group housed calves for one hour during feeding.

Changes in animals and rearing systems

Although calves and their requirements have not changed greatly over the past fifty years, there has been a widespread change from more traditional dairy breeds towards the larger Holstein breed. Roy (1980) reported that in 1975 approximately 115,000 calves were slaughtered in the UK because they were "extreme dairy breed animals (i.e. Ayrshire, Jersey and Guernsey)". He predicted that the increasing contribution of the Friesian breed to the National Dairy Herd in England and Wales (81 per cent in 1973-74) would increase the number of calves reared for beef production. This was because the Friesian breed is dual-purpose; not only do the females have moderately high milk yields, but also the males have good growth rates and carcass conformation. The trend towards the Friesian breed did continue in the UK, so that in 1990 approximately 90% of UK dairy cows were nominally Friesians and male calves could all be reared for beef production. However, within the commercial dairy herd there was much crossing with the Holstein breed, starting in the 1970s, followed by almost complete breed substitution by the end of the millennium. The Holstein is an even more extreme dairy breed than the Ayrshire, Jersey and Guernsey breeds, which has greatly improved milk yield and efficiency in the dairy herd, but has resulted in a surplus of purebred male calves that are unsuitable for beef production.

Purebred male calves from the dairy herd used to be exported from the UK to other EU countries for veal production. The value of calves exported annually was £20 million in the early 1980s, £40 million in the late 1980s, and £80 million in the early 1990s; the trade was banned by the EU in March 1996 because of fears over Bovine Spongiform Encephalopathy (BSE Inquiry, 2000). A calf processing aid scheme was introduced in April 1996 to compensate dairy farmers for loss of income from male calves and to remove calves from the human food chain. The scheme ran until 1999, and during its operation nearly 2 million calves were slaughtered. After the scheme was withdrawn, the market value of purebred male dairy calves for beef production in the UK was negligible so most were, and still are, euthanased on farms. Therefore, the majority of calves now reared artificially are either dairy heifer replacements or beef cross dairy calves of both sexes.

The Ayrshire, Friesian and Friesian-cross cows that traditionally formed the dominant part of the UK dairy herd (typically 450-550 kg body weight) were smaller than their modern Holstein and Holstein-Friesian counterparts (typically 600-725 kg body weight). This has implications for calf and heifer rearing. There is a potential problem of increased dystocia that might arise from increased birth weight of calves (Sieber et al., 1989). Hansen et al. (1999) reported that in a long-term selection study for body size in Holsteins, cows selected for large body size weighed 50 to 90 kg more than cows selected for small body size and produced calves that weighed 2.5 kg more at birth. Larger cows, however, are likely to have larger pelvises, which should reduce the incidence of dystocia and offset the potential problem of increased birth weight (Sieber et al., 1989). This was found to be true in the study of Hansen et al. (1999), where no significant difference was found in the number of cows culled due to calving complications in the small (8.8 %) and large (7.6 %) body size lines.

More important than slightly increased birth weight of calves produced by modern Holsteins is the higher target for live weight at calving, which has increased by 13 – 22 % since 1956 (Figure 1.1). This necessitates increases in targets for growth rate and live weight at different stages of the rearing period (Figure 1.2). Over the past 100 years, there has been a general tendency for target daily gains to increase during the calf rearing phase (0 - 3 months of age) and to decrease during the following phase (3 - 6 months of age). For the period from 6 to 12 months of age, target live-weight gain is considerably higher today than it was in 1956 or 1973; this is the period when much of the additional growth needed to reach a higher target calving weight is made. Targets for daily live-weight gain over the second year (12 – 24 months of age) were higher in 1973 than in either 1956 or 2004. In the early 1970s, it was recommended that cows and heifers should be in high body condition at calving to compensate for reduced feed intake in early lactation, which explains the higher target live-weight gain even though target live weight at calving was lower than today. It was subsequently discovered that high body condition at calving has a negative effect on feed intake in early lactation (Garnsworthy and Topps, 1982), and targets for body condition score at calving were reduced (Garnsworthy, 1988). In addition, rapid growth during the final phase before calving is now known to have detrimental effects on mammary tissue development (Sejrsen, 2005).

For many years, the recommended target age for heifer replacements at first calving has been 24 months (e.g. Russell, 1985). In commercial practice, however, first calving is sometimes delayed to allow heifers to attain heavier

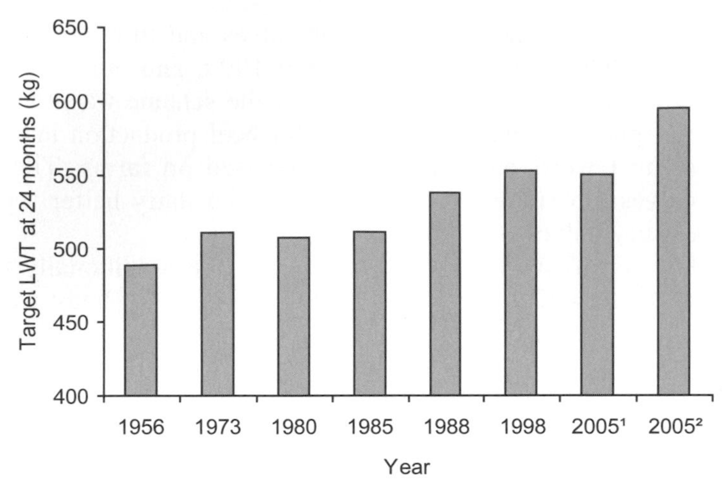

Figure 1.1 Target live weight at 24 months of age for Holstein heifer replacements, as given by Morrison (1956), Smith and Davies (1973), Roy (1980), Russell (1985), Johnsson (1988), Park (1998), [1]Dawson and Carson (2005), [2]Margerison and Downey (2005).

weights or sizes. The objective of this strategy is to reduce the requirement for continued growth during the first lactation, which diverts nutrients away from milk production and is a feature of cows selected for large body size (Hansen et al., 1999). Kossaibati and Esslemont (1997) calculated that the total costs of rearing a heifer to calve at a particular age was £1,037 for 24 months, £1,340 for 30 months, and £1,638 for 36 months. The increased cost of delayed calving can be offset partially by increased milk yield in the first lactation (Johnsson, 1988), but the economic benefit is never likely to exceed 20 % of the extra costs. Higher age at calving is always associated with a lower lifetime economic performance (Strandberg, 1992) and the optimal economic age is 22.6 months (Mourits et al., 1999). In terms of lactation performance, body composition at calving is more important than body weight or size, although body composition is difficult to assess under commercial conditions (Hoffman, 1997). In addition to economic considerations, strategic decisions on age at first calving have resource implications because the number of replacements present on the farm at any one time is greater when calving is delayed (Table 1.1); the greater numbers of youngstock will compete with productive cows for land, feed and labour.

Table 1.1 EFFECTS OF AGE AT FIRST CALVING AND REPLACEMENT RATE ON NUMBER OF HEIFER REPLACEMENTS[1] PRESENT ON A FARM PER 100 COWS

	Age at first calving (months)		
Replacement rate (%)	*24*	*30*	*36*
20	44	55	66
25	55	69	83
30	66	83	99

[1] Including a 10 % allowance for losses during calf and heifer rearing.

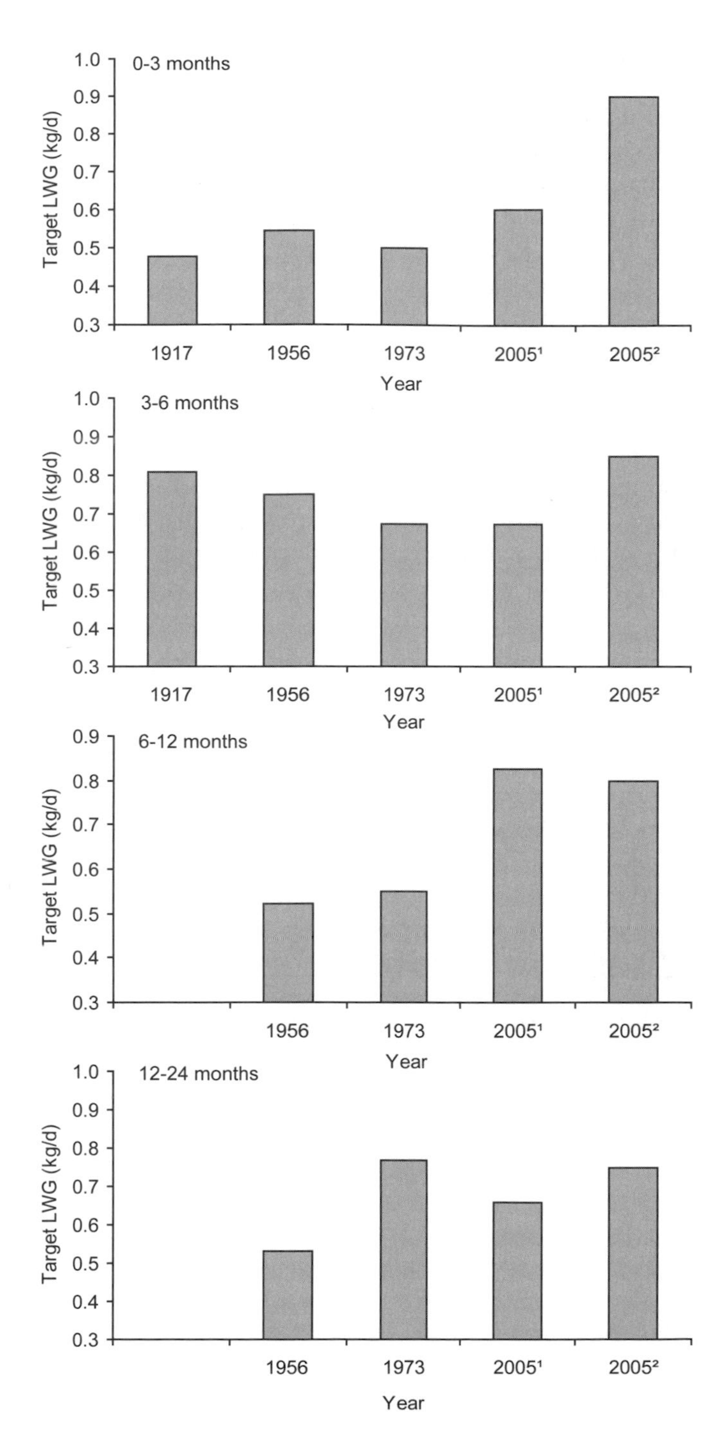

Figure 1.2 Target live-weight gain (LWG) for Holstein heifer replacements destined to calve at 24 months of age during different phases of the rearing period, as given by Henry and Morrison (1917), Morrison (1956), Smith and Davies (1973), [1]Dawson and Carson (2005), [2]Margerison and Downey (2005).

Changes in dairy systems

There have been dramatic changes in the structure of the dairy industries in the UK and most other developed countries of the world. In many countries, there have been national moves towards reduced numbers of dairy cows, fewer dairy herds, larger herd sizes, and greater milk yields per cow. Since 1965, the total number of cows in the UK has decreased from 3.1 to 2.1 million, the number of dairy herds has decreased from 125,000 to 28,000, average annual milk yield per cow has increased from 3,520 to 6,531 litres, and average herd size has increased from 25 to 81 cows (Figure 1.3). Similar trends have been observed in the EU and the USA, although timescales and rate of change differ between countries (Figure 1.4). The trend most likely to have an impact on calf rearing is the increase in herd size, which has direct effects on labour availability. The distribution of herd sizes has changed since 1970, when 80% of herds had less than 50 cows and very few had more than 100 cows. In 2002, over 80% of herds had more than 50 cows, with 50% of herds having more than 100 cows and many herds having over 200 cows (MMB, 1981; MDC, 2003).

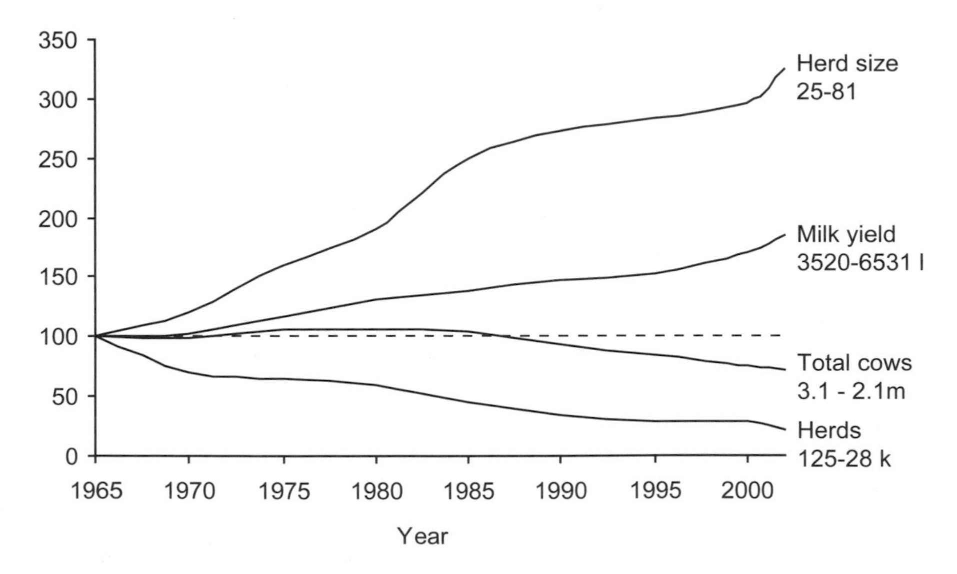

Figure 1.3 Changes in average dairy herd size, average milk yield per cow, total number of cows, and total number of herds in the UK from 1965 to 2002 (1965 = 100). Source: MMB (1981); Defra (2004)

Labour requirements for dairy herds have been estimated by Nix (2003) as 36 hours per cow per year for a 60-cow herd, and 28 hours per cow per year for a 100-cow herd, excluding fieldwork such as hay and silage making. Labour requirements for calf and heifer rearing were estimated at 2.9 hours per month per "replacement unit" (calf + yearling + in-calf heifer) in winter and 1.2 hours

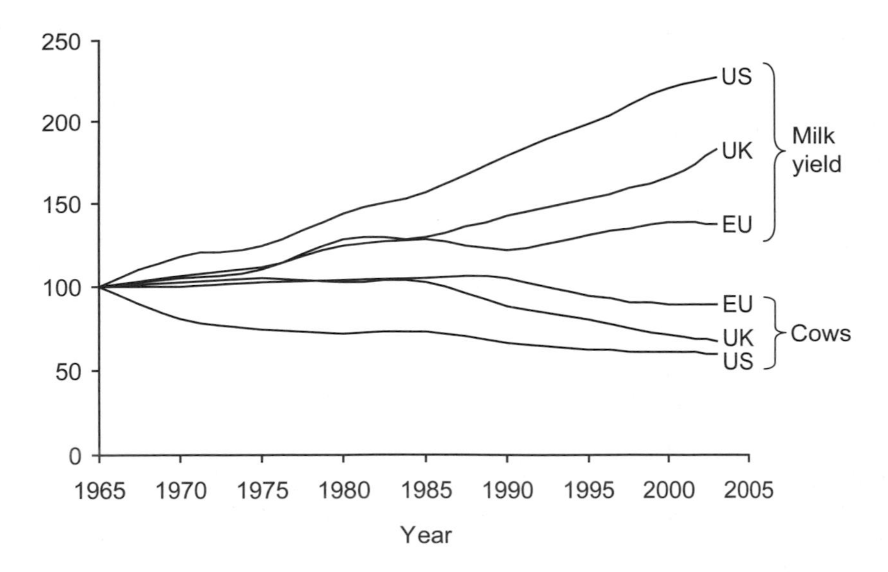

Figure 1.4 Changes in average milk yield per cow and total number of cows in the US, UK and EU from 1965 to 2003 (1965 = 100). Source: FAO (2004)

in summer, with requirements during the calf-rearing phase being 2.3 hours per calf per month. Thus, a 100-cow herd with a replacement rate of 30% would have a total labour requirement of 3,500 hours per year, of which 20% would be required for rearing replacements. Average hours per week would range from 61 in summer to 84 in winter. These figures are considerably in excess of a standard 39-hour working week, even though 100 cows is generally considered to be a "one-man unit".

In recent years there have been severe economic problems in the dairy industry. The price of milk has been low and net profits have been severely reduced. To save fixed costs, labour inputs have been reduced, particularly on larger units. In a survey of 704 UK dairy farms in 2001/2002, Defra (2004) found an average herd size of 78 cows with an annual labour input of 2.14 units, of which the farmer and spouse accounted for 1.25 labour units. This is close to the estimated annual requirement of 2.04 labour units, calculated by interpolation of figures from Nix (2003). Using data from the University of Nottingham Rural Business Research Unit (Seabrook and Johnson, 2000; Seabrook, 2004), it was calculated that the average dairy farm (over 100 ha) used 3.11 units of labour in 1999/2000, and 2.43 units in 2002/2003; the reduction was mainly in employed labour, which was reduced from 2.01 units to 1.26 units. However, annual requirements mask the seasonal nature of labour requirements for calf and heifer rearing. Although the quantity of labour required for dairy cows housed in winter is only slightly greater than when they are grazing in summer, housed youngstock require approximately 2.5 times as much labour as grazing youngstock. This difference is particularly important in an autumn calving herd, where the majority of calves are raised during the housing period.

The type of labour available on dairy farms has also changed in recent years. On small family farms, the farmer's wife would historically have been responsible for rearing the calves, and she would lavish a great deal of care and attention on them. The survey statistics from Defra (2004) show that the farmer's spouse contributes 0.25 labour units to the average UK dairy farm. This implies either that only one spouse in four works on the farm, or that spouses now spend only a quarter of their time working on the farm. The observation that 50% of English dairy farms have some form of diversification (enterprises on or off the farm that are not related to the main farm business), supports the latter probability (Centre for Rural Research, 2002). Nowadays, calf and heifer rearing is often a secondary activity that comes low down on the list of priorities for busy employees.

To reduce labour requirements for calf rearing, various systems have been developed, and their merits are discussed by Thickett et al. (2003). The traditional twice-daily provision of milk replacer in buckets allows precise control of individual calf nutrition without a large investment in equipment; this system, however, requires the greatest labour input. In order to reduce labour requirements, the bucket system was modified so that, after the first week, calves were fed a higher-strength milk replacer in only one feed per day. As well as reducing labour, the once-daily bucket system also reduced feed costs because calves consumed slightly less milk replacer and slightly more dry feed over the rearing period, without significantly altering performance. It should be noted, however, that the Welfare of Farmed Animals (England) Regulations 2000 state that "All calves shall be fed at least twice daily."

Systems where intake of milk replacer is not restricted can also reduce labour requirements. These are usually based on either cold milk, which is acidified to act as a preservative and to aid digestion, or warm milk, which is freshly mixed by a machine every time a calf suckles. These ad libitum systems produce much faster live-weight gains, and require less labour, but the use of milk replacer is approximately twice as great as the restricted twice-daily bucket rearing system. The optimum system that is currently available is a computer-controlled system that allocates milk replacer to individual calves, which are identified by electronic transponders. Although initial equipment costs are greater than other systems, precise control of nutrition is possible, whilst still using minimal labour for feeding calves. Whichever system is practiced, it is essential to ensure that welfare of calves is not compromised when labour is in short supply.

The dramatic increases in genetic potential for milk yield seen in the UK, Europe and USA over the past 20 years have been accompanied by a trend for reduced fertility in dairy cows. In the UK, for example, conception rate to first service has declined by about 1% per year over the past 20 years, to the current level of 38 % (Royal et al., 2000). This means that involuntary culling rates have increased, so more dairy heifers are required to maintain herd size because replacement rate is the reciprocal of the number of years that cows spend in the herd. Therefore, a greater proportion of cows have to be bred to a dairy bull and a greater number of heifer replacements have to be reared. The effect of replacement rate on number of heifers is shown in Table 1.1. These factors represent increased costs, or reduced profits, for dairy producers. In addition, there is an increased environmental impact from the extra replacement heifers. Garnsworthy (2004) calculated that restoring fertility levels to those observed in 1995 would reduce methane and ammonia emissions by approximately 10 %, principally through a reduction in emissions by replacements.

A further consequence of reduced fertility and increased replacement rate is the reduced opportunity for genetic selection within a dairy herd. The current average replacement rate in UK dairy herds is approximately 30 %, so the average cow only survives for three lactations. Therefore, assuming an average pregnancy rate of 80 %, 5 % calf mortality, and 5 % other losses during rearing, she will only produce 2.16 reared offspring during her lifetime. The natural sex ratio means that only 1.08 heifer calves will be reared. In other words, with current levels of fertility the dairy animal is only just breeding enough female offspring to replace herself, leaving 8 % "spare" for genetic selection within the herd. This 8 % has to include selection against animals with poor conformation, mastitis, and other undesirable characteristics, leaving little scope for positive genetic selection, and virtually no scope for producing crossbred calves from a beef bull. In the future, to overcome these restrictions, the use of sexed semen is likely to become more widespread in dairy herds.

Challenges for modern rearing systems

The objectives of heifer rearing are to obtain good animal performance with minimal losses from disease, death and infertility, whilst maintaining a high standard of welfare, optimum (low) input costs and minimum labour inputs. As can be seen from the foregoing discussions, calves and heifers have not changed greatly over the past fifty years, except that mature size has increased necessitating increased targets for live weight and live-weight gain. More influential changes have occurred in the dairy systems within which replacements are reared. The principle changes have resulted from economic pressure, increased herd size and reduced availability of labour. It is these factors that have introduced the biggest challenges for modern rearing systems.

In conclusion, the challenges for modern rearing systems are as follows:

- Firstly, we need to ensure that calves are provided with the highest standards of nutrition, health, welfare and housing.

- Secondly, we need to ensure that heifer rearing programmes post-weaning lead to optimum rates of growth to meet target live weights and levels of fertility.
- Thirdly, we need to remember the constraints on modern dairy farms, especially labour quantity and quality.
- Finally, all these challenges have to be addressed in ways that are most economical for the producer and profitable for the feed supplier.

References

BSE Inquiry (2000) The export of live cattle, beef, and bovine serum and embryos. Volume 10, Section 6, Report of the BSE Inquiry. www.bseinquiry.gov.uk.

Centre for Rural Research (2002) *Farm Diversification in England 2002*: Final Report. Centre for Rural Research, University of Exeter; Rural and Tourism Research Group, University of Plymouth. http://statistics.defra.gov.uk/esg/reports/farmdiv

Dawson, L.E.R and Carson, A.F. (2005) Grazing systems for dairy herd replacements. In *Calf and Heifer Rearing* (Ed P.C. Garnsworthy) pp 253-276. Nottingham University Press, Nottingham.

Defra (2004) *Defra Statistics*. www.defra.gov.uk

FAO (2004) FAOSTAT data, 2004. http://apps.fao.org/faostat/default.jsp

Garnsworthy, P.C. (1988) The effect of energy reserves at calving on performance of dairy cows. In *Nutrition and Lactation in the Dairy Cow*, (Ed. P.C. Garnsworthy), 157-170, Butterworths, London.

Garnsworthy, P.C. (2004) The environmental impact of fertility in dairy cows: a modelling approach to predict methane and ammonia emissions. *Animal Feed Science and Technology*, **112**, 211-223.

Garnsworthy, P.C. and Topps, J.H. (1982) The effect of body condition of dairy cows at calving on their food intake and performance when given complete diets. *Animal Production*, **35**, 113-119.

Hansen, L.B., Cole, J.B., Marx, G.D. and Seykora, A.J. (1999) Productive life and reasons for disposal of Holstein cows selected for large versus small body size. *Journal of Dairy Science*, **82**, 795-801.

Henry, W.A. and Morrison F.B. (1917) *Feeds and Feeding*. The Henry Morrison Company, Madison, WI, USA.

Hoffman, P. C. (1997) Optimum body size of Holstein replacement heifers. *Journal of Animal Science*, **75**, 836–845.

Johnsson, I.D. (1988) The effect of prepubertal nutrition on lactation performance by dairy cows. In *Nutrition and Lactation in the Dairy Cow*, (Ed. P.C. Garnsworthy), pp. 171-192, Butterworths, London.

Kossaibati, M.A. and Esslemont, R.J. (1997) *Understanding the Rearing of Dairy Heifers. A Stockman's Guide.* National Milk Records, Chippenham, Wiltshire.

Margerison, J.K. and Downey, N. (2005) Guidelines for optimal dairy heifer rearing and herd performance. In *Calf and Heifer Rearing* (Ed P.C. Garnsworthy) pp 307-338. Nottingham University Press, Nottingham.

MDC (2003) *Dairy Facts and Figures.* MDC Datum, Cirencester.

MMB (1981) *UK Dairy Facts and Figures.* Federation of UK Milk Marketing Boards, Thames Ditton, Surrey.

Morrison, F. B. 1956. *Feeds and Feeding*, 21st ed. Morrison Publishing Company, Ithaca, NY, USA.

Mourits, M. C. M. Huirne, R. B. M. Dijkhuizen, A. A. Kristensen, A. R. Galligan, D. T. (1999) Economic optimization of dairy heifer management decisions. *Agricultural Systems.* **61**, 17-31.

Nix, J. (2003) *Farm Management Pocketbook* 34th ed.. The Andersons Centre, Melton Mowbray.

Park, C.S. (1998) Heifer rearing for optimum lifetime production. In *Recent Advances in Animal Nutrition - 1998* (Eds P.C. Garnsworthy and J. Wiseman) pp 165-180. Nottingham University Press, Nottingham.

Reindhart, V. (2002) Artificial weaning of calves: benefits and costs. *Journal of Applied Animal Welfare Science*, 5, 251-255.

Roy, J.H.B. (1980) *The Calf* 4th ed. Butterworths, London

Royal, M.D., Darwash, A.O., Flint, A.P.F., Webb, R., Woolliams, J.A. and Lamming, G.E. (2000) Declining fertility in dairy cattle: changes in traditional and endocrine parameters of fertility. *Animal Science*, **70**, 487-501.

Russell, K. (1985) The Principles of Dairy Farming. 10th ed. Farming Press, Ipswich.

Seabrook, M.F. (2004) Farming in the East Midlands Financial Results 2002 – 2003. 52nd Annual Report, Rural Business Research Unit, University of Nottingham, Nottingham.

Seabrook, M.F. and Johnson C.A. (2000) Farming in the East Midlands Financial Results 1997-1998. 47th Annual Report, Rural Business Research Unit, University of Nottingham, Nottingham.

Sejrsen, K. (2005) Mammary development and milk yield potential. In *Calf and Heifer Rearing* (Ed P.C. Garnsworthy) pp 237-252. Nottingham University Press, Nottingham.

Sieber, M. Freeman, A.E. and Kelley, D.H. (1989) Effects of body measurements and weight on calf size and calving difficulty of Holsteins. *Journal of Dairy Science*, **72**, 2402-2410.

Smith, W.R. and Davies, A.J. (1973) Rearing Friesian dairy heifers. *ADAS Profitable Farm Enterprise Booklet 1.* MAFF, Pinner, Middlesex.

Strandberg, E. (1992) Lifetime performance in dairy cattle: definition of traits and influence of systematic environmental factors. *Acta Agriculturae Scandinavica. Section A, Animal Science*, **42**, 71-81.

Thickett, W., Mitchell, D. and Hallows, B. (2003) *Calf Rearing.* Farming Press, Ipswich.

2

PHYSIOLOGY OF GROWTH

JOHN M. BRAMELD
Division of Nutritional Biochemistry, University of Nottingham, School of Biosciences, Loughborough, Leics. LE12 5RD

Introduction – classical concepts of growth

Growth is a simple concept and relates to an increase in size and weight of an animal. However the underlying mechanisms for the regulation of growth are proving to be very complex. Classical studies by Hammond and others in the 1930s and 1940s (see Lawrence and Fowler, 2002) demonstrated the changes in size and form of the various organs and tissues with increasing age. Thus, for example, it was realised that not all tissues grow at the same rate or at the same time. The brain and nervous system were shown to grow and develop at an early stage, followed by bone, skeletal muscle and later fat (see Lawrence and Fowler, 2002). The growth of these various tissues can be represented mathematically by sigmoid curves (e.g. Gompertz equation), indicating changes in the rate of growth with age, eventually reaching a plateau. We now realise, however, that skeletal muscle and adipose tissue are not single tissues; they contain a number of different muscles or depots that also grow and develop at different stages of life.

Growth is also associated (in mammals) with development. Development relates to changes in shape, morphology and function of the various tissues, and also relates to differentiation of cell types. For example, the brain grows and develops quite early, hence the head accounts for a large proportion of body size and weight at birth in most mammals. As the animal subsequently grows and develops, the head stays roughly the same size, but the rest of the body increases proportionately, so that the proportions change dramatically as an animal gets older (see Lawrence and Fowler, 2002).

How do tissues grow?

Growth and development actually take place at the cellular level. For a tissue to get bigger, either the number of cells within the tissue increases (hyperplasia) or the size of those cells increases (hypertrophy). Hyperplasia relates to cell proliferation (also referred to as cell division and mitosis), which is the formation of two daughter cells from one parent cell. At the molecular level this is a complex process involving progression through the cell cycle, which is regulated by combinations of mitogens, including various hormones and polypeptide growth factors (see Lawrence and Fowler, 2002). Hypertrophy, on the other hand, relates to changes in size of the cell, normally as a result of increased deposition or storage of macromolecules. For example, hypertrophy of skeletal muscle fibres is due to increased protein deposition as a consequence of either an increase in protein synthesis or a decrease in protein degradation (proteolysis), or both (see Buttery, Brameld and Dawson, 2000). In comparison, hypertrophy of fat cells (adipocytes) is due to increased fat (triacylglycerol) deposition, again as a consequence of either increased fat synthesis (lipogenesis) or decreased fat degradation (lipolysis), or both. Interestingly, the same factors appear to be involved in regulating both hyperplasia and hypertrophy in many tissues.

Two more terms often used in describing growth and development are cell proliferation and differentiation. Proliferation has already been mentioned, but cell differentiation relates to development and the changes in cell shape, morphology and function that take place in various tissues at different stages of life (see Lawrence and Fowler, 2002). Differentiation is often dependent upon expression of tissue-specific transcription factors, which in turn induce transcription of the tissue-specific proteins that make a cell a particular cell type. For example, the process of differentiation of skeletal muscle cells is referred to as myogenesis (see Brameld, Buttery, Dawson and Harper, 1998; Buttery *et al.*, 2000). This involves the conversion of an embryonic mesenchymal precursor cell to a mature muscle fibre. The differentiation stage mainly relates to the alignment and fusion of mononuclear myoblasts to form a multinuclear myotube or myofibre that expresses particular transcription factors (e.g. myogenin) and proteins (e.g. desmin, creatine kinase).

Endogenous hormones and growth factors as regulators

A variety of hormones (both peptide/protein and steroid) and polypeptide growth factors have been demonstrated to alter rates of growth and development, and their mechanisms of action have been identified to varying degrees.

(I) THE GH-IGF AXIS

The growth hormone (GH) – insulin like growth factor (IGF) axis has been studied extensively, due to its important role in regulating growth (see Brameld,

1997). Although termed growth hormone, endogenous GH doesn't actually have much direct influence on regulation of the growth processes (hyperplasia and hypertrophy). In most mammals (except rodents), high growth rates are associated with relatively low blood concentrations of endogenous GH, for example in well-fed animals. High growth rates are associated more with increased blood concentrations of IGF1, and also with increased numbers of GH-receptors in the liver; whereas reduced growth rates (e.g. during fasting/ starvation or poor nutrition) are associated with reduced blood concentrations of IGF1, and reduced numbers of GH-receptors in the liver, but increased circulating concentrations of GH (see Brameld, 1997). Tissue-specific gene knockout (via cre-lox technology) studies in mice (see Butler and Leroith, 2001) indicate that the liver is the major source of circulating IGF1, but that loss of hepatic IGF1 has no effect on growth rate, suggesting that other tissue sources are a) more important or b) can compensate for the loss. There are actually two IGFs; IGF1, which is positively regulated by GH, and IGF2, which isn't (Brameld, 1997). Both are very potent mitogens and inhibitors of apoptosis, and both also stimulate the differentiation of a variety of cell types, making them unique as growth factors. They also have insulin-like effects on protein, glucose and fat metabolism, and therefore play an important role by regulating both hypertrophy and hyperplasia in various tissues. There are a family of IGF-binding proteins (Brameld, 1997) that appear to have a variety of functions, including (i) acting as a storage site/form for IGFs (binding increases the half-life); (ii) helping shuttle IGFs to their receptors (i.e. enhancing the effects of IGFs); (iii) preventing IGFs from binding to their receptors (i.e. inhibiting the effects of IGFs); and (iv) having direct (IGF-independent) effects on cells. The IGFs bind to two types of IGF receptor (see Brameld, 1997). The type 1 receptor (IGF1R) binds IGF1 with greater affinity than IGF2, which is bound with much greater affinity than insulin, and IGF1R is thought to be responsible for the majority of growth-related effects of the IGFs. The type 2 receptor (IGF2R) binds IGF2 with much greater affinity than IGF1, and is thought to be a mechanism for removing IGF2 (see Butler and Leroith, 2001). The type 1 receptor is very similar to the insulin receptor and consequently insulin can bind to the IGF1R and the IGFs can bind to the insulin receptor, with differing binding affinities. Interestingly, the main regulator of the IGF1R appears to be IGF1 itself, such that high levels of IGF1 are associated with lower receptor numbers, presumably limiting the response (Brameld, 1997).

(II) POLYPEPTIDE GROWTH FACTORS

There are a number of other polypeptide growth factors that, by definition, regulate cell proliferation and differentiation. Growth factors tend to stimulate cell proliferation and to inhibit cell differentiation, because when a cell differentiates it has to withdraw from the cell cycle (and therefore stop proliferating). Examples of such growth factors include the fibroblast growth factors (FGF) and platelet derived growth factor (PDGF). There are exceptions, however, such as the IGFs, which appear to stimulate both proliferation and differentiation, although this action is dose-dependent and possibly cell-dependent as well (see Brameld *et al.*, 1998). Epidermal growth factor (EGF) has also been shown to increase both proliferation and differentiation, at least in skeletal muscle cells; whilst the transforming growth factor beta (TGFß) family of growth factors tend to inhibit both proliferation and differentiation (of skeletal muscle cells at least). An important example of the latter would be myostatin (see Bass, Sharma, Oldham and Kambadur, 2000), where loss of functional protein can lead to double muscling in cattle (e.g. Belgian Blue cattle). Consequently, growth factors have a major role in the regulation of cell hyperplasia, but some of these growth factors can also influence hypertrophy of tissues. For example, the IGFs and EGF have been shown to increase protein synthesis and decrease protein degradation in skeletal muscle cells (see Brameld *et al.*, 1998; Buttery *et al.*, 2000), presumably resulting in hypertrophy of those cells.

(III) INSULIN

Insulin is another hormone that has an important role in regulating growth and development. As indicated above, insulin can bind to the IGF1R as well as to the insulin receptor, which might explain its effects. In vitro studies indicate effects similar to those of the IGFs, although the insulin concentrations required are often higher and might be supraphysiological (see Brameld *et al.*, 1998). Insulin is also thought to regulate expression of IGF1 by the liver and therefore any effects might be indirect via this mechanism (see Brameld, 1997). Insulin is extremely important in regulating global translation of mRNA into protein, since it increases ribosomal RNA and the numbers of ribosomes, and it regulates initiation factors for translation within cells (see Dawson, 1999; Grizard, Picard, Dardevet, Balage and Rochon, 1999; Nieto and Lobley, 1999). Hence insulin is a potent regulator of protein synthesis, which will also relate to hypertrophy, particularly in skeletal muscle.

(IV) THYROID HORMONES

Thyroid hormones play an important role in growth and development, with low levels (hypothyroidism) being associated with reduced growth rates (see Lawrence and Fowler, 2002). The two biologically active thyroid hormones are tri-iodothyronine (T3) and thyroxine (T4), with conversion of T4 to T3 being catalysed by iodothyronine deiodinase enzymes within various tissues (see St Germain, 2001). They function as ligands for the thyroid hormone receptors (THR), which are part of the nuclear hormone receptor family, with T3 having a ten-fold higher binding affinity than T4, resulting in T3 being more active than T4 (Lawrence and Fowler, 2002). Hence conversion of T4 to T3 results in an increase in activity. THR are able to bind to specific sequences of DNA and thereby regulate gene transcription, with binding of the ligand (e.g. T3) changing the effect from inhibition to stimulation of gene transcription (see Jameson, 2001). This regulation of gene expression means that thyroid hormones have important effects in the regulation of cell differentiation. There are interactions between the thyroid hormones and the GH-IGF axis, since T3 has been shown to be a potent inducer of GH-receptor gene expression in the liver (see Brameld, 1997) and may be involved in the switch from prenatal to postnatal growth (see later). Hence effects of thyroid hormones on growth might be indirect via an increase in GHR and then GH-stimulated IGF1 expression.

As indicated, low levels of thyroid hormones are associated with reduced growth rates, but high levels (hyperthyroidism) can also be associated with reduced growth rates and even muscle wasting. This probably relates to the role of thyroid hormones in regulating basal metabolic rate, particularly mitochondrial metabolism (see Jameson, 2001), and therefore availability of energy for growth. Hyperthyroidism is associated with increased energy expenditure and consequent reduction in adipose tissue. Prolonged hyperthyroidism results in depletion of adipose tissue stores, so the body starts to mobilise skeletal muscle as a means of providing an alternative energy source (providing amino acids for gluconeogenesis).

(V) GLUCOCORTICOIDS

Glucocortocoids are associated with reduced growth rates or even reduction in body weight. However, the role of glucocorticoids is very complex. Glucocorticoids, like thyroid hormones, are found in both active and inactive forms; in most mammals the inactive form is cortisone, whereas the active form is cortisol (see Munck and Naray-Fejes-Toth, 2001). Conversion of the inactive to the active form is catalysed by the enzyme 11ß hydroxy steroid dehydrogenase type 1 (11ß HSD1), which is present in a variety of tissues. Like thyroid hormones, differences in activity relate to binding affinities to the receptor protein called the glucocorticoid receptor, which again is a member of the nuclear hormone receptor family that includes receptors for all the steroid hormones. In some skeletal muscles, cortisol inhibits DNA synthesis and thereby cell proliferation (see Lawrence and Fowler, 2002). However, in some situations, dexamethasone (a synthetic glucocorticoid) enhances responses to insulin (which stimulates proliferation and differentiation) via regulation of the insulin receptor. Similarly, dexamethasone also increases GH receptor expression, and thereby GH-stimulated IGF1 expression, in liver (Brameld,

1997). Indeed, the natural cortisol surge that is seen prior to birth of all mammals has been shown to induce GHR expression in the liver of late gestation foetal sheep (see Breier, Oliver and Gallaher, 2000). This results in an increase in hepatic IGF1 expression and a decrease in IGF2 expression, apparently "switching" from a foetal, GH-independent type of growth regulation to a postnatal, GH-dependent regulation. The cortisol surge is also responsible for maturation of a number of tissue types (e.g. lungs), which relates to the later stages of differentiation and development of tissues. It might also be responsible for induction of iodothyronine deiodinase enzymes and therefore increased activity of thyroid hormones. Indeed, the combination of glucocorticoid and thyroid hormone results in the greatest increase in GHR expression in cultured hepatocytes (see Brameld, 1997).

(VI) SEX STEROIDS

Males have less fat, more skeletal muscle and higher growth rates than females. These differences are thought to relate partly to the sex steroids. Indeed, plasma testosterone, oestradiol and dehydroepiandrosterone levels in the mid-gestation foetus can be equivalent to those of the pubertal male (see Rosenfield and Cuttler, 2001). The actions of androgens are suggested to account for the increased birth weight in males. Maximal growth rates are seen at puberty in most mammals, which again relates to the sex steroids. The actions of the various sex steroids are via binding to their intracellular receptors, which are members of the nuclear hormone receptor family, like the glucocorticoid and thyroid hormone receptors. As for the glucocorticoid receptor, binding of the ligand (i.e. the steroid hormone) results in binding of the complex to DNA response elements and stimulation of gene transcription. The sex steroids have been shown to have both direct and indirect effects on growth processes. For example, oestrogens have been shown to stimulate cell proliferation in some tissues (e.g. mammary cells), while testosterone has been shown to stimulate muscle cell proliferation, inhibit muscle cell differentiation and to stimulate muscle fibre protein synthesis (see Buttery *et al.*, 2000). Hence testosterone appears to stimulate both proliferation and hypertrophy of muscle cells. However, the main role of sex steroids in growth appears to be via their interactions with the GH-IGF axis. Circulating GH concentrations are increased by androgens in most species, although in some species (e.g. humans) oestrogens have an equal or greater effect (see Gatford, Egan, Clarke and Owens, 1998). GH production is regulated by 2 hypothalamic factors, GH-releasing hormone (GHRH) and somatostatin, which have stimulatory and inhibitory effects respectively. Pituitary sensitivity to GHRH is increased in male rats, cattle and horses (see Gatford *et al.*, 1998). In rats, these effects are mimicked by exogenous androgen and oestrogen administration and exposure to androgens during neonatal life increased the numbers of GHRH neurones in the hypothalamus. Circulating IGF1 concentrations are also higher in male rats, sheep, cattle and pigs (see Gatford *et al.*, 1998), presumably as a consequence of the increased blood GH levels, although there may also be effects of sex steroids on GHR and/or IGF1 expression/production.

The role of nutrition and nutrients – substrates or regulators?

Animals do not grow if they are not provided with adequate nutrition. Obviously, animals cannot deposit protein in their skeletal muscles without being provided with amino acids to form that protein. Likewise, animals cannot deposit fat in their adipose tissue if they are not provided with glucose and fatty acids. These arguments simply relate to nutrients being substrates or "building blocks" to allow growth. However, a rapidly developing field within nutritional sciences is the idea that these same nutrients might actually regulate the processes for which they are being used as substrates. If you consider the role that vitamins and minerals (micronutrients) play in metabolism, then this may seem obvious; we know that vitamins A and D work in a similar manner to steroid hormones, by binding to nuclear receptors and thereby regulating gene transcription. It is now becoming apparent that macronutrients (e.g. glucose, amino acids and fatty acids) can have similar effects, although the mechanisms are still being elucidated. Thyroid hormones and catecholamines are simply converted amino acids; therefore it might be expected that amino acids would act as regulators of metabolism, although until recently this has been a poorly investigated and even controversial idea. Indeed in pig and poultry nutrition, diets are now formulated on the basis of amino acid composition, not just protein content, realising that some amino acids are more important than others. It is well known that an animal requires sufficient energy before there is any effect of increasing the protein content of the diet. Recent studies have demonstrated a molecular basis for this need for energy and individual amino acids for optimum growth. Research at Nottingham has identified effects of glucose on expression of the GH receptor gene in cultured pig hepatocytes (Brameld, Gilmour and Buttery, 1999a), such that maximal mRNA expression is obtained via a combination of high glucose, T3 and dexamethasone in the culture medium. This then results in an enhanced response of IGF1 mRNA to any added GH. We have also demonstrated, via radiolabelled GH-binding studies, that these effects on GHR mRNA are reflected in levels of GHR on the cell membrane (Brameld and Buttery, 2002). Hence energy in the form of glucose is required for optimal GHR mRNA expression. Similar studies have identified dose-dependent stimulatory effects of specific amino acids (arginine, proline, threonine, tryptophan and valine) on GH-stimulated IGF1 expression (Brameld *et al.*, 1999a). Hence reduction in any one of these amino acids results in reduced hepatic IGF1 mRNA. This has been shown to be due to decreased numbers of GHR on the cell membrane (Brameld and Buttery, 2001).

Hence glucose appears to regulate transcription of the GHR gene, while certain amino acids regulate its translation into functional protein; this matches the observed requirements for dietary energy prior to amino acids/protein. Similar studies in sheep hepatocytes (Wheelhouse, Stubbs, Lomax, MacRae and Hazlerigg, 1999) indicate a similar reduction in IGF1 secretion in media containing low levels of all amino acids, with methionine appearing to be particularly important (Stubbs, Wheelhouse, Lomax and Hazlerigg, 2002).

The studies described specifically relate to growth-regulatory genes, but there are now a number of groups investigating the effects of glucose (see Foufelle and Ferre, 2003) and amino acids (see Fafournoux, Bruhat and Jousse, 2000; Kilberg, Leung-Pineda and Chen, 2003) on gene expression, particularly genes relating to fat and amino acid metabolism respectively. A number of genes appear to be regulated in similar ways. For example, a number of glucose-regulated genes in the liver are also regulated by insulin, thyroid hormones and glucocorticoids, as is the GHR gene.

Since nutrients appear to be both substrates and regulators of the growth processes, then their rate of supply (i.e. food intake) will have an impact upon growth. Nutrients and many of the hormones described earlier (e.g. insulin) also play a role in regulating food intake or appetite (Langhans, 1999).

Pre- and post-natal growth

(I) ENDOCRINE CHANGES

One interesting aspect of growth is the comparison between pre- and post-natal growth. Pre-natal or foetal growth has long been considered to be GH-independent. One of the main growth factors was thought to be IGF2, with IGF1 apparently having little effect on foetal growth and being more important in post-natal, GH-dependent growth (see Breier *et al.*, 2000). However, there is increasing evidence that this isn't strictly correct. Loss of either IGF1 or IGF2 via gene knockout technologies results in impaired foetal growth and growth retarded offspring at birth (see Butler and LeRoith, 2001); while only the loss of IGF1 has an effect on post-natal growth. As indicated earlier, the growth-regulatory effects of the two IGFs is via IGF1R and loss of IGF1R by gene knockout results in extreme growth retardation and death. The role of IGF1 in post-natal growth has been investigated further and loss of liver-specific IGF1 expression (via cre-lox technology) results in reduced circulating IGF1 levels, but has no effect on post-natal growth rate (see Butler and LeRoith, 2001). Interestingly, the loss of hepatic, and thereby circulating, IGF1 results in increased plasma GH levels, presumably due to the loss of negative feedback of IGF1 on GH production by the pituitary, and this results in mice developing insulin-resistance (Yakar, Liu, Fernandez, Wu, Schally, Frystyk, Chernausek, Mejia and LeRoith, 2001). Hence, the role of hepatic-derived IGF1 is being questioned and brings in the possibility that other tissue sources (e.g. adipose tissue or skeletal muscle) may either compensate for the lack of circulating IGF1 or may be more important.

The apparent GH-independence of foetal growth might also be questioned, since this has normally been assumed because of the lack of GHR expression by the foetal liver, but we do not know what the levels of GHR expression in many other tissues are (Breier *et al.*, 2000). Plenty of GH and IGF2 is certainly found in foetal blood, as well as some IGF1. We have studied changes in IGF1 and IGF2 expression in developing skeletal muscle, both in vivo (Fahey, Brameld, Parr and Buttery, 2003a) and in vitro (Brameld, Smail, Imram, Millard and Buttery, 1999b). Both genes are expressed at low levels when myoblasts are proliferating. IGF2 expression increases and peaks at the same time as early markers of myoblast differentiation (myogenin or creatine kinase), suggesting that it might be involved in inducing early differentiation (Brameld *et al.*, 1999b; Fahey *et al.*, 2003a). IGF1 expression, on the other hand, increases slightly later and appears to be as a consequence of fibre formation rather than as an inducer of it. Once fibres are formed, the level of IGF1 expression remains constant thereafter, including during post-natal life (Brameld, Weller, Pell, Buttery and Gilmour, 1995). This skeletal muscle IGF1 expression might then have an impact on postnatal growth rates, since the numbers of muscle fibres correlates strongly with post-natal growth rates in a variety of breeds of pig, and we have found reduced skeletal muscle IGF1 expression in the slower growing breeds (Brameld, Atkinson, Budd, Saunders, Pell, Salter, Gilmour and Buttery, 1996).

There is increasing evidence that the GH-IGF axis is affected by nutritional manipulation in a similar manner in both the foetus and the pregnant mother. Hence, reduced levels of nutrition are associated with increased GH and decreased IGF1 in the blood of both, at least during mid-to-late pregnancy in sheep (see Brameld, Fahey, Langley-Evans and Buttery, 2003). We have also shown reduced expression of both GHR and IGF1 mRNA in mid-gestation foetal sheep liver after a period of reduced nutrition to the mother (see Brameld, Fahey, Langley-Evans and Buttery, 2003). Hence, there appears to be a functional GH-IGF axis in foetal life; major changes do take place, however, during late foetal and early neonatal life that appear to result in the switching from relative GH-independence to GH-dependence. The cortisol surge has been shown to result in a co-ordinated induction of GHR and IGF1 and a reduction in IGF2 expression in the foetal liver (see Breier *et al.*, 2000). This may also involve thyroid hormones, since the cortisol surge also induces hepatic expression of the iodothyronine deiodinase activating enzymes. This also agrees with our in vitro studies that demonstrated stimulatory effects of T3 and dexamethasone on GHR mRNA, which resulted in an increased response of IGF1 to GH (see earlier). At the time of birth there are relatively low levels of GHR and IGF1 expression in the liver, while blood GH concentrations are high. There is then a rapid decrease in GH production by the pituitary and a gradual increase in both GHR and IGF1 expression by the liver. These gradual changes in hepatic GHR are matched by increased responsiveness in GH-stimulated IGF1 (see Breier *et al.*, 2000). Just prior to weaning in pigs, hepatic IGF1 expression starts to plateau (Brameld *et al.*, 1995), whereas GHR expression and plasma IGF1 levels continue to rise, the latter probably as a result of increased IGFBP-3 production by the liver, which increases the half-life of IGF1 in the blood.

(II) PRE-NATAL GROWTH AND FOETAL PROGRAMMING

Foetal life is associated with very rapid increases in cell number, via cell proliferation. Then at different times during gestation, certain cells stop proliferating and differentiate into functional cell types. It is becoming apparent that the number of functional cell types within some tissues is actually determined at or close to the time of birth. Hence, foetal life is associated mainly with hyperplasia of cell types, with the numbers of cells being fixed at the time of birth. This is certainly true for the numbers of nephrons found in kidneys, the numbers of gonadotrophs (and probably somatotrophs and thyrotrophs) found in the pituitary and the numbers of muscle fibres found in skeletal muscles. In skeletal muscle for instance, proliferating myoblasts exit the cell cycle, align and fuse together to form first myotubes and later muscle fibres (see Brameld *et al.*, 1998; 2003). There are 2 (or 3) rounds of this, which result firstly in the formation of primary muscle fibres that develop along the bones; with secondary fibres then forming around the primary fibres. In some (larger) species, tertiary fibres have also been described that form between secondary fibres. Initially, it has been shown that primary fibres tend to form type I (slow oxidative) fibres, whereas secondary fibres tend to form type II (fast glycolytic or fast oxidative glycolytic) fibres, although a degree of switching of fibre types is observed later on (see Brameld *et al.*, 2003). The timings for these fibre formations vary between species (Table 2.1) but, in all farm animal species, take place during foetal life (tertiary fibres are a very minor population in pigs). Hence the numbers of muscle fibres are fixed at birth for cattle, sheep and pigs. It is becoming increasingly apparent that the numbers of muscle fibres can be altered by such factors as genetics, exogenous hormone administration and nutrition.

The genetic component is illustrated by double muscling in cattle. Double muscling has been shown to relate to various genetic mutations that result in production of a non-functional form of myostatin (Bass *et al.*, 2000). This is a member of the TGFß family of growth factors that inhibits myoblast proliferation (and differentiation). Hence its loss results in greater myoblast proliferation for a longer period, a delay in differentiation and thereby an increase in the numbers of muscle fibres formed. However, myostatin cannot be the only factor involved, since other breeds of cattle have been found to have similar mutations, but they do not have a double-muscled phenotype (Smith, Lewis, Wiener and Williams, 2000). It should be pointed out that due to the increased foetal growth observed in the double-muscled cattle, the offspring have to be delivered by caesarean section.

Table 2.1 STAGE OF APPEARANCE OF THE DIFFERENT GENERATIONS OF MUSCLE FIBRES IN VARIOUS SPECIES (ADAPTED FROM BRAMELD *et al.*, 2003).

Species	*Length of Gestation*	*Primary fibres*	*Secondary fibres*	*Tertiary fibres*
Poultry	21 days	3-7 df	8-16 df	-
Rat	22 days	14-16 df	17-19 df	-
Guinea pig	68 days	30 df	30-35 df	-
Pig	114 days	35 df	55 df	0-15 dpn
Sheep	145 days	32 df	38 df	62-76 df
Bovine	278-283 days	60 df	90 df	110 df
Human	280 days	56 df	90 df	110-120 df

df: days of foetal life, dpn: days of postnatal life

Administration of exogenous GH to pregnant sows has also been shown to result in increased numbers of secondary muscle fibres in the resulting pigs, but only when administered in early pregnancy (10-24d gestation), prior to secondary fibre formation (see Brameld *et al.*, 2003; Buttery *et al.*, 2000). Administration after (80-94d gestation) the time of fibre formation had no effect on the numbers of the fibres but resulted in increased fibre diameters, indicating a hypertrophic effect similar to that seen post-natally. GH appeared to result in more pigs within a litter having the higher number of fibres rather than all pigs having increased numbers of fibres. This suggests that more reach their genetic potential and that GH administration prevented or reduced the incidence of runting (foetal growth retardation).

Nutrition has been shown to affect the numbers of secondary muscle fibres in a variety of species, including rats, guinea pigs, pigs and sheep (see Brameld *et al.*, 2003). A reduced level of nutrition is thought to result in runting in pigs, with runts having fewer skeletal muscle fibres than their larger littermates. Likewise, maternal nutrient restriction results in reduced numbers of muscle fibres in the offspring of rats and guinea pigs; whereas doubling the nutrient intake of pregnant sows (which are normally restricted) has a similar effect to GH administration in terms of increasing the average numbers of fibres in the piglets and reducing the incidence of runting (see Brameld *et al.*, 2003). We have recently demonstrated reduced proportions of fast fibres and increased proportions of slow fibres in sheep whose mothers were restricted immediately prior to (but not during or after) secondary/tertiary fibre formation (i.e. 30-75d gestation), indicating a reduction in the number of secondary fibres (Fahey,

Brameld, Parr and Buttery, 2003b). All these observed effects of nutrition are dependent upon the nutritional manipulation taking place prior to major fibre formation, which does differ between species.

Hence the pre-natal period of growth can have a tremendous impact upon the numbers of functional, differentiated cell types found within a tissue. This can then impact upon post-natal growth, development and metabolism resulting in altered growth rates, carcass composition and even disease-susceptibility in later life. The latter is probably not so important in farm animals, which tend to have a relatively short lifespan, but this is obviously of great interest for humans and is often termed foetal/nutritional programming or the foetal origin of adult disease hypothesis.

(III) POST-NATAL GROWTH

Postnatal growth includes both hyperplasia and hypertrophy of tissues, although hypertrophy predominates in some tissues. Two contrasting tissues would be skeletal muscle and adipose tissue. As has already been stated, the number of muscle fibres is fixed at the time of birth for all farm animal species. Hence post-natal growth involves an increase in size of those existing fibres. This principally involves an increase in the amount of protein deposited in those fibres, resulting in increased width of fibres via the addition of more myofibrils and increased length of fibres via the addition of more sarcomeres to the ends of the myofibrils (see Lawrence and Fowler, 2002). Such changes are induced via increased protein synthesis and/or decreased proteolysis. While some "growth promoters" work via reducing proteolysis (e.g. beta agonists), the majority increase protein synthesis (e.g. insulin and IGFs). Increased protein synthesis involves satellite cells, which are myoblast-like cells found beneath the basement membrane of muscle fibres (see Brameld *et al.*, 1998). These cells are able to increase in number (proliferate) and later fuse with the existing fibres (i.e. differentiate) to increase the number of nuclei within the muscle fibre and thereby increase RNA and protein synthesis. A limit in terms of maximum protein deposition in skeletal muscle will therefore depend upon the numbers of both muscle fibres and satellite cells present.

In contrast, the numbers of fat cells (adipocytes) are not limited in any way. Instead, preadipocytes are able to proliferate and then differentiate into adipocytes at all stages of life (see Lawrence and Fowler, 2002). Those adipocytes are then able to increase or decrease in size in response to the needs of the body. They can store more fatty acids and glucose (via lipogenesis) and thereby increase in size (hypertrophy) when necessary. They can also release fatty acids and glycerol (via lipolysis) for use by other tissues when nutrients are scarce, thus resulting in a reduction in cell size. Adipocytes can only reach a certain size though, and therefore if more energy substrates need storing, then more preadipocytes are recruited into adipocytes to meet the demand. Hence both hyperplasia and hypertrophy of adipocytes can take place at all stages of life, but particularly once mature body weight is reached and energy requirements plateau or even fall. Importantly, the question of why different fat depots grow at different times and at different rates has still to be answered.

It is relevant to point out that skeletal muscle mass and adipose tissue mass appear to be interrelated, as is postnatal growth rate. Animals with increased muscle mass tend to have less adipose tissue at similar body weights (see Brameld *et al.*, 2003; Lawrence and Fowler, 2002). For example, it is well known in pigs that increased numbers of muscle fibres are associated with increased growth rates and reduced fatness. The same is true in cattle where breeds with increased muscle mass are known to have higher growth rates and reduced fat mass (Lawrence and Fowler, 2002). It might be the case, therefore, that factors affecting the numbers of muscle fibres formed in utero might also alter post-natal growth rates and body composition. This obviously has implications for calf and heifer producers. For meat production, high growth rates, increased muscle and reduced fat are desirable. However, for milk production, increased fat may be more desirable, since this can have implications in terms of time to reach puberty and first pregnancy and therefore first lactation. A higher proportion of adipose tissue might also be beneficial in terms of recovery for second and subsequent pregnancies and lactations. However, all these possibilities are dependent upon post-natal nutrition as well, since an animal will not grow at all if not provided with the right nutrition.

(IV) COMPENSATORY GROWTH.

Compensatory or catch-up growth relates to the increased growth rates observed in animals switched from a low level of nutrition to a high level (see Dawson, 1999; Lawrence and Fowler, 2002). Obviously, the low level of nutrition is associated with decreased growth rates, whereas the high level of nutrition is associated with increased growth rates. This relates to the interactions between hormones and nutrients in regulating the processes of hyperplasia and/or hypertrophy described earlier. However, it has been demonstrated that alternate low then high levels of nutrition can result in growth rates higher than if the animal is simply kept at a high level of nutrition. There are also cost implications, since the low level of nutrition will be cheaper. The great interest in this field is whether an animal will reach a certain weight (and/or body composition) quicker than a control animal fed at the higher level throughout; with the possible advantage of reduced feed costs. This isn't clear from the literature (see Lawrence and Fowler, 2002). However, the degree of catch-up that takes place will depend upon (i) when the nutritional restriction takes place in terms of stage of life (e.g. pre- or post-natal); (ii) the degree of restriction (e.g. 30 or 50% of requirements); (iii) the length of time the restriction is imposed (e.g.

days, weeks or months), and the equivalent criteria (stage, amount and time) for the "catch-up" diet. As seen earlier, a restriction during muscle development (foetal life) might result in a permanent effect, so the animal will never catch-up. It has been suggested that real catch-up is only seen in older animals, with restrictions not greater than 50% of requirements for a limited period of time. The one area where true "catch-up growth" has been observed is in dairy heifers, where a step-stair feeding regime during very specific stages of development has been shown to both increase growth rates and milk yields (Choi, Han, Woo, Lee, Jang, Myung and Kim, 1997). However, in a number of studies the restricted animals do not catch-up, even though they have increased growth rates for a period after restriction. The composition of the "catch-up" growth might also vary. Hence more work is needed to try and identify the important stages of development when a compensatory effect might be observed, although this will probably also depend upon previous nutritional restrictions and body composition at the start. In terms of mechanisms for catch-up growth, it has been suggested that it relates to reduced maintenance (and physical activity) energy expenditure and reduced energy content of the gain (see Dawson, 1999). The periods of rapid growth rates are associated with increased plasma concentrations of both IGF1 and insulin and increased tissue responsiveness due to increased IGF1R levels. As indicated, low nutrition and reduced growth rates are associated with low circulating IGF1 concentrations. IGF1 has an inhibitory effect on IGF1R, therefore low IGF1 will lead to higher IGF1R. Following realimentation, circulating insulin and IGF1 will both increase, but insulin will increase faster than IGF1. Therefore, the early period after realimentation will be associated with high IGF1R and higher insulin and possibly IGF1. Hence there may be a bigger response in terms of growth, since both insulin and IGF1 can bind and activate the IGF1R (see earlier).

What limits growth? Nature vs nurture

We now understand a lot more in terms of the factors involved in regulating growth. Obviously the genetics of the animal are important, since a genotype that results in a small animal will never be overcome by increased nutrition. On the other hand, optimal nutrition is required for a "large-genotype" animal to grow to its full potential. The latter relates to lifetime nutrition and might even include nutrition of the parent prior to conception. It might therefore be suggested that genetics set the limits or boundaries, while nutrition allows the animal to reach those limits. Thus, both genetics and nutrition of the animals involved (parents and offspring), and their interactions, are what limit growth. Previous studies, including my own, have always investigated either genetic or nutritional components separately; very few have investigated interactions between the two.

The future?

With increased understanding of nutrimental regulation of gene expression, and identification of the mechanisms for these effects, it is becoming increasingly evident that interactions between nutrition and genetics are important. I therefore suggest that a likely development in the future may be to select animals for "responsiveness" to dietary interventions. This might even be possible via simple genotyping once the "response elements" within the transcription regulatory regions of DNA are identified in the genes important for growth. This, however, is far into the future. More immediate will be studies into the interactions of restriction during different stages of life. For example, does a restriction in foetal life alter the response to restriction or over-feeding in later life?

References

Bass, JJ, Sharma, M, Oldham, J and Kambadur, R (2000) Muscle growth and genetic regulation. In: *Ruminant Physiology: Digestion, Metabolism, Growth & Reproduction.* Edited by PB Cronje. CABI Publishing, 227-236.

Brameld, JM (1997) Molecular mechanisms involved in the nutritional and hormonal regulation of growth in pigs. *Proceedings of the Nutrition Society* **56**, 607-619.

Brameld, JM, Atkinson, JL, Budd, TJ, Saunders, JC, Pell, JM, Salter, AM, Gilmour, RS and Buttery, PJ (1996) Expression of insulin-like growth factor-I (IGF-I) and growth hormone receptor (GHR) mRNA in liver, skeletal muscle and adipose tissue of different breeds of pig. *Animal Science* **62**, 555-559

Brameld, JM and Buttery, PJ (2002) Regulation of growth hormone (GH) binding to cultured pig hepatocytes by certain amino acids. *Proceedings of the Nutrition Society* **61**, 48A.

Brameld, JM, Buttery, PJ, Dawson, JM and Harper, JMM (1998) Nutritional and hormonal control of skeletal muscle cell growth and differentiation. *Proceedings of the Nutrition Society* **57**, 207-217.

Brameld, JM, Fahey, AJ, Langley-Evans, SC and Buttery, PJ (2003) Nutritional and hormonal control of muscle growth and fat deposition. *Archives of Animal Breeding* **46** (special issue), 143-156.

Brameld, J.M., Gilmour, R.S. and Buttery, P.J (1999a) Glucose and amino acids interact with hormones to control expression of insulin-like growth factor-I (IGF-I) and growth hormone receptor (GHR) mRNA in cultured pig hepatocytes. *Journal of Nutrition* **129**, 1298-1306.

Brameld, JM, Smail, H, Imram, N, Millard, N and Buttery, PJ (1999) Changes in expression of IGF-I, IGF-II and GH-receptor (GHR) mRNA during differentiation of cultured muscle cells derived from adult and foetal sheep. *South African Journal of Animal Science* **29 (ISRP)**, 307-310.

Brameld, JM, Weller, PA, Pell, JM, Buttery, PJ and Gilmour, RS. (1995) Ontogenic study of insulin-like growth factor-I and growth hormone receptor mRNA expression in porcine liver and skeletal muscle. *Animal Science* **61**, 333-339.

Breier, BH, Oliver, MH and Gallaher, BW (2000) Regulation of growth and metabolism during postnatal development. In: *Ruminant Physiology: Digestion, Metabolism, Growth & Reproduction.* Edited by PB Cronje. CABI Publishing, 187-204.

Butler, AA and LeRoith, D (2001) Minireview: Tissue-specific versus generalized gene targeting of the igf1 and igf1r genes and their roles in insulin-like growth factor physiology. *Endocrinology* **142** (5), 1685-1688.

Buttery, PJ, Brameld, JM and Dawson, JM (2000) Control and manipulation of hyperplasia and hypertrophy in muscle tissue. In: *Ruminant Physiology: Digestion, Metabolism, Growth & Reproduction.* Edited by PB Cronje. CABI Publishing, 237-254.

Choi YJ, Han IK, Woo JH, Lee HJ, Jang K, Myung KH and Kim YS (1997) Compensatory growth in dairy heifers: The effect of a compensatory growth pattern on growth rate and lactation performance. *Journal of Dairy Science* **80** (3), 519-524.

Dawson, JM (1999) Variation in nutrient supply and effects on whole body anabolism. In *Protein metabolism and nutrition: Proceedings of the VIIIth International Symposium on Protein Metabolism and Nutrition.* Edited by G.E. Lobley, A. White and J.C. MacRae. EAAP publication No. 96, 1999, Wageningen Pers, The Netherlands, 101-126.

Fafournoux, P, Bruhat, A and Jousse, C (2000) Amino acid regulation of gene expression. *Biochemical Journal* **351**, 1-12.

Fahey, AJ, Brameld, JM, Parr, T & Buttery, PJ (2003a) Ontogeny of factors thought to control the development of ovine muscle in utero. Proceedings of the British Society for Animal Science, 95.

Fahey, AJ, Brameld, JM, Parr, T & Buttery, PJ (2003b) The effect of maternal undernutrition on muscle fibre type in the newborn lamb. Proceedings of the British Society for Animal Science, 60.

Foufelle, F and Ferre, P (2003) Glucose regulation of gene expression in mammals. In *Molecular Nutrition* edited by J Zempleni and H Daniel, CABI publishing, 91-104.

Gatford, KL, Egan, A, Clarke, IJ and Owens, PC (1998) Sexual dimorphism of the somatotrophic axis. *Journal of Endocrinology* **157**, 373-389.

Grizard, J, Picard, B, Dardevet, D, Balage, M and Rochon, C (1999) Regulation of muscle growth and development. In *Protein metabolism and nutrition: Proceedings of the VIIIth International Symposium on Protein Metabolism and Nutrition.* Edited by G.E. Lobley, A. White and J.C. MacRae. EAAP publication No. 96, 1999, Wageningen Pers, The Netherlands, 177-201.

Jameson, JL (2001) Mechanisms of thyroid hormone action. In *Endocrinology*, 4th edition, volume 2. Edited by LJ DeGroot and JL Jameson. WB Saunders Company, USA, 1327-1344.

Kilberg, MS, Leung-Pineda, V and Chen, C (2003) Amino acid-dependent control of transcription in mammalian cells. In *Molecular Nutrition* edited by J Zempleni and H Daniel, CABI publishing, 105-119.

Langhans, W (1999) Appetite regulation. In *Protein metabolism and nutrition: Proceedings of the VIIIth International Symposium on Protein Metabolism and Nutrition.* Edited by G.E. Lobley, A. White and J.C. MacRae. EAAP publication No. 96, 1999, Wageningen Pers, The Netherlands, 225-251.

Lawrence, TLJ and Fowler, VR (2002). Growth of farm animals, 2nd edition. CABI publishing.

Munck, JL and Naray-Fejes-Toth, A (2001) Glucocorticoid action: physiology. In *Endocrinology*, 4th edition, volume 2. Edited by LJ DeGroot and JL Jameson. WB Saunders Company, USA, 1632-1646.

Nieto, R and Lobley, GE (1999) Integration of protein metabolism within the whole body and between organs. In *Protein metabolism and nutrition: Proceedings of the VIIIth International Symposium on Protein Metabolism and Nutrition.* Edited by G.E. Lobley, A. White and J.C. MacRae. EAAP publication No. 96, 1999, Wageningen Pers, The Netherlands, 69-99.

Rosenfield, RL and Cuttler, L (2001) Somatic growth and maturation. In *Endocrinology*, 4th edition, volume 1. Edited by LJ DeGroot and JL Jameson. WB Saunders Company, USA: 477-502.

Smith, JA., Lewis, AM, Wiener, P and Williams, Jl. (2000) Genetic variation in the bovine myostatin gene in UK beef cattle: Allele frequencies and haplotype analysis in the south devon. *Animal Genetics* **31**, 306-309.

St Germain, DL (2001) Thyroid hormone metabolism. In *Endocrinology*, 4th edition, volume 2. Edited by LJ DeGroot and JL Jameson. WB Saunders Company, USA, 1320-1326.

Stubbs, AK, Wheelhouse, NM, Lomax, MA and Hazlerigg, DG (2002) Nutrient-hormone interaction in the ovine liver: methionine supply selectively modulates growth hormone-induced IGF-1 gene expression. *Journal of Endocrinology* **174**, 335-341.

Wheelhouse, NM, Stubbs, AK, Lomax, MA, MacRae, JC and Hazlerigg, DG (1999) Growth hormone and amino acid supply interact synergistically to control insulin-like growth factor-I production and gene expression in cultured ovine hepatocytes. *Journal of Endocrinology* **163**, 353-361.

Yakar, S, Liu, JL, Fernandez, AM, Wu, Y, Schally, AV, Frystyk, J, Chernausek, SD, Mejia, W and LeRoith, D (2001) Liver-specific igf-1 gene deletion leads to muscle insulin insensitivity. *Diabetes* **50** (5), 1110-1118.

3

BOVINE GUT DEVELOPMENT

J. W. BLUM
Division of Nutrition and Physiology, Institute of Animal Genetics, Nutrition and Housing, VetSuisse Faculty, University of Berne, CH-3012 Berne, Switzerland

Introduction

The development of the bovine gastrointestinal tract (GIT) is a topologically and temporally well organized process. The ontogenetic development of the GIT can be divided in several phases:

1. morphogenesis
2. cytodifferentiation
3. preparation for the postnatal period
4. ingestion of colostrum and milk or milk replacers, and
5. possibly weaning and development of forestomachs.

An important goal of this process is the development of a specialized epithelium that fulfills digestive and (or) absorptive functions.

This paper summarizes some elementary aspects of the development of the GIT during embryonic and foetal periods and during postnatal stages. Studies on GIT development, especially on the epithelium and more recently on gut-associated lymphoid tissues (GALT) in the small intestine (SI) of neonatal calves will be reviewed, particularly with respect to feeding effects (colostrum feeding intensity and components), effects of glucocorticoids, and the importance of the growth hormone (GH) – insulin-like growth factor (IGF) axis. This adds to previous reviews on nutritional, gastrointestinal, metabolic and endocrine aspects in calves (Guilloteau, Le Huërou-Luron, Toullec, Chayvialle and Blum, 1995; Guilloteau, Le Huërou-Luron, Toullec, Chayvialle, Zabielski and Blum, 1997; Blum and Hammon, 1999 and 2000; Blum, 2002; Blum and Baumrucker, 2002; Blum, Hammon, Georgieva and Georgiev, 2002). Development of the forestomachs and of gut regulatory substances are addressed in other chapters of this book (Lesmeister and Heinrichs, 2005; Guilloteau and Zabielski, 2005).

Development of the bovine gastrointestinal tract (general overview)

As outlined by Rüsse and Sinowatz (1991) and Trahair and Sangild (2002), during embryonic stages the GIT develops from the end(t)oderm lining of the yolk sack. During early embryonic stages the ectoderm invaginates cranially and caudally. Ruptures of the cranial and caudal membranes between endo- and ectoderm allow access of the GIT to the amniotic fluid. Endoderm cells are the origin of epithelia, intrinsic glands, accessory glands and splanchnic organs. Mesenchyme cells are the origin of muscles, connective tissue and lymphoid tissues. Cells from the neural crest lead to the formation of the neuronal network. The origin of entero-endocrine cells is not fully clear. Blood supply is important for GIT development: the coeliac artery supplies the foregut, the superior mesenteric artery supplies the midgut and the inferior mesentery artery supplies the hindgut. The foregut devlops into pharyngeal pouches (that develop into liver, gall bladder, pancreas, lung, inner ear, pharynx, tonsils, thymus, thyroid and parathyroids), oesophagus, stomach and upper duodenum; the midgut develops into the lower duodenum, jejunum, ileum, caecum and upper colon; and the hindgut develops into the lower colon and allantois.

All parts of the GIT increase in weight and size from embryonic stages to postnatal full maturity, but at different rates, resulting in marked form differences of the various GIT sites (Church, 1971). Growth rates of the GIT, especially of the SI during late foetal development, are greater than body weight (BW) growth (allometric growth). Postnatal growth rates of the different GIT sites in solid-fed calves are as follows: rumen > reticulum > omasum > caecum > colon and rectum > SI > abomasum > oesophagus. Importantly, the GIT during foetal development is turned around its own axis (Rüsse and Sinowatz, 1991), leading to marked positional changes of GIT sites within the abdominal cavity. Relative masses and volume capacities of the forestomachs, SI and colon are greatly modified by nutrition, especially dependent on whether calves are fed liquid diets (as in veal calves) or solid diets.

In ruminants, development of the forestomachs is of special interest. According to Church (1971) in the 4-week bovine embryo a spindle-shaped stomach primordium is present. In the 6-week embryo the primordia of the four stomach compartments are visible. In the 8-week embryo the epithelial stratification of the oesophagus and forestomachs is initiated. In the 9-week foetus, rumen sacs and pillars are present, primordia of omasal laminae are visible, and the cardia and pyolorus of the abomasum are established. From the 12-week foetus until birth, progressive rumen epithelial cell differentiation and papillae development take place. Forestomach development progressively decreases (from about 80 to about 55%), while abomasal development progressively increases (from about 20 to about 45%) up to birth. Forestomach development is inhibited in calves fed only milk or milk replacers for months, so veal calves can, therefore, be considered as (pseudo-)monogastrics (Guilloteau *et al.*, 1995). However, forestomachs develop markedly, and more than the abomasum (that relatively decreases in size), when calves start to ingest solid foods (Church, 1971; Davis and Drackley, 1998). The development of specific metabolic functions (for instance ketogenesis) of ruminal keratinocytes is initiated and modified by microbial fermentation products, and is influenced by hormones and growth factors (Galfi, Neogrady and Sakata, 1991; Baldwin, 1991, 2000; McLeod and Baldwin, 2000; Lane and Jesse, 1997; Lane, Baldwin and Jesse, 2000 and 2002; Zanming, Seyfert, Löhrke, Schneider, Žitnan, Chudy, Kuhla, Hammon, Blum, Martens, Hagemeister and Voigt, 2004). Size of the SI increases in late foetal periods and during 8-10 weeks postnatally, in length and diameter (primarily due to a rise of the mucosal mass), but decreases relative to the size of other components of the GIT, especially in later postnatal growth periods (Church, 1971; Davis and Drackley, 1998). The SI mass in solid-fed calves is influenced by energy intake. The increase of SI length (8-15 m in the 50 kg calf; 29-43 m in the 140 kg calf) in milk-fed calves is greater than in solid-fed calves. The caecum and colon (especially the ascending loop) also increase steadily in size, and the colon forms loops.

The GIT wall consists of several layers composed of a mixture of diverse cells. The SI consists of epithelium, lamina propria, muscularis mucosae, submucosa, muscularis externa and serosa. The GALT is particularly localized in the lamina propria, in part (in the ileum and colon) concentrated in the form of Peyer's patches (PP), in villi and in the epithelium. The different components of the GIT, and differentiation of different cells, develop successively during the embryonal and foetal development. GIT development and the onset of different functions, such as digestive enzyme and transport activities, start early in mammals with relatively long gestations (such as in cattle), but not until about 80% in mammals with short gestations (such as rats, mice and rabbits) (Buddington, 1994); i.e., there are marked differences between precocial and altricial species with respect to the maturity of the SI. The different cellular compartments in the mucosa and submucosa during pre- and postnatal stages do not grow at the same rate or at the same time. Thus, the increase of mucosal weight immediately after birth represents the greatest proportional rise of any intestinal tissue (Widdowson, 1985). Because any variation in the GIT lumen is encountered first by the epithelium, development and differentiation of the epithelium are of primary importance for GIT function and health. In pre-ruminant calves the epithelium of the SI is particularly important because it is the primary site of digestion, absorption and (or) exclusion from absorption of substances in the SI lumen. The epithelium of the SI during early foetal stages consists first of a simple layer of undifferentiated cells, then becomes stratified, followed by the formation of nascent villi, and finally by the establishment of a columnar epithelium with crypts and villi (Pacha, 2000). The formation of villi depends on invagination of the mesenchyme. Villi are formed in the SI, but villus-like structures, that may have similar functions as villi of the SI (Xu, 1996), can also be found in the colon in calves shortly after birth (Bittrich,

Philipona, Hammon, Romé, Guilloteau and Blum, 2004). Whereas crypt cells during foetal stages are polyclonal, crypt cells in the mature intestine are monoclonal, but crypts can be populated by the clonal progeny of a single, pluripotent or initially committed master crypt stem cell (Pacha, 2000). During differentiation the epithelial cells become polarized by the formation of apical and basolateral membranes that are sepratued by tight junctions. Stem cells at the base of crypts in the SI give rise to differentated epithelial cells: Paneth cells, goblet cells, entero-endocrine cells and columnar absorptive cells (Smith, 1985; Pacha, 2000). In calves the epithelium of the SI during later foetal stages and at birth contains vacuoles that allow intracellular digestion, especially of proteins, but soon after birth and after ingestion of colostrum, vacuoles disappear and digestion occurs extracellularly, i.e. at the brush border membrane and in the GIT lumen (Bainter, 2002). The presence of mesenchymal tissue is necessary for epithelial cell differentiation. Brush border enzyme distribution in calves during development and weaning has been studied by Le Huërou, Guilloteau, Wicker, Mouats, Chayvialle, Bernard, Burton, Toullec and Puigserver (1992). Their production and activities increase postnatally with the well-known exception of lactase, as confirmed for lactase also at the mRNA level (Ontsouka, Korczack, Hammon and Blum, 2004a). Goblet cells produce mucin that protects the epithelium (Montagne and Lallès, 2000). Epithelial cells are lost into the GIT lumen, in the SI, from villi after apoptosis by exfoliation (Shibahara, Sato, Waguri, Iwanaga, Nakahara, Fukumoti and Uchiyama, 1995). There is continuous regeneration of epithelial cells whose half-life is short; in the SI the time for migration of cells from crypts to villus tips is only 2-3 days. To maintain a stable epithelial cell mass, crypt cell mitotic rates and villus exfoliation rates must be coordinated. Crypt depths and villus lengths are negatively correlated, i.e. there is a negative feedback regulation that allows the epithelial cell mass to stabilize within normal limits (Johnson, 1981; Lipkins, 1981). Thus, in the suckling lamb villus sizes decrease, whereas crypt depths increase (Attaix and Meslin, 1991). However, villus heights and crypt depths remain rather stable, despite marked postnatal increases in the diameter of the SI, as shown in sheep (Trahair and Sangild, 2002), indicating that enterocyte development is governed by a well controlled program, i.e. homeostasis can be established by maintaining a balance among proliferating, differentiating and apoptotic cells (Smith, 1985).

During the first days of life there is a marked increase in the diameter and length (cell proliferation rate > apoptotic rate), in protein turnover (synthesis > degradation) and in energy utilization (glutamine > butyrate, long-chain fatty acids, glucose) of the SI. Epithelial cells of the SI differentiate and there is also maturation of other digestion-associated organs (such as of the pancreas and the liver. Various GIT functions (such as the site of glucose absorption, bile acid reabsorption, leakiness, etc.) of the GIT are different in neonates and adults, as shown in humans, but this has barely or not been studied in neonatal calves (Potter, 1990).

The GIT development is regulated by gene activation and expression ("developmental clock"), and modulated by nutritional factors, by microbes and their products, by hormones, growth factors, cytokines and neurotransmitters, and depends on vascularisation (Johnson, 1981; Lipkin, 1981; Veereman-Wauters, 1996; Koletzko, Aggett, Bindels, Bung, Gil, Lentze, Roberfroid and Strobel, 1998; Pacha, 2000; Sanderson and Naik, 2000; Thiesen, Wild, Keelan, Clandinin, McBurney and Van Aerde, 2000; Trahair and Sangild, 2002). Food in the GIT has direct and indirect effects (Buddington, 1994). Direct effects of ingested food depend on the supply of nutrients, energy, vitamins, minerals, trace elements, polyamines, nucleotides, and bioactive substances (hormones, growth factors, cytokines and other ingested peptides, such as lactoferrin), but nutritional effects also include the desquamation of epithelial cells. In calves, as in pigs, there is a very marked GIT increase within hours after ingestion of colostrum, due to pinocytosis and storage of milk proteins (especially immunoglobulin G, IgG) within the SI wall, due to enhanced protein synthesis, and due to an increased cell number as a consequence of enhanced cell proliferation (Buddington, 1994; Xu, 1996). Nutritional effects are in part indirectly mediated by hormones that are produced outside the GIT, and in part mediated by hormones, growth factors and cytokines that are synthesized within the GIT. Hormones, growth factors and cytokines exert their effects locally in an autocrine, paracrine and (or), after release into the blood circulation, in an endocrine manner, and there are juxtacrine effects by cell-to-cell communication. Some of the bioactive factors may exert their effect also after release into the GIT lumen as so-called luminokines (Kuwahara and Fujimiya, 2002). Of GIT regulatory peptides, the physiology of gastrin, cholecystokinin, secretin, vasointestinal polypeptide, motilin, pancreatic polypeptide, somatostatin and glucose-dependent insulinotropic polypeptide have been intensively studied in calves (Guilloteau *et al.*, 1995, 1997; Guilloteau, Biernat, Wolinski and Zabielski, 2002). Of GIT regulatory peptides, gastrin, glucagon-like peptide-2 are recognized as being particularly known for their trophic GIT effects (Carver and Barness, 1996; Guilloteau *et al.*, 2002; Burrin, Stoll and Guan, 2003). Hormones and some growth factors (such as IGFs) that are produced outside the GIT and that exert effects on the GIT are circulating in blood, are swallowed with amniotic fluid or with saliva or are ingested with colostrum or milk. Amniotic and placental supply both contribute to preterm GIT growth (Buddington, 1994; Trahair and Harding, 1995). Bioactive substances are in part produced during digestion (such as caseinomacropeptide that derives from casein) within the GIT lumen and have, too, been shown to affect SI development (Froetschel, 1996). There are important interactions among the endo-, para- and autocrine sytems of the GIT, and the autonomous nervous system (Guilloteau *et al.*, 1995). Food ingestion also modulates the activity of nerves, especially of the autonomous nervous system, which releases acetycholine, noradrenaline, serotonine, nitric oxide and other neurotransmitters (Ekblad and Sundler, 2002). The effects of ingested hormones and growth factors on the GIT depend on the localization and availability of receptors from the luminal and (or) basolateral sites, on regional versus overall control, and on the coordination of endocrine and neuronal regulation, as for example shown for cholecystokinin in calves (Guilloteau *et al.*, 2002).

In ruminants and pigs, cortisol is considered to play a key role in GIT development and cell differentiation, particularly in the promotion of digestive enzymes in the pre- and perinatal period and in rats around weaning (Johnson, 1981; Sangild, Xu and Trahair, 2002; Schmidt, Sangild, Blum, Andersen and Greve, 2004) and in neonatal humans (Thiesen *et al.*, 2000). Enhanced pre-, peri- and early postnatal cortisol or corticosterone release in sheep and pigs, and at weaning in rats, accelerates GIT maturation. In the SI the proliferation of crypt cells, the production of brush border enzymes and the development of carrier systems for glucose and amino acid transport in neonates are advanced in time by cortisol or corticosterone. However, there are catabolic effects at high doses (Burrin, Wester, Davis, Fiorotto and Chang, 1999).

Growth hormone is well known to enhance GIT growth (Johnson, 1981). We have shown that the GH-IGF-1 axis is functional, but not fully mature in neonatal calves (Hammon and Blum, 1997a). However, injections of bovine GH for 7 days in neonatal calves even seemed to reduce SI mucosal growth (Bühler *et al.*, 1998). IGF-1 is important for growth regulation in the pre- and postnatal period, wheras IGF-2 is primarily important in the prenatal period. As shown in sheep, IGF-1 stimulates GIT growth during the foetal period (Trahair, DeBarro, Robinson and Owens, 1997). Swallowed IGF-1 is obviously important for GIT grwoth in the foetal period (Kimble, Breier, Gluckman and Harding, 1999). Effects of IGF-1 (and IGF–2) are mediated by specific receptors that are present on epithelial cells, fibroblasts, endothelia and smooth muscles in the GIT (Howarth, 2003). Receptor numbers are different among different GIT sites and change in calves during the first week of life dependent on age and nutrition (Baumrucker, Hadsell and Blum, 1994; Georgiev, Georgieva, Pfaffl, Hammon, and Blum, 2003; Georgieva, Georgiev, Ontsouka, Hammon, Pfaffl and Blum, 2003; Hammon and Blum, 2002). Parenteral IGF-1 administration enhances mucosal (epithelial), submucosal und muscularis thickness, longitudinal and crossectional GIT growth, sodium absorption and sodium-dependent glucose absorption (Howarth, 2003; Burrin, 1997; Burrin, Davis, Ebner, Schoknecht, Fiorotto, Reeds and McAvoy, 1995; Burrin, Davis, Fiorotto and Reeds, 1997). The regulation of the effects on the GIT of ingested IGFs, of IGFs produced in the GIT and of IGFs produced outside the GIT and circulating in blood, is complex and further complicated by interactions with effects of other endocine systems (GH, insulin and cortisol), by modulation of IGF effects with IGFBPs, and because IGFBP3 may exert effects via lactoferrin (Lf), as shown in the bovine mammary gland (Baumrucker and Erondu, 2000; Blum and Baumrucker, 2002).

Bovine colostrum, besides cells (lactocytes, erythrocytes, leukocytes) and nutrients (including essential fatty and amino acids), minerals, trace elements and (pre-)vitamins, contains non-nutrient components (immunoglobulins and biologically active substances) (Blum and Hammon, 2000). The greatest mass of most colostral components (except lactose) is available to newborns, including calves, in first colostrum. The following non-nutritional substances can be found in bovine colostrum: hormones, releasing factors, growth factors, cytokines, prostaglandins, enzymes, Lf, transferrin, breakdown products of milk proteins, nucleotides, polyamines, and gangliosides. Oligosaccharides and glucoconjugatesare present in bovine milk, but concentrations are relatively low even in colostrum (Gospal and Gill, 2000). In bovine colostrum concentrations of IGFs are high, but epidermal growth factor can barely be detected, whereas the inverse situation is true for human and rat colostrum (Campana and Baumrucker, 1995). Concentrations of bioactive petides, such as IGF-1, insulin, prolactin and GH decline rapidly after the onset of lactation in cows (Campana and Baumrucker, 1995; Blum and Hammon, 2000). Bioactive substances are distributed among the casein fraction (caseinopeptides), the whey fraction (most peptide hormones, hormone releasing factors, growth factors, cytokines and other bioactive peptides, including protein breakdown products, nucleotides, polyamines, oligosaccharides) and the fat fraction (steroid hormones, triiodothyronine, gangliosides). As a consequence there is variable availability during milking and suckling. Thus, during milking the greatest amounts of IGF-1, IGF-2, insulin, prolactin and other bioactive peptides are in the cisternal fraction, followed by transiently low amounts in ensuing fractions and then by a gradual rise toward the end of milking (Ontsouka, Bruckmaier and Blum, 2003). There is now ample evidence that non-nutritional factors affect GIT development. Widdowson in 1976 first described that colostrum intake provokes drastic morphological and functional changes of the GIT in neonates (Widdowson, 1985) and Klagsbrun (1978) first demonstrated that milk contains mitogens whose concentration is highest in colostrum and that they are of different nature in bovine and human colostrum. Various researchers then found that colostrum contains more non-nutrient (bioactive) factors than later milk, and that these factors have trophic effects on the GIT, modulate the immune system and in part exert systemic effects (Grosvenor, Picciano and Baumrucker, 1992; Koldovski, 1994; Burrin, 1997; Burrin *et al.*, 1995 and 1997). Studies in various species have shown that non-nutritional factors in neonates modulate the GIT microbial population, the development (cell proliferation, migration, differentiation and apoptosis), protein synthesis and degradation, digestion, absorption and motility of the GIT, and immune functions in the GIT. In the SI there are important interactions between ingested and processed antigens in the GIT lumen, with the epithelium and with the GALT. Luminal molecules or antigens can exert effects on lymphocytes and lymphocyte production after they have passed the epithelial layer transcellularly (as is the case for IgG in neonatal calves), by paracellular transport, after recognition and selection by M-cells, and through epithelial signalling. It is well known that various of the ingested proteins, such as IgG, glutamyltransferase, transferrin, Lf and others are absorbed early postnatally by calves, and some of them may even reach the brain and thus can exert GIT effects following absorption (Michanek, Ventorp and Westrom, 1989; Hadorn and Blum, 1997; Talukder, Takeuchi, and Harada, 2002). However, for some peptides (such as for IGF-1, insulin) this could not been shown to occur, or only to a very small extent, in calves (Vacher, Bestetti and Blum, 1995), suggesting that these substances may exert their effects at the luminal site.

Studies on the small intestine in neonatal calves: epithelial effects and effects on gut-associated lymphoid tissues

In fullterm and in part in preterm neonatal calves we have conducted a series of studies on effects of feeding colostrum (that contained many bioactive substances besides nutrients, minerals, vitamins, enzymes, nucleotides, enzymes and cells), or only formulas, or milk replacers (that contained no or only negligible amounts of bioactive substances). In some studies IGF-1, long-R3-IGF-1, vitamin A or Lf were added to formulas or milk replacers to study effects on the SI epithelium (and partly the colon). Additional studies have been performed on effects of dexamethasone (DEXA) on the SI and colon epithelium and GALT to simulate effects of cortisol.

Intestinal diseases in neonatal calves may be due to morphological and functional immaturity. We have tested the hypothesis that this is particularly the case in preterm calves (Bittrich *et al*., 2004). We have therefore studied histomorphology, crypt cell proliferation rates (based on incorporation of 5-bromo-2'-deoxyuridine into DNA), presence of apoptotic cells (based on terminal deoxynucleotidyl transferase-mediated X-dUTP nick end labeling), and brush border enzyme activities in preterm calves (277 days of gestation), euthanased on Day 1 (P0) or Day 8 (P8) and in fullterm calves (290 days of gestation), euthanased on Day 1 (F0) or Day 8 (F8). Vacuolated epithelial cells were present in the ileum of P0 and F0, but not in P8 and F8. Therefore, the absence of vacuolated cells on day 8 of life suggests that in this respect the differentiation stage of epithelial cells in both groups was similar. Villus-like structures were present in the colon only in preterm calves immediately after birth, but not in fullterm calves and they were absent in preterm calves on day 8 of life. Their presence in preterm calves at birth expressed a premature state of colon morphology and function. However, their absence in preterm calves on day 8 of life indicated that colonic absorption through villus structures had probably ceased, and that colon morphology and function had matured during the first week of life. During the first 8 days of life villus sizes, crypt depths, and proliferation rates of crypt cells in the SI of preterm calves did not change significantly. In contrast, in fullterm calves during the first 8 days of life, villus sizes in the jejunum decreased, crypt depths increased in the small intestine and colon, and crypt cell proliferation increased in the duodenum and jejunum. Submucosal thickness in the jejunum was highest in P0, but in the ileum it increased with gestational age and feeding. Gestational age × feeding interactions indicated increased activities of aminopeptidase N and there were reduced lactase activities only in F8, and reduced dipeptidylpeptidase IV activities only in P8. This study showed that the epithelial system of the SI of preterm calves was unresponsive to feeding, in contrast to full term calves. The study showed that the SI epithelium in preterm calves is immature and that brush border protease activities differed in part from those in fullterm calves. Differences between pre- and fullterm calves in digestive capacities could therefore be expected. Constitutional factors were probably responsible for these different responses. Morphological and digestive dysfunctioning of the SI might be important causes for greater GIT problems, such as diarrhoeas, and might contribute to higher morbidity and mortality rates in preterm calves than in fullterm calves.

In preliminary studies, we have demonstrated enhanced xylose absorption in calves fed colostrum six times, rather than only once or only milk replacer, indicating enhanced absorptive capacity of the SI in calves fed colostrum for a prolonged time (Hammon and Blum, 1997a). Other studies have shown that increased intake of colostrum or feeding of colostrum (that contained bioactive substances) as such rather than feeding a milk-derived formula (that did not contain bioactive substances) enhanced absorptive capacity, based on absorption studies with xylose (Rauprich, Hammon, and Blum, 2000a,b). These data are in line with studies in other species, showing that early lactation products are important for early GIT development (Grosvenor *et al.*, 1992; Kelly, 1994; Koldovsky, 1994; Xu, 1996). This prompted us to test whether morphological changes in the small intestine of calves fed different amounts of colostrum or only milk replacer after birth were possibly responsible for previously found differences in absorptive capacities (Bühler, Hammon, Rossi, Blum, 1998). Effects of feeding different amounts of colostrum or only milk replacer on SI mucosal morphology in newborn calves were studied by histomorphometry. Neonatal calves fed colostrum six times exhibited significantly greater villus circumferences, areas and heights in total SI, especially in the duodenum, than calves fed only milk replacer. Furthermore, villus circumferences and areas in total small intestine were significantly greater in calves fed colostrum once than in calves fed no colostrum. The study showed that prolonged feeding of colostrum was beneficial for small gut development. Based on that, prolonged colostrum feeding is recommended because of enhanced villus growth. Absorptive capacities of ingested components were expectedly enhanced, as shown by increased absorption of xylose.

Effects of amounts of colostrum fed on intestinal morphology and proliferation and digestive enzyme activities were studied in fullterm neonatal calves (Blättler, Hammon, Morel, Philipona, Rauprich, Romé, Le-Huërou-Luron, Guilloteau and Blum, 2001). Group C_{max} was fed first-milking colostrum undiluted from day 1 to day 3, and diluted with 25, 50, 75 and 75 parts of a milk replacer from day 4 to day 7. Group C_{1-3} was fed colostrum of milkings 1 to 6 up to day 3 and then a milk replacer up to day 7. Group F_{1-3} was fed a milk-based formula (containing only traces of growth factors and hormones) up to day 3 and then a milk replacer up to day 7. Calves were euthanased on day 8. Differences in feeding influenced villus sizes and villus height : crypt depth ratios in the duodenum ($C_{max} > C_{1-3}$), villus areas and villus heights : crypt depth ratios in the jejunum ($C_{1-3} > F_{1-3}$), and crypt depths in the colon ($F_{1-3} > C_{1-3}$). Furthermore, different feeding regimes affected proliferation rates of epithelial cells in the duodenum ($C_{1-3} > C_{max}$; $C_{1-3} > F_{1-3}$) and in the jejunum ($F_{1-3} > C_{1-3}$). The data showed that colostrum intake differentially affected intestinal epithelial surface and proliferation, and enzyme activities. Maximal compared with "normal" colostrum intake (C_{max}) did not affect the proliferation rate of crypt cells, but villus size was increased. This suggested that increased villus size was the consequence of reduced apoptosis rate, increased survival rate or enhanced migration rate of epithelial cells. Lack of colostrum appeared to decrease epithelial growth.

In neonatal calves we have investigated effects on intestinal epithelial cell morphology, proliferation and absorption of feeding a bovine colostrum whey extract that contained bioactive substances, such as IGF-1 and insulin (Roffler, Fäh, Sauter, Hammon, Brem and Blum, 2003). Calves were fed a milk-based formula containing amounts of nutrients comparable to those in colostrum for the first 3 days and a milk replacer from day 4 onwards. Formula and milk replacer contained only trace amounts of non-nutritional factors. The extract of first-milked bovine colostrum that provided physiological amounts of IGF-1 (0.50, 0.15 and 0.09 mg IGF-1/L formula on day 1, 2 and 3, respectively and 0.09 mg IGF-1/L milk replacer on day 4) was added to formula or milk replacer. Plasma xylose concentration, after addition to the feed as an absorption marker in the control group, was transiently higher than in calves fed the colostrum extract. On day 5 villus circumferences and heights in the SI were significantly higher, and epithelial cell proliferation rate tended to be higher, in calves fed the colostrum extract than in controls. Feeding the bovine colostrum extract enhanced intestinal villus size. Enhancement of mucosal growth by colostrum intake was probably not caused by one single growth promoting peptide, but by a pool of several growth promoting substances in colostrum.

We have studied whether DEXA treatment from day 1 to day 5 of life affects villus size, crpyt cell proliferation, digestive enzyme activities, and xylose absorption in the SI of calves, and whether different feeding (colostrum or a formula) modifies the effects of DEXA on SI development (Sauter, Roffler, Philipona, Morel, Guilloteau, Blum and Hammon, 2004). Calves of groups $GrFD^-$ and $GrFD^+$ were fed a milk-based formula, whereas calves of groups

GrCD$^-$ and GrCD$^+$ received colostrum. DEXA (30 µg/kg BW per day) was injected at feeding times to calves of GrFD$^+$ and GrCD$^+$. On day 3, xylose absorption was measured. Calves were euthanased on day 5 of life. Colostrum feeding increased villus sizes in the jejunum and ileum, enhanced xylose absorption capacity and increased peptidase activities in the ileum. DEXA administration diminished sizes and cell proliferation rates in PP in the ileum, but increased cell proliferation of crypt cells in the ileum of formula-fed calves. DEXA reduced amino peptidase N activities in the jejunum of formula-fed calves, but increased peptidase activities, mainly in the ileum, of colostrum-fed calves. Thus, there were DEXA x feeding interactions because cell proliferation was stimulated by DEXA treatment mainly in formula-fed calves and activities of peptidases were highest in calves treated with DEXA and fed colostrum. DEXA effects therefore depended on intestinal site and on different feeding. This resulted in stimulation of crypt cell proliferation in the less mature ileum of formula-fed calves and in stimulation of peptidase activities in the more mature ileum of colostrum-fed calves. These observations suggest that effects of DEXA were modified by the developmental stage of the SI and promoted SI development depending on the maturity stage.

Neonatal calves need passive immunoprotection by the ingestion of immunoglobulins in colostrum, thereby reducing the need for active immunity and thus delaying the postnatal activation of lymphoid organs. This might be particularly so in preterm calves. Because no quantitative studies have been published on cell proliferation and apoptosis rates, nor on B- and T-lymphocyte numbers in the GALT in pre- and fullterm calves and at the end of the neonatal period, we have conducted corresponding studies. In full-term calves there were marked ontogenetic changes during the first 5 days of life of lymphocytes (increased proliferation rate, reduced apoptotic rates and number of T-lymphocytes) in PP and of T-lymphocytes within epithelia, but there were no significant changes of B-lymphocyte numbers in PP of the ileum, possibly due to effects of colostrum feeding (David, Norrman, Hammon, Davis and Blum, 2003). Feeding calves colostrum (containing bioactive substances), compared with a milk-derived formula (containing no bioactive substances), for 5 days reduced the number of proliferating cells in follicles and of B-lymphocytes in PP, but did not change apoptotic rates and the number of T-lymphocytes in the ileum, suggesting that colostrum feeding spared active immune responses or may thus also inhibit immune system overreactions (Norrman, David, Sauter, Hammon and Blum, 2003). In addition, effects of DEXA administration were studied to simulate an enhanced glucocorticoid status that can be seen in stressed neonatal calves and might negatively affect immune status, despite the fact that glucorticoids have basically an important role in the development of the GIT and other organs (Norrman *et al.*, 2003). There were significant effects of DEXA treatment (decrease in cell proliferation rates in follicles of PP, increase of apoptotic rate in follicles of PP, decrease of B-lymphocyte numbers in follicles of PP, increase of B-lymphocyte numbers in domes of PP, increase of T-lymphocyte numbers in follicles of PP and a decrease of intraepithelial T-lymphocyte numbers). The DEXA treated calves had a markedly decreased size of PP. In another study with the same calves, the submucosa thickness of the ileum was therefore greatly reduced (Sauter *et al.*, 2004). Due to a high glucocorticoid status, mimicked by DEXA treatment, neonatal calves under adverse conditions are expected to become more susceptible to infections. Thus, while glucocorticoids might have favourable effects on GIT epithelia, they might have adverse effects on lymphoid cells in this and other tissues.

We have shown that mRNAs for the GH receptor (GHR), the IGF type-1 and -2 receptors (IGF-1R, IGF-2R) and for the insulin receptor (InsR), of IGF-1 and IGF-2, and of IGFBP-1, -2 and -3 are present in all parts of the GIT of neonatal calves (Cordano, Hammon and Blum, 1998; Cordano, Hammon, Zurbriggen and Blum, 2000; Pfaffl, Georgieva, Georgiev, Hageleit, and Blum, 2002; Georgiev *et al.*, 2003; Georgieva *et al.*, 2003; Ontsouka *et al.*, 2003, 2004a; Ontsouka, Hammon and Blum, 2004b; Ontsouka, Sauter, Blum and Hammon, 2004d). There were marked differences in mRNA levels between different GIT sites (and between the GIT and liver), indicating variable mRNA synthesis and (or) turnover rates and variable importance of the traits for GIT growth and maturation. Higher mRNA levels of the GH-IGF system and for insulin (except IGF-2R) on day 5 than on day 1, mainly in the SI, suggest that the SI may be a main target for colostral nutritional and/or growth promoting factors. The mRNA levels and maximal binding (B_{max}) of IGF-1R, IGF-2R and InsR were negatively associated, indicating different receptor turnover (Georgiev *et al.*, 2003). The B_{max} of the IGF-1R, IGF-2R and InsR as a measure of receptor numbers was additionally determined in the SI and colon of neonatal calves (Baumrucker *et al.*, 1994, Hammon and Blum, 2002, Georgiev *et al.*, 2003). These studies showed that B_{max} of IGF-2R decreased, B_{max} of InsR increased, whereas B_{max} of IGF-1R did not change during the first week of life; B_{max} of IGF-1R, IGF-2R and InsR were modified by differences in feeding, and B_{max} of IGF-2R and InsR at birth were lower in pre-term calves (born on d 277 of gestation) than in full-term calves (born on d 290 of gestation). We have additionally measured by real-time PCR the mRNA levels of IGF-1 and -2, IGFBP -2 and -3, and GHR, IGF-1R, IGF-2R and InsR in the oesophagus, rumen, fundus, pylorus, duodenum, jejunum, ileum and colon of calves on days 1 and 5 of life (Ontsouka *et al.*, 2004b). Levels of mRNA of measured traits were significantly different among different GIT sites. Furthermore, mRNA levels of IGF-1 and -2, IGFBPs and of IGF-1R, IGF-2R and InsR differed significantly on days 1 and 5. Differences in mRNA abundance of IGF-1 and -2, IGFBP-2 and -3 and of GHR, IGF-1R, IGF-2R and InsR among GIT sites on days 1 and 5 of life suggest site-specific functional importance and demonstrate that changes are the consequence of ontogenetic development and (or) are due to feeding. We have, furthermore, studied by real-time PCR, whether there are differences in the abundance of mRNA coding for IGF-1, IGF-2, IGFBP-2, IGFBP–3, IGF-1R, IGF-2R, GHR and InsR in fractions of jejunum and ileum consisting mainly of villus tips, crypts, and PP of 5-day-old calves fed colostrum (Ontsouka *et al.*, 2004c). In the jejunum, mRNA levels of IGF-1, IGF-2, IGF-1R, IGF-2R, InsR, GHR, IGFBP-2 and IGFBP-3 were significantly higher in villus than in crypt fractions. In the ileum mRNA levels of IGF-1 were significantly higher in PP fractions than in villus and crypt fractions, and mRNA levels of GHR tended to be higher in crypts than in the PP fraction. Thus, members of the somatotropic axis were variably expressed in different jejunal and ileal layers of neonatal calves. The high IGF-1 mRNA abundance in the PP fraction suggests that IGF-1 is important for lymphocyte functions. Feeding Long-R^3-IGF-I had no significant influence on SI morphology (Bühler *et al.*, 1998). Feeding a formula (containing only traces of non-nutritional components) plus oral IGF-1 (milk-derived from transgenic rabbits; amounts corresponding to those present in colostrum, i.e. 0.4 mg IGF-1/L formula) for 7 days had no effects on proliferation of small intestinal crypt cells nor on villus growth, i.e. ingested IGF-1 in physiological amounts alone is not responsible for intestinal growth and development in neonatal calves (Roffler *et al.*, 2003). In previous studies, feeding recombinant human IGF-1 (that had the same amino acid sequence as bovine IGF-1) for 7 days enhanced the incorporation of (^{3}H)-thymidine into the DNA of isolated enterocytes ex vivo (Baumrucker, Hadsell and Blum, 1994), but it has not been checked whether this had led to morphological changes. We have also measured mRNA levels of members of the somatotropic axis to test the hypothesis that colostrum intake and dexamethasone treatment affect respective gene expression in the GIT (Ontsouka *et al.*, 2004d). Calves were fed either colostrum or an isoenergetic milk-based formula, and in each feeding group, half of the calves were treated with dexamethasone (DEXA; 30 mg/kg body weight per day). Individual parameters of the somatotropic axis differed among different GIT sections and formula feeding significantly increased mRNA levels of individual parameters at various GIT sites. DEXA effects on the somatotropic axis in the GIT partly depended on different feeding. Thus, in colostrum-fed calves, DEXA decreased mRNA levels of IGF-1 (oesophagus, fundus, duodenum, and ileum), IGF-2 (fundus), IGFBP-2 (fundus), IGFBP-3 (fundus), IGF-1R (oesophagus, ileum, and colon), IGF-2R (fundus), GHR (fundus), and InsR (oesophagus, fundus), but in formula-fed calves DEXA increased mRNA levels of IGF-1 (oesophagus, rumen, jejunum, and colon). Furthermore, DEXA increased mRNA levels of IGF-2 (pylorus), IGFBP-3 (duodenum), IGF-2R (pylorus), and GHR (ileum), but decreased mRNA levels of IGFBP-2 (ileum), and IGF-1R (fundus). Whereas formula feeding had stimulating effects, effects of DEXA treatment on mRNA levels of parameters of the somatotropic axis varied among GIT sites and depended in part on feeding.

It is well known that retinoic acid, a metabolite of vitamin A (retinol), is an important morphogen and has effects on epithelial tissues and exerts pre-and postnatal effects on GIT growth (DeLuca, 1991). Recent studies in the bovine mammary gland have demonstrated essential interactions between IGFBP-3 (and thus IGFs), retinoic acid and Lf (Baumrucker and Erondu 2000; Baumrucker and Blum, 2002). Based on that, we have studied effects of feeding a milk-based formula that contained just traces of vitamin A or Lf, or we added to this formula vitamin A alone, Lf alone, or vitamin A plus Lf (Muri, Schottstedt, Hammon and Blum, unpublished observations). Calves that were not fed colostrum and fed a formula without vitamin A addition had a very poor vitamin A status that rapidly deteriorated during the first days of life, as expected (Blum, Hadorn, Sallmann, and Schuep, 1997), and plasma Lf concentrations also remained low, in contrast to colostrum-fed controls. In calves that were fed the formula plus vitamin A, plasma Lf concentrations increased slightly and those of retinol increased markedly. Calves were euthanased on day 5 of life to study effects on the GIT. Vitamin A supplementation alone significantly affected, and vitamin A plus Lf (but not Lf alone) tended to affect, villus lengths and circumferences, i.e. villus sizes were greater in calves fed the formula with vitamin A, but only in the ileum. Furthermore, vitamin A negatively affected the proliferation rates of crypt cells in the jejunum, ileum and colon and vitamin A plus Lf (but not Lf alone) negatively affected proliferation rates in ileum and colon, i.e. proliferation rates were reduced in calves fed the vitamin A supplemented formula. These data suggest that vitamin A in the SI enhanced villus sizes by stimulation of the migration rate or by reduction of the apoptosis rate.

Conclusion

In conclusion, there is much to be learned about effects of feed components and other factors that regulate GIT development in cattle. The high complexity of the GIT makes it necessary to enhance studies in single cell systems, although this will only mirror the real situation to a small extent. There are few answers, and there remain many questions.

Acknowledgements

Studies have been supported by Swiss National Science Foundation (grants nbr. 32-30188.90, 32-36140.92, 32-051012.97, 32-56823.99, 32-59311.99, 32-67205.01), CH-Berne; by Schaumann Foundation, D-Hamburg; by F. Hoffmann-La Roche, CH-Basle; by Novartis (formerly Ciba Geigy AG), CH-Basle, Switzerland; by Gräub AG, CH-Berne, and by the Swiss Federal Veterinary Administration, CH-Liebefeld-Berne.

References

Attaix, D. and Meslin, J.C. (1991) Changes in small intestinal mucosa morphology and cell renewal in suckling, prolonged suckling, and weaned calves. *American Journal of Physiology*, **261**, R811-R818.

Bainter, K. (2002) Vacuolation in the young. *In* Biology of the Intestine in Growing Animal, pp 55-110. Edited by R. Zabielski, P.C. Gregory and B. Weström. Elsevier Science, Amsterdam.

Baldwin, R.L. (1999) The proliferative actions of insulin, insulin-like growth factor-I, epidermal growth factor, butyrate and propionate on ruminal epithelial cells in vitro. *Small Ruminant Research,* **32**, 261-268.

Baldwin, R.L. (2000) Sheep gastrointestinal development in response to different dietary treatments. *Small Ruminant Research,* **35**, 39-47.

Baumrucker, C.R., Hadsell, D.L. and Blum, J.W. (1994) Effects of dietary rhIGF-I in neonatal calves: Intestinal growth and IGF receptors. *Journal of Animal Science,* **72**, 428-433.

Baumrucker, C.R. and Erondu, N.E. (2000) Insulin-like growth factor (IGF) system in the bovine mammary gland and milk. *Journal of Mammary Gland Biology and Neoplasia*, **5**, 53-64.

Buddington, R.K. (1994) Nutrition and ontogenic development of the intestine. *Canadian Journal of Physiology and Pharmacology,* **72**, 251-259.

Bittrich, S., Philipona, C., Hammon, H.M., Romé, V., Guilloteau, P. and Blum, J.W. (2004) Preterm as compared with fullterm neonatal calves are characterized by morphological and functional immaturity of the small intestine. *Journal of Dairy Science*, **87**, 1786-1795.

Blättler, U., Hammon, H.M., Morel, C., Philipona, C., Rauprich, A., Romé, V., Le-Huërou-Luron, I., Guilloteau, P. and Blum, J.W. (2001) Feeding colostrum, its composition and feeding duration variably modify proliferation and morphology of the intestine and digestive enzyme activities of neonatal calves. *Journal of Nutrition*, **131**, 1256-1263.

Blum, J.W., Hadorn, U., Sallmann, H.-P. and Schuep, W. (1997) Delaying colostrum intake by one day impairs the plasma lipid, essential fatty acid, carotene, retinol and a-tocopherol status in neonatal calves. *Journal of Nutrition*, **127**, 2024-2029.

Blum, J.W. and Hammon, H.M. (1999) Endocrine and metabolic aspects in milk-fed calves. *Domestic Animal Endocrinology*, **17**, 219-230.

Blum, J.W. and Hammon, H.M (2000) Colostrum effects on the gastrointestinal tract, and on nutritional, endocrine and metabolic parameters in neonatal calves. *Livestock Production Science*, **66**, 151-159.

Blum, J. W. (2002) Nutrition of the milk-fed calf: metabolism, hematology and endocrinology. In *Recent Developments and Perspectives in Bovine Medicine,* pp 116-124. Edited by M. Kaske, H. Scholz and M. Hölderschinken. Proceedings of the 22nd World Buiatrics Congress, Hanover, August 18-23, 2002,

Blum, J.W. and Baumrucker, C.R (2002) Colostral insulin-like growth factors and related substances: mammary gland, and neonatal (intestinal and systemic) targets. *Domestic Animal Endocrinology,* **23,** 101-110.

Blum, J.W., Hammon, H.M., Georgieva, T.M. and Georgiev, I.P. (2002) Role of colostrum and milk components in the development of the intestine structure and function in calves. In *Biology of the Intestine in Growing Animals*, pp 193-202. Edited by R. Zabielski, P.C. Gregory and B. Weström. Elsevier Science, Amsterdam.

Bühler, C., Hammon, H.M., Rossi, G. and Blum, J.W. (1998) Small intestinal morphology in 8-d old calves fed colostrum for different duration or only milk replacer and treated with Long-R3-IGF-I and growth hormone. *Journal of Animal Science,* **76**, 758-765.

Burrin, D.G., Davis, T.A., Ebner, S., Schoknecht, P.A., Fiorotto, M.L., Reeds, P.J. and McAvoy S. (1995) Nutrient-independent and nutrient-dependent factors stimulate protein synthesis in colostrum-fed newborn. *Pediatric Research,* **37,** 593-599.

Burrin D.G. (1997). Is milk-borne insulin-like growth factor-I essential for neonatal development ? *Journal of Nutrition,* **127,** 975S-979S.

Burrin D.G., Davis T.A., Fiorotto M.L. and Reeds P.J. (1997) Role of milk-borne vs. endogenous insulin-like growth factor I in neonatal growth. *Journal of Animal Science,* **75**, 2739-2743.

Burrin, D.G., Wester, T.J., Davis, T.A., Fiorotto, M.L. and Chang, X. (1999) Dexamethasone inhibits small intestinal growth via increased protein catabolism in neonatal pigs. *American Journal of Physiology*, **276**, E269-E277.

Burrin, G.D., Stoll, B. and Guan, X (2003) Glucagon-like peptide 2 function in domestic animals. *Domestic Animal Endocrinology*, **24**, 103-122.

Campana, W.M. and Baumrucker, C.R. (1995) Hormones and growth factors in bovine milk. In *Handbook of Milk Composition,* pp 476-494. Edited by R.G. Jensen. Academic Press, New York.

Carver, J.D., Barness, L.A. (1996) Trophic factors for the gastrointestinal tract. *Neonatal Gastroenterology*, **23**, 265-285.

Cordano, P., Hammon, H.M. and Blum, J.W. (1998) Tissue distribution of insulin-like growth factor-I mRNA in 8-day old calves. In *Symposium on Growth in Ruminants: Basic Aspects, Theory and Practice for the Future,* p 288. Edited by J.W. Blum, T. Elsasser and P. Guilloteau. University of Berne, Berne.

Cordano, P., Hammon, H.M., Zurbriggen, A. and Blum, J.W. (2000) mRNA of insulin-like growth factor (IGF) quantification and presence of IGF binding proteins, and receptors for growth hormone, IGF-I and insulin, determined by reverse-transcribed polymerase chain reaction, in the liver of growing and mature male cattle. *Domestic Animal Endocrinology*, **19**, 191-208.

Church, D.C. (1971) The Ruminant Animal. Digestive Physiology and Nutrition. Prentice Hall, Englewood Cliffs, NJ.

David, C. W., Norrman, J., Hammon, H.M., Davis, W.C. and Blum, J.W. (2003) Cell proliferation, apoptosis, and B- and T-lymphocytes in Peyer patches of the ileum, in thymus and in lymph nodes of pre-term calves and in full-term calves at birth and on day 5 of Life. *Journal of Dairy Science*, **86**, 3321-3329.

Davis, C.L. and Drackley, J.K. (1998) The Development, Nutrition, and Management of the Young Calf. Iowa State University Press, Ames, IA.

DeLuca, L.M. (1991) Retinoids and their receptors in differentiation, embryogenesis and neoplasia. *FASEB Journal,* **5**, 2924-2933.

Ekblad, E. and Sundler, F (2002) Innervation of the small intestine. *In* Biology of the Intestine in Growing Animal, *pp 235-270.* Edited by R. Zabielski, P.C. Gregory and B. Weström. Elsevier Science, Amsterdam.

Froetschel, M.A. (1996) Bioactive peptides in digesta that regulate gastrointestinal function and intake. *Journal of Animal Science,* **74**, 2500-2508.

Galfi, P., Neogrady, S. and Sakata, T. (1991) Effects of volatile fatty acids on the epithelial cell proliferation of the digestive tract and its hormonal mediation. In: *Physiological Aspects of Digestion and Metabolism in Ruminants,* pp 49-59. Edited by T. Tsuda, Y. Sasaki and R. Kawasaki. Academic Press, San Diego, CA.

Georgiev, I.P., Georgieva, T.M., Pfaffl, M., Hammon, H.M. and Blum, J.W. (2003) Insulin-like growth factor and insulin receptors in intestinal mucosa of neonatal calves. *Journal of Endocrinology*, **176**, 121-132.

Georgieva, T. M., Georgiev, I.P., Ontsouka, E., Hammon, H.M., Pfaffl, M. and Blum, J.W (2003) Expression of insulin-like growth factors (IGF)-I and -II and of receptors for growth hormone, IGF-I, IGF-II, and insulin in the intestine and liver of pre- and full-term calves. *Journal of Animal Science*, **81**, 2294-2300.

Gopal, P.K., Gill, H.S. (2000) Oligosaccharides and glycoconjugates in bovine milk and colostrum. *British Journal of Nutrition,* **84**, 69-74.

Grosvenor, C.E., Picciano, M.F. and Baumrucker, C.R. (1992) Hormones and growth factors in milk. *Endocrine Reviews*, **14**, 710-728.

Guilloteau, P., Le Huërou-Luron, I., Toullec, R., Chayvialle, J.A. and Blum, J.W. (1995) Regulatory peptides in young ruminants. In *Ruminant Physiology – 1995, Proceedings 8. International Symposium of Ruminant Physiology,* pp 519-537. Edited by W. v. Engelhardt, S. Leonhard-Marek, G. Breves and D. Giesecke, D. Ferdinand Enke Verlag, Stuttgart

Guilloteau, P., Le Huërou-Luron, I., Toullec, R., Chayvialle, J.-A., Zabielski, R. and Blum J.W. (1997) Gastrointestinal regulatory peptides and growth factors in young cattle and sheep. *Journal of Veterinary Medicine A,***44**, 1-23.

Guilloteau, P., Biernat, M., Wolinski, J. and Zabielski, R. (2002). Gut regulatory peptides and hormones of the small gut. In *Biology of the Intestine in Growing Animals*, pp 325-362. Edited by R. Zabielski, P.C. Gregory and B. Weström. Elsevier Science, Amsterdam.

Guilloteau, P. and Zabielski, R. (2005). Digestive secretions in pre-ruminant and ruminant calves and some aspects of their regulation. In *Calf and Heifer Rearing*, pp 159-190. Edited by P.C. Garnsworthy. Nottingham University Press, Nottingham.

Hadorn, U. and Blum, J.W. (1997) Effects of colostrum, glucose or water on the first day of life on plasma immunoglobulin G concentrations and γ-glutamyltransferase activities in calves. *Journal of Veterinary Medicine A*, **44**, 531-537.

Hammon, H.M. and Blum, J.W. (1997a) The somatotropic axis in the neonatal calf can be modulated by nutrition, growth hormone and long-R3-insulin-like growth factor-I. *American Journal of Physiology*, **273**, E130-E138.

Hammon, H.M. and Blum, J.W. (1997b) Enhanced xylose absorption in neonatal calves by prolonged colostrum feeding. *Journal of Animal Science*, **75**, 2915-2919.

Hammon, H. M. and Blum, J.W. (2002) Feeding different amounts of colostrum or only milk replacer modify receptors of intestinal insulin-like growth factors and insulin in neonatal calves. *Domestic Animal Endocrinology*, **22**, 155-168.

Howarth, G. S. (2003) Insulin-like growth factor-1 and the gastrointestinal system: therapeutic indications and safety implications. *Journal of Nutrition*, **133**: 2109-2112.

Johnson, L.R. (1981) Regulation of gastrointestinal growth. In *Physiology of the Gastrointestinal Tract*, Vol. 1, pp 169-196. Edited by L. R. Johnson. Raven Press, New York, NY.

Kelly, D. (1994) Colostrum, growth factors and intestinal development in pigs. In: *Schriftenreihe Forschungsinstitut für die Biologie landwirtschaftlicher Nutztiere*, pp 151-166. Edited by W.-B. Souffrant and H. Hagemeister, Dummerstorf.

Kimble, R.M., Breier, B.H., Gluckman, P.D. and Harding, J.E. (1999) Enteral IGF-I enhances foetal growth and gastrointestinal development in oesophageal ligated foetal sheep. *Journal of Endocrinology*, **162**, 227-235.

Klagsbrun, M. (1978) Human milk stimulates DNA synthesis and cellular proliferation in cultured fibroblast. *Proceedings of the National Academy of Science*, USA, **75**, 5057-5061.

Koldovsky, O. (1994) Hormonally active peptides in human milk. *Acta Paediatrica*, **402,** 89-93.

Koletzko, B, Aggett, P.J., Bindels, J.G., Bung, P., Gil, A., Lentze, M.J., Roberfroid, M. and Strobel, S. (1998) Growth, development and differentiation; a functional food science approach. *British Journal of Nutrition*, **80**, S5-S45.

Kuwahara, A. and Fujimiya, M. (2002) Luminal release of regulatory peptides and amines: waste or physiological message ? In *Biology of the Intestine in Growing Animals*, pp 363-407. Edited by R. Zabielski, P.C. Gregory and B. Weström. Elsevier Science, Amsterdam.

Lane, M.A. and Jesse, B.W. (1997) Effect of volatile fatty acid infusion on development of the rumen epithelium in neonatal sheep. *Journal of Dairy Science*, **80**, 740-746.

Lane, M.A., Baldwin, R.L. and Jesse, B.W. (2000) Sheep rumen metabolic development in response to age and dietary treatments. *Journal of Animal Science*, **78**, 1990-1996.

Lane, M.A., Baldwin, R.L. and Jesse, B.W. (2002) Developmental changes in ketogenic enzyme gene expression during sheep rumen development. *Journal of Animal Science,* **80**, 1538-1544.

Le Huërou, I., Guilloteau, P., Wicker, C., Mouats, A., Chayvialle, J.-A., Bernard, C., Burton, J., Toullec, R. and Puigserver, A. (1992) Activity distribution of seven digestive enzymes along the small intestine in calves during development and weaning. *Digestive Diseases and Science*, **37**, 40-46.

Lesmeister K. E. and Heinrichs A. J. (2005) Rumen development in the dairy calf. In *Calf and Heifer Rearing*, pp 53-66. Edited by P.C. Garnsworthy. Nottingham University Press, Nottingham.

Lipkin, M. (1981) Proliferation and differentiation of gastrointestinal cells in normal and disease states. In *Physiology of the Gastrointestinal Tract*, Vol. 1, pp 145-168. Edited by L. R. Johnson. Raven Press, New York.

McLeod, K.R. and Baldwin, R.L. (2000) Effects of diet forage: concentrate ratio and metabolizable energy intake on visceral organ growth and in vivo oxidative capacity of gut tissues in sheep. *Journal of Animal Science*, **78**, 760-770.

Michanek, P., Ventorp, M. and Westrom, B. (1989) Intestinal transmission of colostral antibodies in newborn calves: initiation of closure by feeding colostrum. *Swedish Journal of Agricultural Research*, **19**, 125-127.

Montagne, L., Lallès J.P. (2000) Digestion des matières azotées végétales chez le veau préruminant. Quantification des matières azotées endogènes et importance des mucines. *INRA Productiones Animales*, **13**, 315-324.

Norrman, J., David, C.W., Sauter, S.N., Hammon, H.M. and Blum, J.W. (2003) Effects of dexamethasone on lymphoid tissue in the gut and thymus of neonatal calves fed with colostrum or milk replacer. *Journal of Animal Science,* **81**, 2322-2332.

Ontsouka, C.E., Bruckmaier, R. M. and Blum, J.W. (2003) Fractionized milk composition during removal of colostrum and mature milk. *Journal of Dairy Science,* **86**, 2005-2011.

Ontsouka, C.E., Korczack, B., Hammon, H.M. and Blum, J.W. (2004a) Bovine lactase mRNA levels determined by RT-PCR. In *Proceedings of the Meeting of the "Fachgruppe für Physiologie und Biochemie"*, Deutsche Gesellschaft für Veterinärmedizin, Berlin, March 26-27, 2004.

Ontsouka, C.E., Hammon, H.M and Blum, J.W. (2004b) Expression of insulin-like growth factors (IGF)-1 and –2, IGF binding proteins-2 and –3, and receptors for growth hormone, IGF type-1 and –2 and insulin in the gastrointestinal tract. *Growth Factors* (in press)

Ontsouka, C.E., Philipona, C.. Hammon, H.M. and Blum, J.W. (2004c) Abundance of mRNA coding for components of the somatotropic axis and insulin receptor in different layers of the jejunum and ileum of neonatal calves. In *Proceedings of the Meeting of the Swiss Society of Animal Production*, Zürich, March 3, 2004.

Ontsouka, C. E., Sauter, S.N., Blum, J.W. and Hammon, H.M. (2004d) Effects of colostrum feeding and dexamethasone treatment on mRNA levels of insulin-like growth factors (IGF)-I and-II, IGF binding proteins -2 and –3, and on receptors for growth hormone, IGF-I, IGF-II, and insulin in the gastrointestinal tract of neonatal calves. *Domestic Animal Endocrinology*, **26**, 155-175.

Pacha, J. (2000) Development of intestinal transport function in mammals. *Physiological Reviews*, **80**, 1633-1667.

Pfaffl, M.W., Georgieva, T.M., Georgiev, I. P., Hageleit, M. and Blum, J.W. (2002) Real-time RT-PCR quantification of insulin-like growth factor (IGF)-1, IGF-1 receptor, IGF-2, IGF-2 receptor, insulin receptor, growth hormone receptor, IGF-binding proteins 1, 2 and 3 in the bovine species. *Domestic Animal Endocrinology*, **22**, 91-102.

Potter, G.D. (1990) Intestinal development and regeneration. *Hospital Practice*, **20**, 131-144.

Rauprich, A.B.E., Hammon, H. M., and Blum, J.W. (2000a) Influence of feeding different amounts of first colostrum on metabolic, endocrine, and health status and on growth performance in neonatal calves. *Journal of Dairy Science,* **78**, 896-908.

Rauprich, A.B.E., Hammon, H. M., and Blum, J.W. (2000b) Effects of feeding colostrum and a formula with nutrient contents as colostrum on metabolic and endocrine traits in neonatal calves. *Biology of the Neonate,* **78**, 53-64.

Roffler, B., Fäh, A., Sauter, S.N., Hammon, H.M., Brem, G. and Blum, J.W. (2003) Intestinal morphology, epithelial cell proliferation, and absorptive capacity in neonatal calves fed milk-born insulin-like growth factor-I or a colostrum extract. *Journal of Dairy Science*, **86**, 1797-1806.

Rüsse, I. and Sinowatz, F. (1991) Lehrbuch der Embryologie der Haustiere. P. Parey, Berlin and Hamburg.

Sanderson, I.R. and Naik, S. (2000) Dietary regulation of intestinal gene expression. *Annual Review of Nutrition,* **20**, 311-338.

Sangild, P.T., Xu, R.J. and Trahair J.F. (2002) Maturation of small intestine: the role of cortisol and birth. In *Biology of the Intestine in Growing Animals*, pp 111-144. Edited by R. Zabielski, P.C. Gregory and B. Weström. Elsevier Science, Amsterdam.

Sauter, S.N., Roffler, B., Philipona, C., Morel, C., Guilloteau, P., Blum, J.W. and Hammon H. M. (2004) Intestinal development in neonatal calves: effects of glucocorticoids and dependence on colostrum supply. *Biology of the Neonate*, **85**, 94-104.

Shibahara, T., Sato, N., Waguri, S., Iwanaga, T., Nakahara, A, Fukumoti, H. and Uchiyama, Y. (1995) The fate of effete epithelial cells at the villus tips of the human small intestine. *Archives of Histology and Cytology,* **58**, 205-219.

Schmidt, M., Sangild, P.T., Blum, J.W., Andersen, J.B. and Greve, T. (2004) combined ACTH and glucocorticoid treatment improves survival and organ maturation in premature newborn calves. *Theriogenology*, **61**, 1729-1744.

Smith, M.W. (1985) Expression of digestive and absorptive function in differentiating enterocytes. *Annual Review of Physiology*, **47**, 247-260.

Talukder, M.J.R., Takeuchi, T. and Harada, E. (2002) Transport of colostral macromolecules into cerebrospinal fluid via plasma in newborn calves. *Journal of Dairy Science*, **85**, 514-524.

Thiesen, A., Wild, G., Keelan, M., Clandinin, M.T., McBurney, M., Van Aerde, J. (2000) Ontogeny of intestinal nutrient transport. *Canadian Journal of Physiology and Pharmacology*, **78**, 513-527.

Trahair, J.F. and Harding, R. (1995) Restitution of swallowing in the foetal sheep restores intestinal growth after midgestation esophageal obstruction. *Journal of Pediatric Gastroenterology and Nutrition*, **20**, 156-161.

Trahair, J.F., Sangild, P.T. (1997) Systemic and luminal influences on the perinatal development of the gut. *Equine Veterinary Journal*, **24**, Suppl., 40-50.

Trahair, J.F., Wing, S.J., Quinn, K.J., and Owens, P.C. (1997) Regulation of gastrointestinal growth in foetal sheep by luminally administered insulin-like growth factor-I. *Journal of Endocrinology*, **152**, 29-38.

Trahair, J.F. and Sangild, P.T. (2002) Studying the development of the small intestine: philosophical and anatomical perspectives. In *Biology of the Intestine in Growing Animals*, pp 1—54. Edited by R. Zabielski, P.C. Gregory and B. Weström. Elsevier Science, Amsterdam.

Vacher, P.-Y., Bestetti, G. and Blum, J.W. (1995) Insulin-like growth factor I absorption in the jejunum of neonatal calves. *Biology of the Neonate*, **68**, 354-367.

Veereman-Wauters, G. (1996) Neonatal gut development and postnatal adaptation. *European Journal of Pediatrics*, **155**, 627-632.

Walsh, J.H. and Dockray, G.J. (1994) Gut Peptides. Comprehensive Endocrinology. Edited by L. Martini. Raven Press, New York, NY.

Widdowson, E. (1985) Development of the digestive system. Comparative animal studies. *American Journal of Clinical Nutrition*, **41**, 384-390.

Xu, R.J. (1996) Development of the newborn GI tract and its relation to colostrum/milk intake: a review. *Reproduction, Fertility and Development,* **8**, 35-48.

Zanming, S., Seyfert, H.-M., Löhrke, B., Schneider, F., •itnan, R., Chudy, A., Kuhla, S., Hammon, H.M., Blum, J.W., Martens, H., Hagemeister, H., and Voigt, J. (2004) Effect of nutritional level on rumen papillae development and IGF-1 and IGF type 1 receptor in young goats. *Journal of Nutrition*, **134**, 11-17.

4

RUMEN DEVELOPMENT IN THE DAIRY CALF

A. J. HEINRICHS AND K. E. LESMEISTER
Dairy and Animal Science Department, The Pennsylvania State University, University Park, Pennsylvania, USA

The degradation of feedstuffs, rumen microbial synthesis, and the various resulting end products have been a subject of investigation for over a century. In 1884, Tappiener (as cited in Phillipson, 1947) attributed cellulose digestion in the rumen to digestive actions of symbiotic organisms in the rumen. Much work followed these early discoveries, and this research was devoted primarily to the digestion of cellulose, the assimilation of various end products of the microbial population, and the various species of rumen bacteria that exist under certain rumen conditions (Bryant, 1951).

Neonatal ruminants are unique in that at birth they are physically and functionally two different types of animals with respect to their gastro-intestinal system. At birth the physical attributes distinguishing a ruminant from a monogastric animal, i.e. the reticulum, rumen, and omasum, are present. However, the rudimentary state of the reticulo-rumen and omasum, presence of the oesophageal groove (Church, 1988), plus the developing abomasal and intestinal enzymatic state forces neonatal ruminants to function as monogastric animals (Longenbach and Heinrichs, 1998), subsisting on milk-based diets, which are digested and assimilated quite efficiently (Davis and Drackley, 1998; Van Soest, 1994). Digestive enzymatic changes (Longenbach and Heinrichs, 1998) coupled with the high daily costs of maintaining a pre-weaned calf (Gabler *et al.*, 2000) result in an ability and need to transition the calf from a monogastric animal to a ruminant animal (Church, 1988; Davis and Drackley, 1998). A smooth transition from a monogastric to a ruminant animal, with minimal loss in growth, requires adequate size and development of the reticulo-rumen for efficient utilization of dry and forage-based diets. Therefore, understanding the factors responsible for initiating cellular growth and maturation of the non-functional rumen tissues, and establishing rumen development and function in the neonatal calf, are important.

Rudimentary reticulo-rumen

At birth, the reticulum, rumen, and omasum are undeveloped, non-functional, small in size compared with the abomasum, and disproportionate to the adult digestive system (Tamate *et al.*, 1962). Papillary growth, rumen muscularization, and rumen vascularization are minimal to nonexistent; the rumen wall is thin and slightly transparent, and reticulo-rumen volume is minimal (Flatt *et al.*, 1958; Harrison *et al.*, 1960; Sander *et al.*, 1959; Tamate *et al.*, 1962; Warner *et al.*, 1956). In addition, Lane *et al.* (2000) indicated that rumen epithelial cells of sheep younger than 42 days of age were incapable of converting butyrate to ß-hydroxybutyrate, suggesting a metabolically inactive epithelium at birth. However, rumen epithelial metabolic activity in calves has been shown to occur at a younger age than reported in lambs, and to increase with age (Nocek *et al.*, 1980; Sutton, 1963). Ruminant animals require a physically and functionally developed rumen to meet the demands of an innate desire to consume forages and dry feeds (Van Soest, 1994). However, the neonatal rumen will remain undeveloped if the necessary diet requirements for rumen development are not provided (Brownlee, 1956; Harrison *et al.*, 1960; Van Soest, 1994). Rumen development appears to be greatly affected by diet and dietary changes (Brownlee, 1956; Harrison *et al.*, 1960). In addition, the influence of dietary factors on rumen development may vary, and development of rumen epithelium, rumen muscularization, and expansion of rumen volume have been found to occur independently (Brownlee, 1956; Flatt *et al.*, 1958; Harrison *et al.*, 1960; Stobo *et al.*, 1966a). These findings suggest that dietary factors influencing papillary growth and development may not affect rumen muscularization or rumen volume.

Changes in rumen epithelium

Proliferation and growth of squamous epithelial cells causes increases in papillae length, papillae width, and thickness of the interior rumen wall (Church, 1988). Figure 4.1 demonstrates the progression of cellular differentiation and growth that occurs during the first few weeks of life in samples taken from the cranial dorsal sac of young calves. Prior to undergoing the transition from a pre-ruminant to a ruminant, growth and development of the ruminal absorptive surface area (papillae) is necessary to enable absorption and utilization of end products of microbial digestion, specifically rumen volatile fatty acids (Church, 1988; Sutton *et al.*, 1963; Warner *et al.*, 1956; Van Soest, 1994). Presence and absorption of volatile fatty acids stimulate rumen epithelial metabolism and may be key in initiating rumen epithelial development (Baldwin and McLeod, 2000; Sander *et al.*, 1959; Sutton *et al.*, 1963; Tamate *et al.*, 1962). However, it has been suggested that rumen epithelial ketogenesis, indicating metabolic activity, may occur independently of volatile fatty acid production (Sutton *et al.*, 1963). Nevertheless, numerous researchers have indicated that ingestion of dry feeds, and the resultant microbial end products, sufficiently stimulates rumen epithelial development (Brownlee, 1956; Flatt *et al.*, 1958; Greenwood

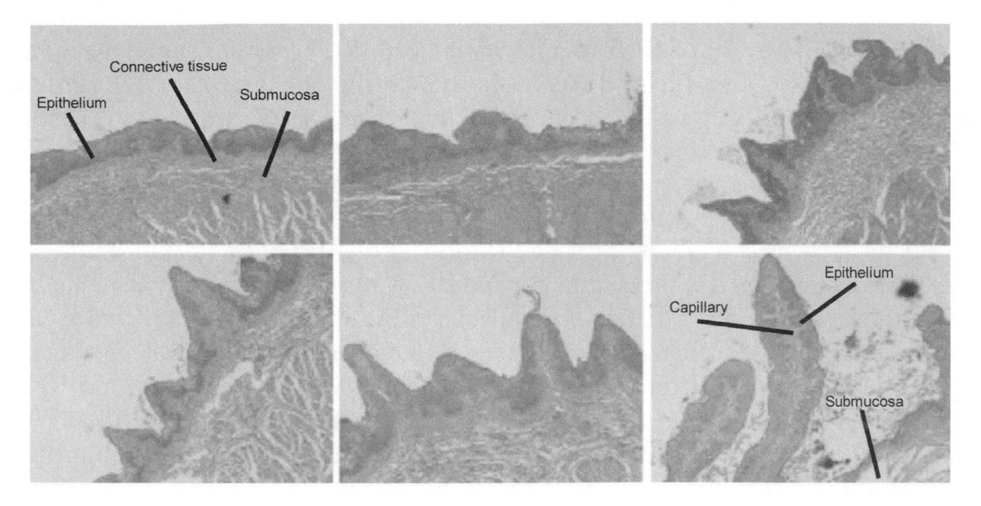

Figure 4.1 The progression of cellular differentiation and growth during the first weeks of life in a grain fed dairy calf. (progression is 3d to 35d of age, upper row L to R, lower row L to R.

et al., 1997; Nocek *et al.*, 1984; Stobo *et al.*, 1966a; Warner *et al.*, 1956). However, the stimulatory effects of different volatile fatty acids are not equal, with butyrate being most stimulatory, followed by propionate (Flatt *et al.*, 1958; Harrison *et al.*, 1960; Sander *et al.*, 1959; Stobo *et al.*, 1966a; Sutton *et al.*, 1963; Tamate *et al.*, 1962). Low activity of the acetyl-CoA synthetase enzyme appears to limit rumen epithelial metabolism of acetate, thereby limiting acetate's ability to stimulate epithelial development (Ash and Baird, 1973; Harmon *et al.*, 1991). Conversely, Baldwin and McLeod (2000) indicated comparable acetate and butyrate metabolism in sheep, stating that animal energy status may influence the metabolism rate of individual volatile fatty acids. However, the concentration of acetate was higher than butyrate in this study, indicating a possible conditioning of the rumen epithelium to acetate use, due to decreased butyrate availability (Baldwin and McLeod, 2000). Rumen epithelial conditioning to utilize specific volatile fatty acids has been reported previously (Rickard and Ternouth, 1965). In addition, epithelial butyrate metabolism appears to increase concomitantly with decreasing rumen pH and increasing butyrate concentrations (Baldwin and McLeod, 2000; Krehbiel *et al.*, 1992; Stevens and Stettler, 1966; Sutton *et al.*, 1963; Weigand *et al.*, 1975). A continuous presence of volatile fatty acids maintains growth, size, and function of rumen papillae (Harrison *et al.*, 1960; Warner *et al.*, 1956). Therefore, it is likely that diets composed of milk, concentrates, or forages affect the rate and extent of rumen epithelial growth differently, and such results have been reported (Harrison *et al.*, 1960; Stobo *et al.*, 1966a; Tamate *et al.*, 1962; Warner *et al.*, 1956).

Papillae length and width are the most obvious factors influencing absorptive surface area, but changes in papillae density should also be considered. Dietary and age differences have been found to alter papillae density of the developing rumen; however, significant differences due to dietary treatment are seldom reported for papillae density in calves (Klein *et al.*, 1987; Lane *et al.*, 2000; Nocek *et al.*, 1984; Zitnan *et al.*, 1998; Zitnan *et al.*, 1999; Lesmeister *et al.*, 2004a). Papillae density is commonly reported as the number of papillae in a fixed area (usually per cm^2) regardless of rumen volume, but rumen volume has been shown to increase with age (Stobo *et al.*, 1966a). Lesmeister *et al.* (2004a) demonstrated a procedure for sampling rumen tissue that was capable of detecting treatment differences for papillae length and width and was moderately capable of detecting treatment differences for rumen wall thickness. Minimal treatment influence on papillae density may be explained by a confounding effect of rumen volume. In addition, McGavin and Morrill (1976) and Lesmeister *et al.* (2004a) reported intra-rumen variation for papillae measurements, demonstrating that papillae growth is not universal in all rumen areas.

The various areas of the rumen are illustrated in Figure 4.2. Typically the right and left sides of the ventral portion of the caudal ventral blind sac are the least developed. The caudal portion of the ventral blind sac, the right and left sides of the caudal dorsal sac, and the left and right sides of the cranial ventral sac are typically more developed and develop faster than other areas of the rumen. Lesmeister *et al.* (2004a) demonstrated that when rumen development is being studied, samples should be taken from the caudal and cranial sacs of the dorsal and ventral rumen to represent papillae growth and development throughout the entire rumen. However, the complete physiological mechanics and genetics of papillae growth have not been thoroughly elucidated in the literature.

Liquid feeds and epithelial development

Milk or milk replacer is initially the primary diet of neonatal dairy calves; however, its chemical composition, in addition to the shunting effect of the oesophageal groove, limits its ability to stimulate rumen development (Brownlee, 1956; Harrison *et al.*, 1960; Lane *et al.*, 2000; Stobo *et al.*, 1966a; Tamate *et al.*, 1962; Warner *et al.*, 1956). Numerous researchers have reported minimal rumen development in calves receiving solely milk or milk replacer (Brownlee, 1956; Stobo *et al.*, 1966a; Warner *et al.*, 1956), even up to 12 weeks of age (Tamate *et al.*, 1962), and others have reported a regression, or stasis, of rumen development when calves were switched from a dry diet to a milk or milk replacer diet (Harrison *et al.*, 1960). In addition, calves receiving only milk or milk replacer exhibit minimal metabolic activity of the rumen epithelium and minimal volatile fatty acid absorption, which again do not increase with age (Sutton *et al.*, 1963). However, ruminal size of the milk-fed calf, regardless of rumen development, has been shown to increase

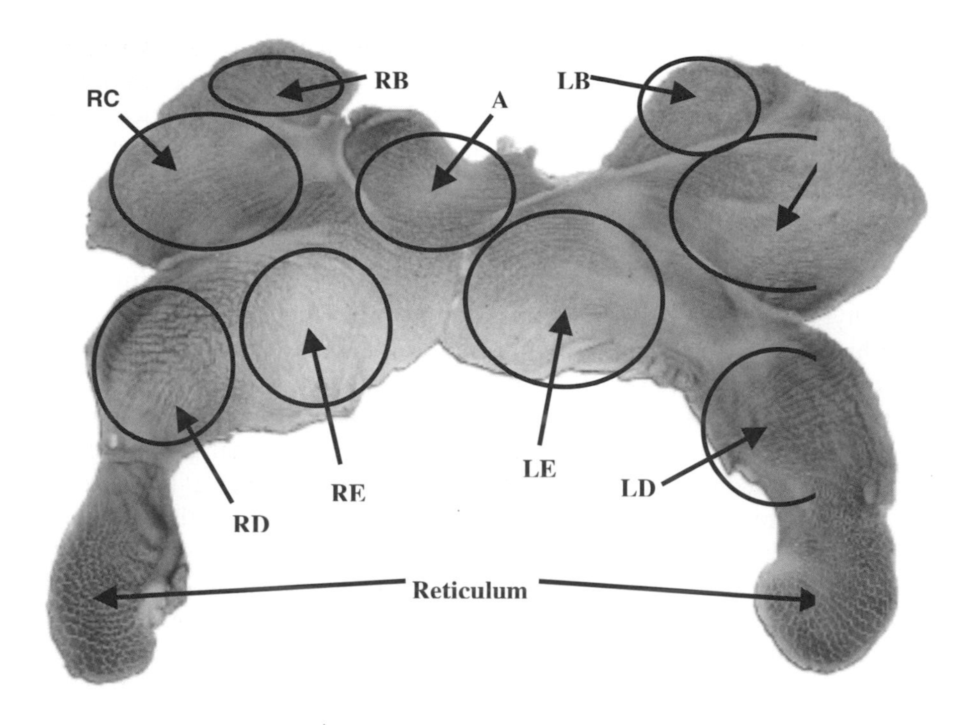

Figure 4.2 Example of a dissected rumen depicting the physical areas of the rumen sampled and corresponding labels. (A) caudal portion of the caudal ventral blind sac; (RB) right side and (LB) left side caudal dorsal sac; (RC) right side and (LC) left side cranial dorsal sac; (RD) right side and (LD) left side cranial ventral sac; and (RE) right side and (LE) left side ventral portion of caudal ventral blind sac.

proportionately with body size (Flatt *et al.*, 1958; Stobo *et al.*, 1966a; Tamate *et al.*, 1962; Vazquez-Anon *et al.*, 1993; Warner *et al.*, 1956). Therefore, although a milk or milk replacer diet can result in rapid and efficient growth, it does little to prepare the pre-ruminant calf for weaning or utilization of grain and forage based diets.

Solid feeds and epithelial development

Solid feeds, unlike liquid feeds, are preferentially directed to the reticulo-rumen for digestion (Church, 1988; Van Soest, 1994). Ingestion of solid feed stimulates rumen microbial proliferation and production of microbial end products, volatile fatty acids, which have been shown to initiate rumen epithelial development (Beharka *et al.*, 1998; Harrison *et al.*, 1960; Hibbs *et al.*, 1956; Pounden and Hibbs, 1948a; Pounden and Hibbs, 1948b; Warner *et al.*, 1956). However, solid feeds differ in their efficacy to stimulate rumen development. Chemical composition of feeds, and the resultant end products of microbial digestion, have the greatest influence on epithelial development (Harrison *et al.*, 1960; Flatt *et al.*, 1958; Nocek *et al.*, 1984; Stobo *et al.*, 1966b; Warner *et al.*, 1956).

Many chemical characteristics of solid feeds appear to influence rumen epithelial growth. Concentrates (Brownlee, 1956; Harrison *et al.*, 1960; Stobo *et al.*, 1966a; Stobo *et al.*, 1966b; Tamate *et al.*, 1962; Warner *et al.*, 1956) and purified diets containing casein, starch, cerelose, and minerals (Flatt *et al.*, 1958) have increased the rate of rumen development when compared with forage sources. When introduced into the rumen as purified sodium salts, sodium butyrate had the greatest influence on rumen epithelial development, followed by sodium propionate; sodium acetate and glucose had minimal effects (Flatt *et al.*, 1958; Sander *et al.*, 1959; Tamate *et al.*, 1962). In addition, research has identified butyrate and propionate as the volatile fatty acids most readily absorbed by rumen epithelium, especially when present at physiological concentrations (Baldwin and McLeod, 2000; Sander *et al.*, 1959). Furthermore, the chemical composition of concentrates causes a shift in the microbial population, subsequently increasing butyrate and propionate production at the expense of acetate (Hibbs *et al.*, 1956; Pounden and Hibbs, 1948a; Pounden and Hibbs, 1948b; Stobo *et al.*, 1966b). Increased microbial production of stronger rumen acids, i.e. lactate, butyrate, and propionate, also decreases rumen pH (Hibbs *et al.*, 1956; Stobo *et al.*, 1966b). Stevens and Stettler (1966) indicated that only free-state volatile fatty acids were absorbed into and across the rumen epithelium, and lower rumen pH increases the free H^+ available, subsequently increasing absorbable free volatile fatty acids (Sutton *et al.*, 1963). Therefore, a lower rumen pH and its effect on volatile fatty acid absorption may be the catalyst driving rumen epithelial growth (Sutton *et al.*, 1963). However, dietary type (Stobo *et al.*, 1966b; Zitnan *et al.*, 1998), microbial population present, and volatile fatty acids produced (Hibbs *et al.*, 1956) greatly influence ruminal pH and cannot be ignored.

Forages on the other hand have the ability to maintain a higher ruminal pH, due to larger particle size and increased fibre content (Hibbs *et al.*, 1956; Stobo *et al.*, 1966b; Zitnan *et al.*, 1998). Larger particle size increases ruminal salivary flow through greater initial mastication and increased subsequent rumination in mature and immature ruminants (Beauchemin and Rode, 1997; Hibbs *et al.*, 1956). Maintenance of a higher ruminal pH supports microbial populations typically associated with forages, which in turn shift volatile fatty acid production from butyrate and propionate to acetate (Hibbs *et al.*, 1956; Van Soest, 1994). Furthermore, there is evidence of higher NH_3 concentrations with forage diets (Sutton *et al.*, 1963; Zitnan *et al.*, 1998), resulting in H^+ sinks and effectively decreasing free volatile fatty acid availability (Stevens and Stettler, 1966). Some studies have indicated minimal differences between concentrates and forages for volatile fatty acid proportions and rumen development (Anderson *et al.*, 1982; Baldwin and McLeod, 2000; Klein *et al.*, 1987; Zitnan *et al.*, 1998). Confounding factors, however, such as sampling of rumen volatile fatty acids via oesophageal tube, age at slaughter, and dietary particle size (Coverdale *et al.*, 2004), might have influenced results in these studies. Furthermore, studies indicating increased rumen epithelial development when concentrates (Anderson *et al.*, 1982; Brownlee, 1956; Stobo *et al.*, 1966a; Warner *et al.*, 1956; Zitnan *et al.*, 1998; Zitnan *et al.*, 1999), purified diets (Harrison *et al.*, 1960), or fatty acid salts of butyrate and propionate (Sander *et al.*, 1959; Tamate *et al.*, 1962) were compared with forages are numerous and cannot be ignored.

Increased absorption and utilization of butyrate and propionate, compared with acetate, provide further evidence that the former volatile fatty acids stimulate epithelial development (Baldwin and McLeod, 2000; Sander *et al.*, 1959). Whether the actual stimulant for epithelial development is increased butyrate and propionate production (Flatt *et al.*, 1958; Harrison *et al.*, 1960; Sander *et al.*, 1959; Sutton *et al.*, 1963; Tamate *et al.*, 1962), decreased ruminal pH concomitant with stronger ruminal acid production (Stevens and Stettler, 1966), or a combination, concentrates appear to result in greater development of the rumen epithelium than forages (Brownlee *et al.*, 1956; Nocek *et al.*, 1984; Stobo *et al.*, 1966a; Stobo *et al.*, 1966b; Warner *et al.*, 1956; Zitnan *et al.*, 1998). This concept is demonstrated in Figure 4.3, which shows marked differences in rumen development of 6-week-old calves fed milk, milk and grain, or milk and forage (dry hay).

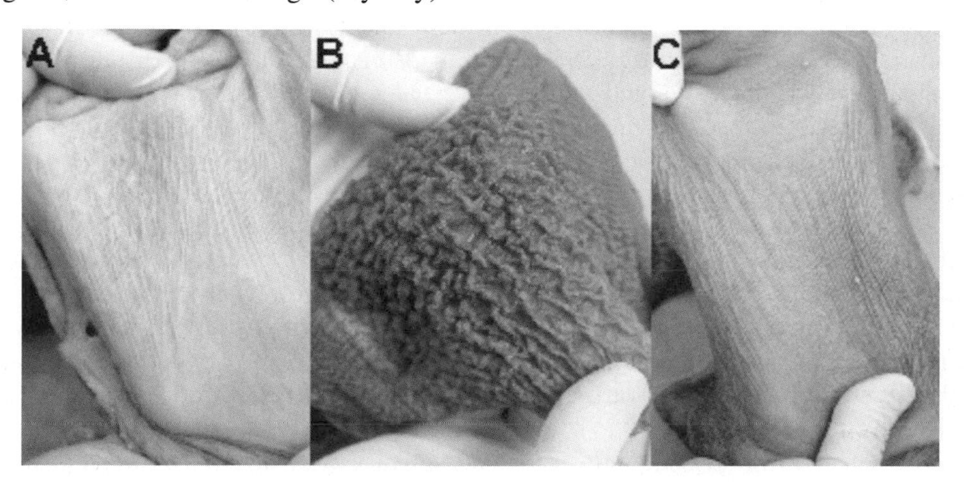

Figure 4.3 Comparison of rumen papillae development at six weeks of age in calves fed milk only (A), milk and grain (B), or milk and dry hay (C).

Recent studies have looked at effects of dietary alterations or additions on rumen development and subsequent effects on rumen microbial end products. Additions of yeast culture increased grain intake by calves, and had marginal affects on rumen development in young calves when added at 2% of the diet (Lesmeister *et al.*, 2004b). Papillae length and rumen wall thickness were significantly greater in 4-week-old calves fed calf starter diets containing steam-flaked maize compared with those fed dry-rolled and whole maize when these maize supplements made up 33% of the calf starter diet (Lesmeister and Heinrichs, 2004). This study showed that type of grain processing can influence rumen development in young calves.

Physical structure and epithelial development

Rumen epithelial development cannot be thoroughly discussed without covering the influence of parakeratosis on papillae development and absorptive ability. Parakeratosis occurs when epithelial squamous cells develop a hardened keratin layer due to a diet's inability to continuously remove degenerating epithelial cells (Bull *et al.*, 1965; Hinders and Owen, 1965). Parakeratosis creates a physical barrier, restricting absorptive surface area and volatile fatty acid absorption, reducing epithelial blood flow and rumen motility, causing papillae degeneration and sloughing in extreme cases (Anderson *et al.*, 1982; Beharka *et al.*, 1998; Brownlee, 1956; Bull *et al.*, 1965; Hinders and Owen, 1965; McGavin and Morrill, 1976; Nocek *et al.*, 1984). Initial evidence of parakeratosis is clumping and branching of papillae, followed by degeneration and sloughing of papillae (Anderson *et al.*, 1982; Beharka *et al.*, 1998; Brownlee, 1956; Bull *et al.*, 1965; Hinders and Owen, 1965; McGavin and Morrill, 1976; Nocek *et al.*, 1984; Zitnan *et al.*, 1998). Concentrate diets, having small particle size and low abrasive value (Beharka *et al.*, 1998; Bull *et al.*, 1965; Greenwood *et al.*, 1997; McGavin and Morrill, 1976), increased volatile fatty acid production, decreased rumen buffering capacity, and subsequently decreased rumen pH (Anderson *et al.*, 1982; Hinders and Owen, 1965), are factors commonly associated with parakeratosis. The abrasive value has been defined as a feed's efficacy in physically removing keratin and/or dead epithelial cells from the rumen epithelium (Greenwood *et al.*, 1997). Therefore, increased feed particle size, especially with forages or coarsely-ground concentrates, maintains epithelial and papillae integrity and absorptive ability via physical removal of the keratin layer (Greenwood *et al.*, 1997; McGavin and Morrill, 1976), increased rumination and rumen motility (Brownlee, 1956), subsequently increased salivary flow and buffering capacity (Bull *et al.*, 1965; Hinders and Owen, 1965), and development of mature rumen function and environment (Hibbs *et al.*, 1956). Factors such as animal susceptibility, intake differences, passage rate, rumination rate, and salivary production, however, may also contribute to occurrences of parakeratosis (Anderson *et al.*, 1982; Zitnan *et al.*, 1999).

Changes in rumen muscularization and volume

Feed physical structure probably has the greatest influence on development of rumen muscularization and volume. Stimulation of rumen motility is governed by the same factors, particle size and effective fibre, in the neonatal ruminant as in the adult ruminant (Beauchemin and Rode, 1997; Mertens, 1997; Van Soest, 1994). In contrast to the advantages of concentrates for epithelial development (Brownlee *et al.*, 1956; Nocek *et al.*, 1984; Tamate *et al.*, 1962; Warner *et al.*, 1956), forages appear to be the primary stimulators of rumen muscle development and increased rumen volume (Hibbs *et al.*, 1956; Stobo *et al.*, 1966a; Zitnan *et al.*, 1998). Large particle size, high effective fibre content, and increased bulk of forages or high fibre sources physically increase stimulation of the rumen wall, subsequently increasing rumen motility, muscularization, and volume (Brownlee *et al.*, 1956; Hibbs *et al.*, 1956; Nocek *et al.*, 1984; Stobo *et al.*, 1966a; Tamate *et al.*, 1962; Vazquez-Anon *et al.*, 1993; Warner *et al.*, 1956; Zitnan *et al.*, 1998). As discussed earlier, increases in rumen muscularization and volume have occurred independently of epithelial development (Brownlee, 1956; Flatt *et al.*, 1958; Harrison *et al.*, 1960; Stobo *et al.*, 1966a). Supporting evidence for independent muscle and epithelial growth is found in studies determining the effects of inert material (sponges, toothbrush bristles, or bedding) on rumen epithelial, muscular, and capacity development (Brownlee, 1956; Flatt *et al.*, 1958; Harrison *et al.*, 1960). Inert materials were ineffective for stimulating papillae growth, but capable of significantly increasing rumen capacity and muscularization (Brownlee, 1956; Flatt *et al.*, 1958; Harrison *et al.*, 1960). However, solid feeds other than forages or bulky feedstuffs can be effective in influencing rumen capacity and muscularization. Coarsely or moderately ground concentrate diets have been shown to increase rumen capacity and muscularization more than finely ground or pelleted concentrate diets, indicating that extent of processing and/or concentrate particle size affects the ability of concentrates to stimulate rumen capacity and muscularization (Beharka *et al.*, 1998; Greenwood *et al.*, 1997). Therefore, concentrate diets with increased particle size might be the most desirable feedstuff for overall rumen development, due to their ability to stimulate epithelial development, rumen capacity, and rumen muscularization.

Although the basics of rumen development have been published, current research in rumen development focuses on dietary manipulation, attempting to optimize the rate and extent of rumen development. Increased availability of feed by-products, development of new feed additives, and differences in particle size of calf starter diets all provide areas for future research. Understanding the cellular biology and physiological changes that occur during rumen development, clarifying digestion kinetics in neonatal calves, and development of low-impact or non-invasive research procedures could be instrumental in advancing this area. Although much is known about rumen development, several areas require additional study. The adoption of newer technologies to stimulate the rate of rumen development may have important economic consequences for dairy and beef producers and warrant further applied studies.

References

Anderson, M. J., Khoyloo, M. and Walters, J. L. (1982) Effect of feeding whole cottonseed on intake, body weight, and reticulorumen development of young Holstein calves. *Journal of Dairy Science,* **65**,764-772.

Ash, R. and Baird, G. D. (1973) Activation of volatile fatty acids in bovine liver and rumen epithelium. *Biochemistry Journal,* **136**,311-319.

Baldwin, R. L. and McLeod, K. R. (2000) Effects of diet forage:concentrate ratio and metabolizable energy intake on isolated rumen epithelial cell metabolism in vitro. *Journal of Animal Science,* **78**,771-783.

Beharka, A. A., Nagaraja, T. G., Morrill, J. L., Kennedy, G. A. and Klemm, R. D. (1998) Effects of form of the diet on anatomical, microbial, and fermentative development of the rumen of neonatal calves. *Journal of Dairy Science,* **81**,1946-1955.

Beauchemin, K. A. and Rode, L. M. (1997) Minimum versus optimum concentrations of fiber in dairy cow diets based on barley silage and concentrates of barley or corn. *Journal of Dairy Science,* **80**,1629-1639.

Bryant, M. P. (1951) Some characteristics of the different bacteria present in the rumen of cattle on a constant ration. *Animal Science,* **10**,1042.

Brownlee, A. (1956) The development of rumen papillae in cattle fed on different diets. *British Veterinary Journal,* **112**,369-375.

Bull, L. S., Bush, L. J., Friend, J. D., Harris, B. and Jones, E. W. (1965) Incidence of ruminal parakeratosis in calves fed different rations and its relation to volatile fatty acid absorption. *Journal of Dairy Science,* **48**,1459-1466.

Church, D. C. (1988) *The Ruminant Animal: Digestive Physiology and Nutrition.* Prentice-Hall, Inc. Englewood Cliffs, New Jersey.

Coverdale, J. A., Tyler, H. D., Quigley, J. D. and J.A. Brown (2004) Effect of various levels of crude fiber and form of diet on rumen development and growth in calves. *Journal of Dairy Science,* **87**, 2554-2562.

Davis, C. L. and Drackley, J. K. (1998) *The Development, Nutrition, and Management of the Young Calf.* Iowa State University Press. Ames, IA.

Flatt, W. P., Warner, R. G. and Loosli, J. K. (1958) Influence of purified materials on the development of the ruminant stomach. *Journal of Dairy Science,* **41**,1593-1600.

Gabler, M. T., Tozer, P. R. and Heinrichs, A. J. (2000) Development of a cost analysis spreadsheet for calculating the costs to raise a replacement dairy heifer. *Journal of Dairy Science,* **83**,1104-1109.

Greenwood, R. H., Morrill, J. L., Titgemeyer, E. C. and Kennedy, G. A. (1997) A new method of measuring diet abrasion and its effect on the development of the forestomach. *Journal of Dairy Science,* **80**,2534-2541.

Harmon, D. L., Gross, K. L., Krehbiel, C. R., Kreikemeir, K. K., Bauer, M. L.

and Britton, R. A. (1991) Influence of dietary forage and energy intake on metabolism and acyl-CoA synthetase activity in bovine ruminal epithelial tissue. *Journal of Animal Science,* **69**,4117-4127.

Harrison, H. N., Warner, R. G., Sander, E. G. and Loosli, J. K. (1960) Changes in the tissue and volume of the stomachs of calves following the removal of dry feed or consumption of inert bulk. *Journal of Dairy Science,* **43**,1301-1312.

Heinrichs, A. J. and Jones, C. M. (2003) Feeding the newborn dairy calf. *Special Circular UD013.* The Pennsylvania State University. University Park, PA.

Hibbs, J. W., Conrad, H. R., Pounden, W. D. and Frank, N. (1956) A high roughage system for raising calves based on early development of rumen function. VI. Influence of hay to grain ratio on calf performance, rumen development, and certain blood changes. *Journal of Dairy Science,* **39**,171-179.

Hinders, R. G. and Owen, F. G. (1965) Relation of ruminal parakeratosis development to volatile fatty acid absorption. *Journal of Dairy Science,* **48**,1069-1073.

Klein, R. D., Kincaid, R. L., Hodgson, A. S., Harrison, J. H., Hillers, J. K. and Cronrath, J. D. (1987) Dietary fiber and early weaning on growth and rumen development of calves. *Journal of Dairy Science,* **70**,2095-2104.

Krehbiel, C. R., Harmon, D. L. and Schnieder, J. E. (1992) Effect of increasing ruminal butyrate on portal and hepatic nutrient flux in steers. *Journal of Animal Science,* **70**,904-914.

Lane, M. A., Baldwin, R. L. and Jesse, B. W. (2000) Sheep rumen metabolic development in response to age and dietary treatments. *Journal of Animal Science,* **78**,1990-1996.

Lesmeister, K. E., Tozer, P. R. and Heinrichs, A. J. (2004a) Development and analysis of a rumen tissue sampling procedure. *Journal of Dairy Science,* **87**, 1336-1344.

Lesmeister, K. E., Heinrichs, A. J. and Gabler, M. T. (2004b) Effects of supplemental yeast (*Saccharomyces cerevisiae*) culture on rumen development, growth characteristics, and blood parameters in neonatal dairy calves. *Journal of Dairy Science,* **87**, 1832-1839.

Lesmeister, K. E. and Heinrichs, A. J. (2004) Effects of corn processing on growth characteristics, rumen development, and rumen parameters in neonatal dairy calves. *Journal of Dairy Science,* **87**, 3439-3450.

Longenbach, J. I. and Heinrichs, A. J. (1998) A review of the importance and physiological role of curd formation in the abomasum of young calves. *Animal Feed Science and Technology,* **73**,85-97.

McGavin, M. D. and Morrill, J. L. (1976) Scanning electron microscopy of ruminal papillae in fed various amounts and forms of roughage. *American Journal of Veterinary Research,* **37**,497-508.

Mertens, D. R. (1997) Creating a system for meeting the fiber requirements of dairy cows. *Journal of Dairy Science,* **80**,1463-1481.

Nocek, J. E., Herbein, J. H. and Polan, C. E. (1980) Influence of ration physical form, ruminal degradable nitrogen and age on rumen epithelial propionate acetate transport and some enzymatic activities. *Journal of Nutrition,* **110**,2355-2364.

Nocek, J. E., Heald, C. W. and Polan, C. E. (1984) Influence of ration physical form and nitrogen availability on ruminal morphology of growing bull calves. *Journal of Dairy Science,* **67**,334-343.

Phillipson, A. T. (1947) Fermentation in the alimentary tract and the metabolism of the derived fatty acids. *Nutrition Abstracts and Reviews.* **17**,12-18.

Pounden, W. D. and Hibbs, J. W. (1948a) The influence of the ration and rumen inoculation on the establishment of certain microorganisms in the rumens of young calves. *Journal of Dairy Science,* **31**,1041-1050.

Pounden, W. D. and Hibbs, J. W. (1948b) The influence of the ratio of grain to hay in the ration of dairy calves on certain rumen microorganisms. *Journal of Dairy Science,* **31**,1051-1054.

Rickard, M. D. and Ternouth, J. H. (1965) The effect of the increased dietary volatile fatty acids on the morphological and physiological development of lambs with particular reference to the rumen. *Journal of Agricultural Science,* **65**,371-377.

Sander, E. G., Warner, R. G., Harrison, H. N. and Loosli, J. K. (1959) The stimulatory effect of sodium butyrate and sodium propionate on the development of rumen mucosa in the young calf. *Journal of Dairy Science,* **42**,1600-1605.

Stevens, C. E. and Stettler, B. K. (1966) Factors affecting the transport of volatile fatty acids across rumen epithelium. *American Journal of Physiology,* **210**,365-372.

Stobo, I. J. F., Roy, J. H. B. and Gaston, H. J. (1966a) Rumen development in the calf. 1. The effect of diets containing different proportions of concentrates to hay on rumen development. *British Journal of Nutrition,* **20**,171-188.

Stobo, I. J. F., Roy, J. H. B. and Gaston, H. J. (1966b) Rumen development in the calf. 2. The effect of diets containing different proportions of concentrates to hay on digestive efficiency. *British Journal of Nutrition,* **20**,189-215.

Sutton, J. D., McGilliard, A. D. and Jacobson, N. L. (1963) Functional development of rumen mucosa. I. Absorptive ability. *Journal of Dairy Science,* **46**,426-436.

Tamate, H., McGilliard, A. D., Jacobson, N. L. and Getty, R. (1962) Effect of various dietaries on the anatomical development of the stomach in the calf. *Journal of Dairy Science,* **45**,408-420.

Van Soest, P. J. (1994) *Nutritional Ecology of the Ruminant.* 2nd ed. Cornell University Press, Ithaca, NY.

Vazquez-Anon, M., Heinrichs, A. J., Aldrich, J. M. and Varga, G. A. (1993) Postweaning age effects on rumen fermentation end-products and digesta kinetics in calves weaned at 5 weeks of age. *Journal of Dairy Science,* **76**,2742-2748.

Warner, R. G., Flatt, W. P. and Loosli, J. K. (1956) Dietary factors influencing the development of the ruminant stomach. *Journal of Agriculture, Food and Chemistry*, **4**,788-792.

Weigand, E., Young, J. W. and McGilliard, A. D. (1975) Volatile fatty acid metabolism by rumen mucosa from cattle fed hay or grain. *Journal of Dairy Science,* **58**,1294-1300.

Zitnan, R., Voigt, J., Schonhusen, U., Wegner, J., Kokardova, M., Hagemeister, H., Levkut, M., Kuhla, S. and Sommer, A. (1998) Influence of dietary concentrate to forage ratio on the development of rumen mucosa in calves. *Archives of Animal Nutrition*, **51**,279-291.

Zitnan, R., Voigt, J., Wegner, J., Breves, G., Schroder, B., Winckler, C., Levkut, M., Kokardova, M., Schonhusen, U., Kuhla, S., Hagemeister, H. and Sommer, A. (1999) Morphological and functional development of the rumen in the calf: Influence of the time of weaning. *Archives of Animal Nutrition*, **52**,351-362.

[illegible] *Biochemistry* [illegible]

5

CURRENT PERSPECTIVES ON THE ENERGY AND PROTEIN REQUIREMENTS OF THE PRE-WEANED CALF

MICHAEL VAN AMBURGH[1] AND JAMES DRACKLEY[2]
[1]*Cornell University, Ithaca, New York, USA;* [2]*Department of Animal Sciences, University of Illinois, Urbana, Illinois 61822 USA*

Introduction

The dairy calf is unusual with respect to how nutrient requirements are calculated. Under most management conditions, the concept of a "requirement" has been passively applied since little evaluation of changes related to age, weight, environment or ingredients and subsequent diet modification has been conducted in calves prior to three months of age. There is a tremendous literature available concerning the growth and developmental responses of calves to various dietary ingredients, amounts fed, environmental challenge, weaning age, rumen development and health. Historically, a system to help frame this knowledge, as is common for more mature cattle, has been unavailable to nutritionists and calf managers. Thus, rearing programmes and nutritional management guidelines have been static and nearly dogmatic.

In the U.S., a substantial move forward in collating knowledge of calf growth and nutrient requirements was accomplished with the release of the Nutrient Requirements of Dairy Cattle, Seventh Ed. (NRC, 2001). An entire chapter was dedicated to the calf, and a section of the software was designed to apply the information with a simple approach that could be used in the field. New data on calf growth and development were being generated during the writing of NRC (2001) and those data will be used to refine what has been published previously.

Guidelines for milk replacer formulation and calf feeding have been developing on a widespread, commercial basis since the 1950s. Roy (1964) examined the origins of commercial milk replacer and clarified the context in which developments like fat concentration, ingredient choice, and feeding practices were made. It is clear from the review of Davis and Drackley (1998) that considerable research has been completed over the last 50 years to elucidate the specific nutrient requirements of the young calf, as well as the potential benefits (or risks) of various feeding practices. It is therefore logical to assume that the advances in this understanding would be subsequently reflected in current feeding practices. However, it appears that nutritional management of the dairy calf, unlike other neonates, is driven more by financial decisions than by optimizing our understanding of their development. Also, aside from the veterinarian's clinical perspective, there is a low level of comfort, or a lack of true expertise, on behalf of many industry professionals in discussing calf management and nutrition on a more dynamic and pragmatic basis. Thus, we revert to a "one size fits all" management scheme.

Our challenge as nutritionists and practitioners is to develop a management system that allows us to make the most use of the available data to improve calf management programmes in a cost effective and practical manner.

RECENT STUDIES DESCRIBING THE COMPOSITION OF GAIN

Unlike the functioning ruminant, the composition of live-weight gain in the pre-weaned calf can be effectively manipulated by diet, intake level and source of energy. The foundation of much of this work was conducted years ago and a reasonable database exists on which to frame nutrient requirements for body gain (Donnelly and Hutton 1976ab; Donnelly, 1983; Toullec, 1989). More recently, additional data sets have been created in an effort to refine our understanding of the composition of gain of growing dairy calves, primarily Holstein (Bartlett, 2001; Blome *et al.*, 2003; Diaz *et al.*, 2001; Tikofsky *et al.*, 2001) and a small set of Jersey calves (Bascom, 2003).

The work conducted by Diaz *et al.* (2001) was conducted to help refine the existing energy prediction equations for light-weight Holstein cattle. Sixty calves were assigned to a comparative slaughter study designed as an energy titration. Calves were slaughtered at birth, 65, 85 and 105 kg BW for determination of body composition. Treatments were designed to achieve three target rates of body-weight gain (BWG): 0.5 0.9 and 1.4 kg/d. The milk replacer was formulated to contain 300g CP/kg DM and 200g fat/kg DM and was an all-milk protein formulation. This dietary CP content was selected based upon previous studies (Donnelly and Hutton, 1976a,b; Gerrits, *et al.*, 1996) that indicated a plateau in daily protein accretion might be achieved at near maximal DMI with a CP concentration of 300g/kg DM. A goal of the diet formulation was to ensure that protein would not be the most limiting nutrient.

Although not unique, the primary findings of this study were the changes in body composition based on level of nutrient intake and diet formulation. Body fat accretion was moderate (5.7, 9.8 and 9.2 g fat/kg EBW) and the protein level of the diet probably affected both protein and fat accretion. Protein content of the gain was similar among treatments (213, 205 and 211 g CP/kg EBW) and we assumed this maintenance of protein content was due to the protein content of the diet. The effect of level of protein intake on body composition was reinforced by the study of Bartlett (2001) and is similar to effects predicted by Gerrits *et al.* (1996). From the perspective of application, the most important finding was the inability of the existing nutrient requirement equations used in the U.S. (the 1989 and 1996 NRC requirement recommendations for dairy and beef cattle) to represent the energy and protein content of the gain (Table 5.1). This inability to predict the composition of gain reinforced one of the hypotheses of the study; that the current equations, based on beef breeds of cattle and not within the weight range of the study, might not represent the composition of light-weight Holstein cattle. The latest edition (2001 Dairy NRC) was also compared, and found to provide more reasonable predictions than the previous editions. Variation still exists, however, as will be discussed later.

Table 5.1 COMPARISON OF OBSERVED ENERGY AND PROTEIN RETAINED FROM THE STUDY OF DIAZ *et al.* (2001) WITH PREDICTION EQUATIONS USED IN THE 1989 DAIRY (NATIONAL RESEARCH COUNCIL, 1989), 1996 BEEF NRC (NATIONAL RESEARCH COUNCIL, 1996) AND THE 2001 DAIRY CALF MODEL (NATIONAL RESEARCH COUNCIL, 2001).

Treatment	*Retained Energy, MJ/d*				*Retained Protein, g/d*			
	Observed	*Predicted 1989 dairy*	*Predicted 1996 beef*	*Predicted 2001 dairy*	*Observed*	*Predicted 1989 dairy*	*Predicted 1996 beef*	*Predicted 2001 dairy*
1	4.89	4.89	3.85	5.86	137	99	130	103
2	10.38	8.87	7.20	12.55	199	161	213	189
3	11.80	10.25	8.41	14.73	244	183	244	227

A study by Bartlett (2001) evaluated the effects of isoenergetic diets at four protein concentrations and two levels of intake (Table 5.2). A group of baseline calves were slaughtered and the remaining calves were fed treatment diets for 35 days and then slaughtered. The protein content of the milk replacers was 140, 180, 220 and 260g/kg DM and the protein source was all milk protein. Intakes were set at approximately 13g/kg BW and 18 g/kg BW on an as-fed basis and milk replacer was reconstituted to 125g DM/kg water.

The data of Bartlett (2001) reinforce the concept that energy intake drives growth rate and the requirement for protein, thus the protein required will be a function of the desired growth rate. Furthermore, growth rate itself appears to be a poor predictor of body composition in calves of this weight, especially given the linear effect of protein concentration on changes in body composition. The data of Bartlett (2001) reinforce the previous observations of Donnelly and Hutton (1976a,b) that indicated body fat deposition decreased with increasing protein content of the diet (111g/kg EB fat at 157g/kg CP level to 75 g/kg EB fat at 315 g/kg CP level) in a linear manner, while the protein content of the EB increased with increasing protein content. It has been suggested that the decrease in body fat concomitant with the increase in body

Table 5.2 INTAKE OF PROTEIN AND METABOLISABLE ENERGY, AVERAGE DAILY GAIN, AND PROTEIN AND FAT TISSUE RETENTION OVER 35 DAYS FOR CALVES FED FOUR LEVELS OF CRUDE PROTEIN AT TWO LEVELS OF INTAKE (BARTLETT, 2001).

	Formulated crude protein			
	140g/kg	*180g/kg*	*220g/kg*	*260g/kg*
Crude protein intake, kg/d				
13 g/kg BW feeding rate	0.089	0.119	0.150	0.162
18 g/kg BW feeding rate	0.138	0.175	0.214	0.245
Metabolisable energy intake, MJ/d				
13 g/kg BW feeding rate	11.87	12.28	12.88	12.47
18 g/kg BW feeding rate	19.38	18.04	18.41	18.89
Average daily gain, kg/d				
13 g/kg BW feeding rate	0.251	0.306	0.409	0.360
18 g/kg BW feeding rate	0.509	0.561	0.690	0.703
Protein tissue accretion[a], kg				
13 g/kg BW feeding rate	0.90	1.28	2.09	1.85
18 g/kg BW feeding rate	2.09	2.61	3.59	3.96
Fat tissue accretion[b], kg				
13 g/kg BW feeding rate	1.87	1.47	1.54	1.13
18 g/kg BW feeding rate	3.69	3.32	2.98	2.80

[a,b]Significant linear effect of diet on protein and fat accretion.

protein accretion was a function of decreasing energy available for gain, due to the lower fat content of the diets with increased protein content. However, the high energetic cost of protein accretion (Thorbek, 1977), and the amount of protein deposited, demonstrate that the ME allowable gain was not decreased with increasing protein content of the diet, and that the protein requirement of the calf under the feeding conditions of this study did not appear to be met, since a plateau in protein accretion was not observed. Coupled with the data of Diaz *et al.* (2001), these data suggest that to optimize protein deposition in calves fed milk-protein based diets, the protein content of the diet would need to be approximately 280g/kg DM, which is not dissimilar to the crude protein content of whole milk. Equations generated from the pooled data of both experiments predict that to meet the energy allowable protein requirement when calves are gaining in excess of 700 g/d, the protein content of the diet must be at least 260 to 280g/kg DM. This is in agreement with levels predicted in a summary of the literature (Davis and Drackley, 1998) primarily from the data of Donnelly and Hutton (1976a,b). The higher protein content would be necessary to achieve high rates of gain without increased fat accretion. Furthermore, these data clearly demonstrate our ability to manipulate the composition of gain by changing diet composition, prior to weaning. This also has implications for veal calf production.

A study by Tikofsky *et al.*, (2001), was conducted to determine the effect of varying levels of dietary fat and carbohydrate for dairy calves fed under isoenergetic and isonitrogenous intake conditions. Furthermore, the study could assess this potential effect under conditions where calculated protein intakes relative to the energy allowable gain were not considered to be limiting growth (Davis and Drackley, 1998; Diaz, *et al.*, 2001). Previous work in this area was confounded by varying fat and or protein concentrations, but allowing intakes to remain similar, thus creating diets that were not isoenergetic or isonitrogenous (Roy *et al.* 1970; Lodge and Lister, 1973). This lack of control confounded interpretation of the primary effect of fat or carbohydrate on the efficiency of use of the energy source, and on growth rate or body composition.

Milk replacer formulations were manufactured according to protein and fat specifications determined by the investigators so that target DMI for each treatment would enable isoenergetic and isonitrogenous intake conditions among treatments (Table 5.3). Treatment diets consisted of three specially formulated MR (Milk Specialties, Co., Dundee, Ill.). The protein of all MR was derived from all-milk sources, and the fat was primarily tallow. Fat and lactose contents of all diets were formulated to deliver treatments defined as low fat, high lactose (LF); medium fat, medium lactose (MF); and high fat, low lactose (HF). Dry matter intake for calves on all treatments was calculated to supply 1.0 MJ/kg $BW^{0.75}$ for treatment days 1 to 14, and then increased to 1.17 MJ/kg $BW^{0.75}$ from day 15 until final slaughter weight was reached. Target energy intakes for individual calves were adjusted every seven days based on changes in animal weight.

Table 5.3 MILK REPLACER COMPOSITION, INTAKE, GROWTH AND BODY COMPOSITION FOR CALVES FED DIFFERENT CONCENTRATIONS OF FAT AND LACTOSE (TIKOFSKY *et al.*, 2001).

	Milk replacer composition		
	Low fat	*Medium fat*	*High fat*
Gross energy, MJ/kg DM	19.33	21.30	24.14
Protein, g/kg	235.4	248.0	270.0
Fat, g/kg	147.9	216.2	306.2
Lactose, g/kg	552.9	466.9	353.6
		Intake and growth	
N	8	8	8
Days on treatment	54.6	56.1	55.1
Dry matter intake, kg	55.2^{x}	52.8^{xy}	46.8^{y}
GE intake, MJ	1077.8	1124.7	1130.9
Protein intake, kg	13.0	13.1	12.6
Fat intake, kg	8.16^{x}	11.41^{y}	14.33^{z}
Empty body gain, kg/d	0.61	0.61	0.65
Empty body protein gain, kg	5.77	5.70	6.08
Empty body fat gain, kg	3.88^{x}	5.11^{xy}	6.12^{z}
Empty body energy gain, MJ	281.58^{x}	308.36^{xy}	358.15^{z}

^{xyz}Significant treatment effects.

The treatments remained isoenergetic and isonitrogenous throughout the study, thus providing better interpretive data than previous studies (Tikofsky *et al.*, 2001). Two observations were of significant interest; firstly, that growth rate was not influenced by changes in the ratio of fat to carbohydrate, and secondly that body composition was significantly altered by fat content of the diet, independently of the rate of growth. This study demonstrated that the source of energy, primarily increasing the fat content of the diet, could affect body composition and that fat as an energy source does not appear to be used more efficiently than carbohydrate in supporting protein deposition. Lodge and Lister (1973) investigated the effect of source of energy in milk-fed calves in an effort to understand the role of dietary energy on the biological value (BV) of protein. Their data indirectly demonstrated that the higher the energy intake, especially from carbohydrate, the higher the BV of protein. Part of the increased BV was due to increased energy decreasing the protein concentration, but it also reinforces the concept that fat as an energy source is not used as efficiently as carbohydrate in promoting protein synthesis and accretion. In the study of Tikofsky *et al.* (2001), the apparent partial efficiency of fat deposition (fat accretion over fat intake) was 0.45 on average across the three treatments. Donnelly (1983) made similar observations concerning energy utilization in calves.

Energy requirements

MAINTENANCE

Maintenance energy requirements and fasting heat production of calves have been extensively researched and reviewed. This section will deal only with the maintenance requirement of the calf, since fasting heat production can be more variable due to weight of the animal, previous level of nutrient intake and composition of the diet. Toullec (1989) summarized the data from pre-ruminant calves and indicated that the range in maintenance requirement was 0.377 to 0.452 MJ/kg$^{0.75}$ per day. Davis and Drackley (1998) revisited the maintenance values in the literature and concluded the range in ME for maintenance (MEm) requirements was 0.39 to 0.47 MJ/kg$^{0.75}$. Thus, when calves are within thermo-neutral conditions, the average maintenance energy requirement for pre-ruminant calves appears to be close to 0.42 MJ/kg$^{0.75}$ per day on a metabolisable energy (ME) basis. The maintenance value published in a previous edition of the NRC (1989) was 0.36 MJ/kg$^{0.75}$ on a NE basis. From the available data the efficiency of use of energy for maintenance for milk fed calves under no stress conditions ranges from 0.83 to 0.86 (Blaxter, 1952; Holmes and Davey, 1976; Agricultural Research Council, 1980; Arieli *et al.*, 1995). Applying the value of 0.83 to the NRC (1989) NE value gives 0.433 MJ/kg$^{0.75}$ on an ME basis, which is similar to the value adopted by the current NRC (2001) and the average observed in the literature.

Using data from 153 individual calves in three separate studies (Bartlett, 2001, Diaz *et al.* 2001, Tikofsky *et al.*, 2001) the maintenance energy value was re-evaluated by regression and the intercept on a NE basis was 0.35 MJ/kg$^{0.75}$. Applying the ARC (1980) efficiency of use (0.83) to convert to a ME basis provides a value of 0.422 MJ/kg$^{0.75}$ when averaged across the data. This supports the value adopted by the current NRC (2001) under no-stress conditions (Table 5.4).

Table 5.4 MAINTENANCE REQUIREMENTS OF CALVES LESS THAN 21 DAYS OF AGE AT DECREASING TEMPERATURE AND INCREASING BODY WEIGHT[a].

Temperature, °C	*15*	*10*	*5*	*0*	*-5*	*-10*	*-15*
Body weight, kg	*MJ metabolisable energy per day*						
30	5.4	7.2	8.1	9.0	9.6	10.8	11.6
40	6.9	8.5	9.4	10.3	11.2	12.1	12.9
50	7.9	9.7	10.6	11.5	12.4	13.3	14.2
60	9.1	10.8	11.7	12.6	13.5	14.4	15.3
70	10.2	12.0	12.8	13.7	14.6	15.5	16.4

[a]Fifteen degrees Celsius is the lower critical temperature for a calf of this age.

Obviously there are other factors, such as age, environmental temperature and exposure, disease and other stressors that affect the maintenance requirement of the neonatal calf. The thermo-neutral range for young adapted calves appears to be 15 to 25°C (Schrama, *et al.*, 1993; Arieli *et al*, 1995) and the lower end of the range decreases with increasing body weight and age. Data from several studies (Gonzalez-Jimenez and Blaxter, 1962; Holmes and Davey, 1976; Scibilia *et al.*, 1987; Shrama *et al.*, 1992, 1993; Arieli *et al.*, 1995) demonstrate that the maintenance requirements of young calves will vary by age and weight, temperature and movement to another environment (transportation combined with new surroundings).

From the study of Gonzalez and Blaxter (1962) the lower critical temperature (LCT) of newborn calves was approximately 13°C and decreased with increasing age to about 6 °C at 30 days of age. Scibilia *et al.* (1987) observed a 32% increase in maintenance energy requirement when calves aged 6 to 28 days were housed at – 4 °C compared with 10 °C (0.56 versus 0.42 MJ/kg$^{0.75}$, respectively) and fed energy at maintenance or slightly above, indicating that the LCT in their study was close to 10 °C and consistent with the data of Gonzalez and Blaxter (1962).

The MEm requirements calculated by Arieli *et al.* (1995) demonstrated a significant age-related adjustment to both energy intake and temperature. Calves approximately seven days of age were transported to the research facility, fed two levels of intake (70 or 110% of MEm assuming 0.46 MJ/kg$^{0.75}$/d requirement) at either 7.5 or 19 °C over a two-week period. Heat production on average was 0.05 MJ/kg$^{0.75}$/d higher during week one compared to the end of week two. This suggests that an adaptation period of approximately 14 days is required when moving calves to a new environment, a new temperature and a modified diet, in addition to the apparent age-related changes.

NRC (2001), using the summary of Davis and Drackley (1998) adopted an adjustment of 0.09 MJ/kg$^{0.75}$ per day per degree to the basal MEm requirement when the temperature falls below the LCT (Schrama *et al.*, 1993). This adjustment was applied to a calf younger than 21 days of age when the temperature was 15°C or lower. When the calf is older than 21 days of age, the adjustment is applied starting at 5 °C. Since the release of NRC (2001), many empirical evaluations of this MEm adjustment have been conducted and it appears, based on field observations and controlled research data, the adjustment accounts for a reasonable amount of the variation in ME available for gain during periods of cold stress.

An additional adjustment for stressed or unadapted calves might be necessary for calves that have been transported with inadequate feed and water intake. Based on work with commercial calf growers where calves are moved reasonable distances to new environments, the additional heat increment observed by Arieli *et al.* (1995) (0.05 MJ/kg$^{0.75}$/d) should be applied for 14 days to the basal maintenance requirement, separately if the temperature is greater than 15°C, or in combination with the LCT adjustment if the calves are also cold stressed. On average, this adjustment would be equivalent to ME contained in approximately 100 g DM of milk or milk replacer for the average calf of 45 to 50 kg BW.

The National Animal Health Monitoring Service (2002) in the U.S. conducts periodic surveys of dairy cattle health. The recent survey indicated that calf mortality from birth to weaning in the U.S. was 8.7% (range 7.7 to 9.4% among three distinct herd sizes). Sickness or morbidity, although not reported in the latest survey, was previously reported at 34% suggesting that over 40% of calves were considered to be ill or unthrifty during the preweaning period. In North America, from October to March, a predisposition for this amount of illness could be something as simple as feeding at or below maintenance requirements. Since many areas of the world experience temperatures that could induce severe cold stress in a calf, adjusting requirements to account for cold stress should help reduce calf deaths and sickness.

GROWTH

Requirements for growth are a function of the composition of the gain. Recent work supports previous studies that indicate the composition of gain is directly influenced by at least three factors; 1) overall nutrient intake; 2) source of energy; and 3) protein content of the diet. These factors have additive effects on how calves partition nutrients into tissue.

Energy conversions

Conversion of intake energy (IE) to energy of useful products requires that indigestible feed, gaseous, urinary and heat energy losses be accounted for as we progress from one energy currency to another. Intake energy is the amount of energy consumed by the animal and is a function of the gross energy (GE) content of the feed. Conversion from gross energy (GE) to digestible energy (DE) requires that faecal losses be accounted for, and in general the digestibility of energy from milk or milk ingredients is high. The published range for the conversion of GE to DE is 0.87 to 0.98 and the average value is 0.97. The data of Diaz *et al.*, (2001) and Blome *et al.*, (2003) fall within this range and are close to the average value (0.95 and 0.94 respectively).

The average conversion of DE to metabolizable energy (ME) chosen by NRC (2001) was 0.96. The value observed by Diaz *et al.* (2001) was 0.96, which agreed with previous findings (Neergard, 1976; Toullec, 1989), although the data of Blome *et al.*, (2003) indicated an efficiency of 0.95 for converting DE to ME. Thus the average value of 0.96 seems appropriate. Summarizing the conversion from GE to ME, using the average efficiency values provides an overall efficiency of 0.93, which was the value adopted by NRC (2001). The average efficiency of conversion of GE to ME reported by Davis and Drackley (1998) was 0.936. Using the values from the Illinois and Cornell data sets gives a slightly lower conversion of GE to ME at 0.91. From the data set of Bartlett (2001) the conversion of GE to ME was 0.90. For the purposes of this chapter, 0.91 will be adopted for the efficiency of converting GE to ME.

A more difficult issue exists when discussing the conversion of ME to net energy for gain (NEg) or retained energy (RE). The calculation must be made after accounting for the amount of ME used for maintenance, thus ME available for growth is prone to some variation due to the accuracy of predicting maintenance requirements. Toullec (1989) suggested that a constant conversion of 0.69 should be used for ME to RE and that the value did not vary with live weight, weight gain or diet composition. Other observations are in the range of 0.63 to 0.79 (Johnson and Elliot, 1972; Holmes and Davey, 1976; Neergard, 1976). The value of 0.69 was adopted by NRC (2001), as suggested by Davis and Drackley (1998), but based on more recent studies, a constant does not seem appropriate. A constant conversion of ME to RE implies that the composition of tissue gain is similar under all conditions, but data from experiments such as Bartlett (2001) and Tikofsky *et al.* (2001) do not support that concept.

Using the data from University of Illinois and Cornell data sets, the conversion of ME to RE varied from 0.47 to 0.74, and based on study and treatment design, these differences were apparently due to nutrient intake, source of energy, and protein content of the diet under isoenergetic intake conditions. Under conditions of high protein content and low nutrient intake (energy limited), fat accretion will be reduced, so the conversion of ME to RE will have a lower value, as illustrated in the data of Diaz *et al*., (2001). In the Diaz *et al.* study, the crude protein content of the diets was fixed (300 g/kg) and the calves had three intake levels resulting in conversion of ME to RE values of 0.482 to 0.74. This is consistent with data from Bartlett (2001) and Tikofsky *et al.* (2001). The average conversion of ME to RE from the eight separate treatments of Bartlett *et al.* was 0.52 (range, 0.47 to 0.59); the conversion tended to decrease with increasing protein content of the diet, even though the diets were isoenergetic on a ME basis, reflecting the increasing protein accretion that paralleled protein supply.

Conversely, diets with high fat content and adequate protein content will result in higher efficiency values because the composition of gain will result from more fat accretion if fed under thermo-neutral conditions. In the study of Tikofsky *et al*., (2001) diets were formulated to vary in source of energy, and fat content ranged from 148 g/kg to 306 g/kg in the milk replacer. The diets were fed at similar $KJ/kg^{0.75}$ (1.0 $KJ/kg^{0.75}$ from day 1 to day 14, and 1.17 $KJ/kg^{0.75}$ from day 15 to slaughter) and the protein content was formulated to be similar per unit of GE. Total energy intake was not different among treatments and averaged 1,111 MJ GE. Total fat intake among the three diets was 8.16, 11.41 and 14.33 kg, while protein intake was similar and averaged 12.9 kg. The conversion of ME to RE among the three diets was 0.60, 0.61 and 0.70, and increased with increasing fat intake and accretion. Using the data from Diaz *et al*. (2001), Bartlett (2001) and Tikofsky *et al.* (2001) the mean conversion of ME to RE was 0.60 ± 0.08; a difference of 0.15 compared to the value suggested by Toullec (1989). The data from these more recent studies suggest that a more dynamic approach to calculating conversion of ME to RE should be applied to the calculation of nutrient requirements.

A potential difference between the predicted energy value of Toullec (1989) and the value derived from the current data sets could be an underestimation of the maintenance requirement. However, Toullec used heavier veal calves fed higher levels of intake, and diets more similar to the highest fat diet in the Tikofsky *et al.* study. Thus, with higher fat intakes, the conversion of ME to RE will be higher, and this might explain most of the difference. Furthermore, many of the studies summarized by Davis and Drackley (1998) fed whole milk, which would provide higher fat and energy intakes than most milk replacers available commercially.

The data suggest that if the fat content of the diet is in the range of whole milk (28 to 40 g/kg), the efficiency of use of ME for RE will be 0.69 or higher. However if the fat content of the diet is in the range of most commercially available milk replacers (120 g/kg to 220 g/kg), and the calves are of lighter body weight (< 85 kg) then the conversion of ME to RE will be in the range of 0.59 to 0.63. These values were used to re-evaluate two sets of predicted energy requirements. In the paper by Diaz *et al.* (2001) a comparison of the actual RE versus the RE predicted by the Toullec equation (1989) and using 0.69 for the efficiency of ME to RE, showed that the predicted RE was overestimated by 25%. Recalculating the Toullec predictions using 0.60 as the conversion of ME to RE results in a 9% over-prediction of RE requirements. A similar re-evaluation of the estimated ME requirements for pre-weaned calves as described by Davis and Drackley (pp. 62 Table 4.7, 1998) was conducted; application of the 0.60 ME to RE conversion accounted for the over-prediction of ME required for growth. To improve the predictive capability of a calf model, a more appropriate approach for ME to RE conversions would be to make the adjustment dynamic with energy intake, energy content of the diet, source of energy and weight of the calf.

A new equation to describe the composition of gain of calves from birth to 105 kg BW was generated from the combined Illinois and Cornell data sets (Bartlett, Diaz and Tikofsky):

$$RE = 0.4431 \times EWG^{1.1684} \times EBW^{0.75}$$

where RE is retained energy (Mcal/d), EWG is empty weight gain (kg/d) and EBW is empty body weight (kg). [1 Mcal = 4.184 MJ].

This equation provides a better estimate of the energy content of the gain of calves up to approximately 100 kg compared to the equation of Toullec (1989) (Table 5.5). Using this Cornell-Illinois equation, coupled with a fixed conversion of ME to RE of 0.60, the equation will predict lower energy accretion for calves at 50 kg consistent with the higher protein deposition and energy retention from the studies and consistent with the body weight of and diets fed to dairy replacement calves. The Cornell-Illinois equation will predict slightly higher energy requirement values (~0.53 MJ ME/d) than the Toullec equation, reflecting an increase in fat deposition as the calves achieve heavier body weights.

Table 5.5 PREDICTED ENERGY REQUIREMENTS USING THE CORNELL-ILLINOIS EQUATION COMPARED TO THE TOULLEC (1989) EQUATION AT 50 AND 80 kg BODY WEIGHT.

Body weight	*50 kg ME, MJ/d*		*80 kg ME, MJ/d*	
Rate of gain, kg/d	*Toullec*	*Cornell-Illinois*	*Toullec*	*Cornell-Illinois*
0.2	2.04	1.86	2.40	2.65
0.4	4.69	4.18	5.50	5.96
0.6	7.62	6.73	9.01	9.57
0.8	10.76	9.41	12.71	13.40
1.0	14.07	12.20	16.38	17.30

Protein requirements

MAINTENANCE AND GROWTH

Using the factorial approach taken by Blaxter and Mitchell (1948) and Roy (1970), as summarized by Davis and Drackley (1998), the estimate of protein requirements for maintenance of a pre-ruminant calf weighing 50 kg and gaining approximately 450 g is 30 g/d. The published protein requirements for maintenance of a calf at this weight range from 21 to 35 g/d; based on the current data, these values appear reasonable. Obviously, protein requirements for maintenance will increase with increasing dry feed intake as endogenous and metabolic faecal losses increase, but that is not considered here.

The equation utilized by NRC (2001) to calculate both maintenance and growth requirements for protein is: $6.25 * [1/BV * (E + G + M * D) - M * D]$, where BV = biological value, E = endogenous urinary N ($0.2\ LW^{0.75}$), G = N in gain (30 g/kg LWG), M = metabolic faecal N (1.9 g/kg dry matter intake) and D = dry matter consumed (Blaxter and Mitchell, 1948; Roy, 1970). The review by Davis and Drackley (1998) was adopted by NRC (2001) and chose an efficiency of use of 80%, which is high for calves fed adequate protein and nutrients. The equation appears to work reasonably well for calves prior to weaning, but the BV adopted by NRC (2001) appears to be too high. Some discussion of the constant value used to describe the N content of gain is warranted.

The biological value of milk proteins is a function of their amino acid profile, the amount of protein fed relative to requirements, and the energy allowable gain (Mitchell, 1924a,b; Blaxter and Mitchell, 1948 and Roy *et al.*, 1970). Milk proteins typically have a biological value that ranges from 0.85 to 0.98 (Mitchell, 1924b) when fed at low to limiting levels. Increasing the protein intake in rats from 0.05 to 0.10 decreased BV by 0.09.

In the study of Diaz *et al.* (2001), protein content of the diet was fixed (300g/kg) and total nutrient intake was increased to achieve higher rates of LWG. Using the factorial equation the BV of protein was 0.36, 0.76, and 0.82 for calves gaining 0.6, 0.98 and 1.29 kg/d respectively and demonstrates that the BV of protein is energy dependent. A response was observed in the data of Donnelly and Hutton (1976a,b), where the dietary protein ranged from 157 to 315 g/kg DM and the BV of the protein ranged from 0.84 to 0.56 as the protein content of the diets increased, demonstrating that BV is also dependent on the protein supply relative to requirements. Furthermore, Blome *et al.* (2003) fed diets that ranged from 160 to 260 g CP/kg DM at levels that allowed approximately 0.38 to 0.62 kg/d gain; the BV of the absorbed protein ranged from 0.68 to 0.72 and was unaffected by level of intake. However, in the study of Blome *et al.* (2003), growth rate nearly doubled from the lowest to the highest dietary protein content, suggesting that protein was limiting gain and further indicating that the BV of protein depends on energy intake, since energy intake was similar among treatments. Based on the current available data, a more reasonable value appears to be in the range of 0.70 to 0.73 for the young non-ruminant calf. Thus, using a BV of 0.80 will under-estimate protein required for gain using the factorial approach. As with the energy transactions, a more dynamic approach to calculating the efficiency of use of absorbed protein should be taken that utilizes energy intake, source of energy, protein intake and weight of the calf.

Protein requirements for growth are a function of the energy allowable gain. Summarizing the composition of protein gain in the literature, Davis and Drackley (1998) adopted a value of 30 g N/kg gain. This approach was also adopted by the Dairy NRC (2001) and assumes that the composition of both protein gain and whole empty body gain is relatively fixed, which is not necessarily true. Although the protein content of the fat free empty body is relatively fixed, there is some variation in the protein content due to the dilution of gain by fat. However, the range in N content of gain in the current data sets was 27 to 35 g N/kg live-weight gain and the variation based on study and diet did not suggest that the average value of 30 g/kg LWG adopted by the NRC (2001) for field application is unreasonable. However, a more dynamic approach, as taken by Gerrits *et al.* (1997), would potentially reduce the variation in predicting protein requirements.

Application

Application of the discussion within this chapter to the calculation of requirements of a calf leads to the values in Table 5.6. These values were derived assuming an efficiency of use of ME for RE of 0.60 and a BV of 0.72. The estimated feed efficiencies are provided as a reminder of the potential of calves at this stage of growth. Note that the efficiencies increase as nutrient intake above maintenance increases and that the estimates are reasonable for this weight of calf and fall within the range of published values. Under controlled conditions, feed efficiency would be a reasonable tool for evaluating diet formulation, environmental conditions and stress.

Table 5.6 NUTRIENT REQUIREMENTS AND ESTIMATED FEED EFFICIENCIES FOR A 50 kg CALF UNDER THERMO-NEUTRAL CONDITIONS, BASED ON MODIFICATIONS DESCRIBED IN THIS CHAPTER OF THE DAIRY NRC (NRC, 2001) USING THE CORNELL AND ILLINOIS DATA SETS.

Rate of gain, kg/d	*Dry matter intake, g/kg body weight*	*Metabolisable energy, MJ/d*	*Crude protein, g/d*	*Crude protein, g/100g DM*	*Estimated feed efficiency, g gain/g feed*
0.2	10.5	9.8	94	18.0	0.38
0.4	13.0	12.1	150	22.4	0.63
0.6	15.7	14.6	207	26.6	0.77
0.8	18.4	17.3	253	27.4	0.86
1.0	23.0	20.1	318	28.6	0.87

Summary

The field-applicable approach taken by NRC (2001) for calculating the requirements of the pre-weaned calf was a tremendous step forward. New data are now available with which to refine that approach and have been partially discussed here. A more dynamic approach would improve the predictive capability of the current system and allow for greater refinements based on source of energy (fatty acid profiles) and amino acids. However, the current approach if appropriately applied will allow nutritionists and dairy producers to make more appropriate nutritional management decisions.

References

Agricultural Research Council. 1980. *The Nutrient Requirements of Ruminant Livestock.* Farnham Royal, Slough, England: Commonwealth Agricultural Bureaux.

Arieli, A., Schrama, J.W., Van Der Hel, W. and Verstegen, M.W.A. (1995) Development of metabolic partitioning of energy in young calves. *Journal of Dairy Science,* **78**,1154-1162.

Bartlett, K. S. (2001) *Interactions of protein and energy supply from milk replacers on growth and body composition of dairy calves.* M. S. Thesis. University of Illinois, Urbana-Champaign.

Bascom, S.S. (2003) *Jersey calf management, mortality, and body composition,* Ph.D. Dissertation, Virginia Polytechnic Institute and State University, Blacksburg, VA.

Blaxter, K.L. and Mitchell, H. H. (1948) The factorization of the protein requirements of ruminants and of the protein value of feeds, with particular reference to the significance of the metabolic fecal nitrogen. *Journal of Animal Science,* **7**,351.

Blaxter, K. L. (1952) The nutrition of the young Ayrshire calf. 6. The utilization of the energy of whole milk. *British Journal of Nutrition,* **6**,12-19.

Blome, R. Drackley, J. K., McKeith, F. K., Hutjens, M. F. and McCoy, G.C. (2002) Growth, nutrient utilization, and body composition of dairy calves fed milk replacers containing different amounts of protein. *Journal of Animal Science,* **81**,1641-1655.

Davis, C. L. and Drackley, J. K. (1998) *The Development, Nutrition, and Management of the Young Calf.* Iowa State University Press, Ames, IA.

Diaz, M. C., Van Amburgh, M. E., Smith, J. M., Kelsey, J. M. and Hutten, E. L. (2001) Composition of growth of Holstein calves fed milk replacer from birth to 105 kilogram body weight. *Journal of Dairy Science,* **84**, 830-842.

Donnelly, P. E. and Hutton, J. B. (1976a) Effects of dietary protein and energy on the growth of Friesian bull calves. I. Food intake, growth, and protein requirements. *New Zealand Journal of Agricultural Research,* **19**, 289-297.

Donnelly, P. E. and Hutton, J. B. (1976b) Effects of dietary protein and energy on the growth of Friesian bull calves. II. Effects of level of feed intake and dietary protein content on body composition. *New Zealand Journal of Agricultural Research,* **19**, 409-414.

Donnelly, P. E. (1983) Effects of dietary carbohydrate:fat ratio on growth and body composition of milk-fed calves. *New Zealand Journal of Agricultural Research,* **26**, 71-77.

Gerrits, W.J.J., Tolman, G. H., Schrama, J. W., Tamminga, S., Bosch, M. W. and Verstegen, M.W.A. (1996) Effect of protein and protein-free energy intake on protein and fat deposition rates in preruminant calves of 80 to 240 kg live weight. *Journal of Animal Science,* **74**, 2129-2139.

Gerrits W. J. J., Dijkstra, J. and France, J. (1997) Description of a model integrating protein and energy metabolism in preruminant calves. *Journal of Nutrition,* **127**, 1229-1242.

Gonzalez-Jimenez, E. and Blaxter, K. L. (1962) The metabolism and thermal regulation of calves in the first month of life. *British Journal of Nutrition,* **16**, 199-212.

Holmes, C. W. and Davey, A. W. F. (1976) The energy metabolism of young Jersey and Friesian calves fed fresh milk. *Animal Production,* **23**, 43-53.

Johnson P. T. C. and Elliot, R. C. (1972) Dietary energy intake and utilization by young Friesian calves. 3. The utilization by calves of energy in whole milk. *Rhodesian Journal of Agricultural Research,* 10, 135-142.

Lodge, G. A. and Lister, E. E. (1973) Effects of increasing the energy value of a whole milk diet for calves. I. Nutrient digestibility and nitrogen retention. *Canadian Journal of Animal Science,* **53**, 307- 316.

Mitchell, H.H. (1924) A method of determining the biological value of protein. *J. Biological Chemistry,* **58**, 873-903.

Mitchell, H.H. (1924) The biological value of proteins at different levels of intake. *J. Biological Chemistry,* **58**, 905 – 922.

National Animal Health Monitoring Service (2002). United States Department of Agriculture/Animal and Plant Health Inspection Service. *Dairy 2002. Part 1: Reference of the dairy health and management in the United States, 2002.* Fort Collins, CO. web access: www.aphis.usda.gov/vs/ceah/cahm

National Research Council (1989). *Nutrient Requirements of Dairy Cattle.* Sixth rev. Ed., Natl. Acad. Sci., Washington, D. C.

National Research Council (1996) *Nutrient Requirements of Beef Cattle.* Seventh rev. Ed., Natl. Acad. Sci., Washington, D. C.

National Research Council (2001) *Nutrient Requirements of Dairy Cattle.* Seventh rev. Ed., Natl. Acad. Sci., Washington, D. C.

Neergard, L. (1976) A comparative study of nitrogen and energy metabolism in young calves fed three liquid diets. In: *Energy Metabolism of Farm*

Animals. (Ed. M Vermorel), pp. 205-208. European Association of Animal Production. Claremont-Ferrand, France.

Roy, J. B. H. (1964) The nutrition of intensively-reared calves. *Veterinary Record,***76**, 511-526.

Roy, J. B. H., Stobo, I. J. F. and Gaston, H. J. (1970) The nutrition of the veal calf. 3. A comparison of liquid skim milk with a diet of reconstituted spray dried skim-milk powder containing 20% margarine fat. *British Journal of Nutrition,* **24**, 459-475.

Roy, J.H. B. (1980) *The Calf* (4th Ed.). Butterworths, London.

Schrama, J.W., van der Hel, W., Arieli, A. and Verstegen, M.W.A. (1992) Alteration of energy metabolism of calves fed below maintenance during 6 to 14 days of age. *Journal of Animal Science,* **70**, 2527-2532.

Schrama, J. W., Arieli, A., van der Hel, W. and Verstegen, M.W.A. (1993) Evidence of increasing thermal requirement in young, unadapted calves during 6 to 11 days of age. *Journal of Animal Science,* **71**, 1761-1766.

Scibilia, L.S., Muller, L. D., Kensinger, R.S., Sweeney, T.F. and Shellenberger, P.R. (1987) Effect of environmental temperature and dietary fat on growth and physiological responses of newborn calves. *Journal of Dairy Science,* **70**, 1426-1433.

Tikofsky, J. N., van Amburgh, M. E. and Ross, D. A. (2001) Effect of varying carbohydrate and fat levels on body composition of milk replacer-fed calves. *Journal of Animal Science,* **79**, 2260-2267.

Thorbek, G. (1977) The energetics of protein deposition during growth. *Nutrition and Metabolism,* **21**, 115-118.

Toullec, R. (1989) Veal calves. *In Ruminant Nutrition – Recommended Allowances and Feed Tables* (edited by R. Jarrige). INRA. John Libby, London.

6

NUTRIENT SOURCES FOR LIQUID FEEDING OF CALVES

K.G. TANAN
Provimi Research and Technology Centre, Brussels, Belgium

Introduction

After birth, liquid feeding represents the primary source of nutrients for the calf and must provide proteins, carbohydrates, fats, minerals and vitamins. The first colostrum, given in the first day of life, apart from being a nutritional source for the newborn, is essential to provide the bioactive substances to strengthen the immune system and develop the digestive tract (Blum and Hammon, 2000, Tanan and Newbold, 2002). The following colostrums, so called transitional milk, are generally given for the first two to three days of life, after which the calf receives whole milk or a milk replacer. The calf can receive from less than 25 kg of dry matter from liquid feed in an early weaning system to more than 50 kg dry matter when intake of liquid feed is maximised.

The digestive tract of the calf after birth presents an enzymatic profile adapted to digest nutrients of dairy origin only. The digestive capacities increase during the first month of life and the secretion of enzymes adapts to digest more complex proteins and carbohydrates; stimulated further by the ingestion of solid feed (Le Heurou-Luron, Guilloteau, Wicker-Planquart, Chayvialle, Burton, Mouats, Toullec and Puigserver, 1992). Consequently, liquid feed should contain highly digestible nutrients adapted to the digestive physiology of the young calf. Producers expect their weaning programme to provide healthy growing calves and the feeding rate and composition of the liquid feed should respect calf health and be compatible with an economical and profitable weaning programme.

The production of milk replacer for rearing calves in the European Union has been quite stable for the last decade: 344 000 MT in 2002, 341 000 MT in 1992 (FEFAC, for 11 countries). The tonnage in 2002 represented about 0.25 of total calf milk replacer production; the rest being used by the veal industry. In 1992, the average inclusion rate of skim milk powder was 0.42, decreasing to 0.28 in 2002. This decrease reflects a shortage induced by the milk quota policy and a higher utilization by the dairy industry to manufacture products such as cheese.

Alternative ingredients have been used to replace skim milk powder as a source of protein and lactose, such as whey products whose production has increased in parallel to that of cheese. In addition, the ban on the use of ruminant fats in calf milk replacer in 2001, and of all animal fats later on, has forced the use of alternative vegetable fat sources. Consequently, the formulation of calf milk replacer has seen a change in the last decade in its source of nutrients.

The objective of this chapter is to review the principal current ingredients available and permitted under European Union (EU) legislation, and to evaluate them as sources of nutrients in calf milk replacers, specifically for rearing the young heifer calf.

Dairy ingredients - process and composition

Whole milk is composed, per kg, of: 87 g water, 33-47 g fat, 32-35 g crude protein (caseins, whey proteins and non protein nitrogen in the proportion 80/15/5), 49-46 g lactose and 7 g minerals (De Wit, 2001; Mahaut, Jeantet, Schuck and Brulé, 2000a) (Table 6.1). These indicative compositions are representative of the composition of Holstein whole milk and vary according to diet and season.

Table 6.1 COMPOSITION OF WHOLE MILK (g/kg DM) AND OF COMMONLY USED INGREDIENTS IN CALF MILK REPLACERS (g/kg AIR DRY MATTER)

	Source	*DM*	*Crude protein (N*6.38)*	*Crude fat*	*Lactose*	*Crude fibre*	*Ash*	*Ca*	*P*
Whole milk	1	130	254	308	377	—	63	10	7.5
	2	130	246-269	254-362	377	—	54	9.5	7.3
Skim milk powder	4	947	341	16	478	—	82	14.7	10.2
Sweet whey powder	1,3	950-965	110-140	10-0	700-50	—	80-90	7.6	6.8
Delactosed whey powder	1,3	950-965	200-250	15-25	480-40	—	150-220	28.9	10.7
Whey protein concentrate (a)	1,3	960-970	340-800	30-80	100-550	—	40-80	4.8	5.1
Soya protein concentrate	4	950	665	5	—	40	65	3	8
Soluble wheat gluten	4	950	840	50	—	11	11	0.5	2.2

1: Toullec, 1989; 2: Mahaut *et al.*, 2000a; 3: de Wit, 2001; 4: INRA-AFZ, 2002 (indicative values)

a: commonly used WPC is 340 g CP, 500 g lactose.

The dairy industry collects milk and processes it to create many different products for direct human consumption or for the food sector (liquid milk, fermented milk, butter, cream, cheese, yoghurt, whey, whole milk powder, skim milk powder, caseinate, lactose..).

The collected whole milk is slightly heated, standardised to adjust its fat content and then pasteurized (de Wit, 2001). At that stage, the standardized milk may be used for the production of cheese or be dried as skim milk powder, for example. The composition of skim milk powder is given in Table 6.1. Its usage in calf milk replacers is strictly regulated by EU legislation and its incorporation rate has to be 0.5 to get the EU subsidies.

The term 'whey' is a very broad term which requires further explanation to better understand its origin and the different possible types. Whey, by definition, is the liquid fraction remaining after the precipitation of the caseins. For the production of semi-hard types of cheese (the major cheese type in Europe), lactic acid bacteria and rennet (a mixture of chymosin and pepsin, enzymes naturally present in the abomasum of the calf) are added to the milk. Under the action of rennet, the caseins form a curd and entrap the fat globules, while the whey proteins remain in the liquid fraction (de Wit, 2001). The renneted whey obtained in this way is called 'sweet whey'. An alternative 'acid' whey is obtained when caseins are precipitated either by a natural acidification (fresh cheese), using only lactic bacteria to produce lactic acid from lactose, or by an artificial acidification when either hydrochloric acid or sulphuric acid are added (de Wit, 2001). The different types of whey have different compositions and final pH (Table 6.2). The major difference between sweet whey and acid whey is the presence of a glycomacropeptide in sweet whey. This is produced during the hydrolysis of k-casein by chymosin (de Wit, 2001). The glycomacropeptide is made of 64 amino acids and represents 0.15 to 0.20 of the protein of sweet whey (Saito, Yanaji and Itoh, 1991).

The different nutrients can be concentrated or discarded by different processes such as demineralisation, crystallisation, isolation and fractionation (De Wit, 2001). In the feed sector, the affordable products are delactosed whey (obtained by crystallisation) and whey protein concentrate from sweet whey (obtained by ultrafiltration).

All these processes occur in liquid phase and the products are unstable. In order to reduce storage and transportation costs, and to provide a stable low moisture product, it is necessary to dry the ingredients. The most common way is spray-drying, where fine droplets of pre-concentrated liquid are atomised in a drying tower (Mahaut, Jeantet, Brulé and Schuck, 2000b). This process was developed in the 1970s to dry large quantities of liquid. Although it is not the most cost effective way to evaporate water, the spray-dried powders remain highly soluble (Mahaut *et al.*, 2000b).

The major milk-derived ingredients used in calf milk replacers are: skim milk powder, whey powder, delactosed whey powders and whey protein concentrate. Their composition is given in Table 6.1. They are sources of proteins, lactose and minerals.

Table 6.2 TYPICAL COMPOSITION OF ACID AND SWEET WHEYS, g/l

	Acid whey		*Sweet whey*	
Constituents	*Casein whey*	*Lactic whey*	*Gouda whey*	*Camembert whey*
Total solids	65.0	65.0	65.0	65.0
True protein	6.0	6.0	6.2	6.2
NPN (N*6.38)	2.0	2.2	2.4	2.4
Lactose	47.0	40.0	47.0	45.0
Lipid	0.3	0.3	0.5	0.5
Ash	7.9	7.9	5.3	5.6
Calcium	1.6	1.6	0.6	0.6
Phosphorus	1.0	1.0	0.7	0.7
Lactic acid	0.2	0.6	0.2	0.3
pH	4.7	4.5	6.4	6.0

adapted from de Wit (2001)

Fat sources are generally incorporated in the liquid whey with an emulsifier before the drying process. During the atomisation, the fat globules, of very small size, are surrounded by the whey proteins. This gives the advantage of creating a very stable emulsion during the preparation of the liquid feed at the farm and a very digestible source of fat for the young animal.

Skim milk powders are classified according to the denaturation of the whey proteins that occurs during the pasteurisation, concentration and drying processes. Whey proteins are very sensitive to heat and the undenatured whey protein nitrogen index (WPNI, expressed in mg of N /g of powder) constitutes an official method to classify the powders according to their historical thermal treatments. four classes are defined: low heat (WPNI >= 6), medium heat (WPNI between 4.5 and 5.99), medium-high heat (WPNI between 1.51 and 4.49) and high heat (WPNI less than 1.5) (Niro Atomizer, 1978, Mahaut *et al.*, 2000a). For feed application, only low and medium skim milk powders are used. The high heat skim milk has specific properties developed for the food sector. The effect of heat treatment was well studied in the 1960s (review of Davis and Drackley, 1998a). It was clearly demonstrated that heat-damaged skim milk powders do not clot properly (Emmons and Lister, 1976), have lower digestibility, can cause diarrhoea (Roy, 1970) and are not suitable for the very young calf of less than one month old (Longenbach and Heinrichs, 1998).

In whey based powders, the proportion of denatured protein can be evaluated by determining the proportion of protein that precipitates when the powder is diluted in a buffer at pH 4.6. Heat-denatured lysine is not detected by a normal amino acid analysis and the determination of the lysine content of the powder can be a good indicator of the thermal treatment applied to the powder.

Crude protein sources

SKIM MILK POWDER

The crude protein (CP) of skim milk powder is highly digestible. Chymosin, which is secreted by the abomasum and is responsible for coagulation and digestion of caseins, is present in large quantities in the gastric secretions of the young calf and its activity tends to disappear only very slowly after weaning (Le Huerou-Luron *et al.*, 1992).

The apparent faecal digestibility of CP increases from 0.87 in the first week of life to 0.94 at three weeks of age and to 0.97 after the first month of life (Grongnet, Patureau-Mirand, Toullec and Prugnaud, 1981) (Table 6.3).

When measured simultaneously, apparent ileal digestibility tends to be lower than faecal digestibility: 0.89 versus 0.92 at three weeks of age (Petit, Ivan and Brisson, 1989); 0.91 and 0.92 versus 0.95 at two months of age (Branco-Pardal, Lallès, Formal, Guilloteau and Toullec, 1995, Tukur, Branco-Pardal, Formal, Toullec, Lallès and Guilloteau, 1995). This indicates further digestion of CP posterior to the ileum.

Guilloteau, Toullec, Grongnet, Patureau-Mirand and Prugnaud (1986) observed, in two-month-old calves, that the CP escaping digestion in the small intestine appears to be almost entirely from endogenous and bacterial origin. From this work, many authors assume that the true digestibility of skim milk powder is 1. An analysis of the amino acid profile at the end of the ileum showed that the digesta protein was similar to a mix containing 0.61 of endogenous protein and 0.39 of gut bacteria protein (Toullec and Formal, 1998).

Caugant, Toullec, Formal, Guilloteau and Savoie (1993a) suspected that the lower value of apparent ileal digestibility they determined (0.90) compared to other authors (0.93, 0.95), was due to the quality of the skim milk powder used. The level of 4.8 mg of soluble N/g of skim milk powder indicated that a 'medium heat' type of skim milk powder was used, which caused a gastric emptying faster than usual. The authors suspected that around 0.09 of the amino acids at the end of the ileum were from dietary origin and concluded that this explained their lower digestibility value.

Caseins have the property of clotting in the presence of chymosin in the abomasum. With raw milk, formation of the clot is very fast and occurs within 10 minutes after feeding (Scanff, Savalle, Miranda, Pellisier, Guilloteau, Toullec, 1990). For skim milk powder, the firmness of the clot may vary depending on the denaturation of the whey proteins. This was demonstrated in vitro (Emmons and Lister, 1976) and in vivo (Roy, 1970, Scanff *et al.*, 1990, Caugant, Petit, Charbonneau, Savoie, Toullec, Thirouin and Yvon, 1992). Denatured whey proteins in high heat skim milk powder interfere with the activity of the chymosin (Longenbach and Heinrich, 1998) and limit the formation of a curd in the abomasum. The firmness of the clot can be decreased two-fold between a low and a medium heat skim milk powder and by a greater amount between a low and a high heat skim milk powder (Emmons and Lister, 1976).

Table 6.3 DIGESTIBILITY OF CRUDE PROTEIN FOR INGREDIENTS COMMONLY USED IN MILK REPLACERS

Ingredients	*Age of calves (weeks)*	*Apparent faecal digestibility*	*Apparent ileal digestibility*	*True ileal digestibility (b)*	*Sources*
Skim milk powder	<1	0.87			1
	2	0.94			1
		0.72			9
		0.91			2
	3	0.92	0.89		3
	4	0.97			1
		0.77			9
	6	0.90			9
	>8	0.95	0.91		7
		0.95	0.92		8
			0.92		10
			0.90		5
			0.92		6
		0.9			9
		0.97	0.95		11
Whey protein concentrate	<1	0.85			1
	2	0.8			1
		0.73			9
	4	0.89			1
		0.83			9
	6	0.91			9
	>8	0.88			9
			0.86	0.95	6
Whey protein concentrate (a)	<1	0.9			1
	2	0.87			1
	4	0.94			1
	>8		0.88	0.96	6
Wheat gluten (native)	>8	0.91	0.84	0.92	7
Wheat gluten (native)		0.94			4
Wheat gluten (soluble)			0.87	>0.92	10
Soya protein concentrate	>8	0.84	0.91	0.98	8 (c)
			0.84	0.92	5 (d)
			0.88	0.93	12 (d)

(a): WPC is denatured (75°C, 30 min for source 1; 72°C 20 min for source 6).
(b): true ileal digestibility is calculated by correcting ileal CP flow for non-specific endogenous CP loss.
(c): soya concentrate obtained by precipitation with HCl.
(d) : soya concentrate obtained by hot aqueous ethanol treatment.
1: Grongnet *et al.*, 1981: 2: Petit *et al.*, 1988; 3: Petit *et al.*, 1989; 4: Toullec and Grongnet, 1990; 5: Caugant *et al.*, 1993a; 6: Caugant *et al.*, 1993b; 7: Branco-Pardal *et al.*, 1995; 8: Tukur *et al.*, 1995; 9: Terosky *et al.*, 1997; 10: Toullec and Formal, 1998; 11: Nunes Do Prado *et al.*, 1989b; 12: Montagne *et al.*, 2001

For raw milk, the clot entraps the caseins in the abomasum, which then undergo an intensive hydrolysis. They leave the abomasum relatively slowly (taking more than 6 hours) in the form of small or large peptides that can be absorbed easily (Scanff *et al.*, 1990). With pasteurised milk (95°C, 45s), coagulation of casein does not occur during the first hour of digestion (Scanff *et al.*, 1990). A lower firmness of the curd leads to a lower extent of proteolysis in the abomasum and a higher passage rate of undigested materials into the duodenum (Scanff *et al.*, 1990). The pH of the gastric effluent remains higher after feeding compared with non heat treated milk (Scanff *et al.*, 1990). With 'severely' heat treated skim milk, 33 to 39% of the ingested casein passes through the abomasum undigested (Roy, 1970).

Recent reviews (Davis and Drackley, 1998a, Longenbach and Heinrichs, 1998) showed that pre-digestion in the abomasum does not seem to be a prerequisite for the caseins to be highly digestible, except for calves of less than one month old.

By using an oxalate-NaOH buffer to avoid the coagulation of caseins in the abomasum, Petit *et al.* (1989) did not detect any difference in the apparent ileal and apparent faecal digestibilities of N when calves of four weeks of age were fed milk replacer containing 0.60 of clotting or of non clotting low heat skim milk powder: 0.892 vs 0.844 of total N intake for apparent ileal digestibility and 0.923 and 0.909 for apparent faecal digestibility.

Similarly, Cruywagen and Horn-Quass (1991) did not observe any difference in the apparent faecal digestibility of N in five-week-old calves receiving a milk replacer containing 0.4 of clotting or of non clotting skimmed milk powder (0.810 vs 0.856). In this study, the animals on the two treatments gained similar body weight in four weeks (+10.4 and +10.7 kg, respectively).

In the cited experiments, calves were at least one month old. At that age, the pancreas has grown by about 1.6 since birth (Le Huerou-Luron *et al.*, 1992) and consequently may secrete enough proteolytic activity to cope with a lack of digestion in the abomasum.

In younger calves, less than one month old, the limited intestinal digestive capacity may be overwhelmed by the flow of undigested protein leaving the abomasum.

A higher pH of the abomasal effluent is observed when the clot is less firm or prevented (Roy, 1970, Strudsholm and Lykkeaa, 1988, Scanff *et al.*, 1990). This may increase the risk of passage of pathogenic microorganisms in the digestive tract, such as E.coli.

In conclusion, it can be said that skim milk powder is a highly digestible source of crude protein for the pre-ruminant calf. Its quality, however, which is related to the denaturation of the whey proteins and its ability to form a firm clot, modifies the kinetics of gastric emptying of the proteins and the biochemical form of the proteins delivered in the intestine. These aspects should be considered when skim milk is included in a milk replacer formulated for calves less than one month old.

WHEY PROTEINS

Whey proteins do not clot under the action of chymosin in the abomasum. ß-lactoglobulin, the major protein in whey (De Wit, 2001) is resistant to gastric enzymes (Scanff *et al.*, 1990) and can be detected intact in the gastric effluent (Scanff *et al.*, 1990, Scanff, Yvon, Pélissier, Guilloteau and Toullec, 1992).

Three hours after feeding, the abomasal content observed in slaughtered animals receiving a milk replacer based on whey proteins did not show any clot, but was very similar to the liquid fed. However, calves receiving a milk replacer containing skim milk power had large and firm clots in the abomasum (Strudsholm and Lykkeaa, 1988). At this time, 0.5 of the total ingested proteins in whey fed calves had gone into the small intestine while 0.7 was still remaining in the abomasum of calves receiving a skim milk based milk replacer (Strudsholm and Lykkeaa, 1988).

Trials that have investigated the digestibility of whey proteins have generally used sweet whey protein concentrate, obtained by ultrafiltration, as a 'skim milk replacer' with a composition similar to skim milk, especially the protein content (350 g/kg) (Table 6.1).

The apparent CP faecal digestibility of whey protein concentrate increases from 0.80-0.85 at one week of age to 0.89 from four weeks of age (Grongnet *et al.*, 1981) (Table 3). The increase with time is slower than for skim milk and occurs only after one month of age (Grongnet *et al.*, 1981). These authors found that a partial heat denaturation (75°C for 30 minutes) of whey protein concentrate increases digestibility compared with a normal pasteurization at 72°C for few seconds. During the first month of life, partially heat denatured whey protein concentrate (0.34 of the CP denatured) had a CP digestibility of 0.88 versus 0.83 for the native whey protein concentrate (0.08 of the CP denatured). At four weeks of age these values were 0.94 and 0.89 respectively. From this trial it can be concluded that native (normal pasteurised) whey protein concentrate had a faecal digestibility equal to 0.92 of that of skim milk powder during the first month of life and equal to 0.95 from one month onwards. In addition, a partial denaturation of the protein increases the digestibility to a value similar to that of skim milk: 0.94 vs 0.97 from one month onwards.

The faecal digestibilities of skim milk powder and whey protein concentrate observed by Terosky, Heinrichs and Wilson (1997) are lower than those observed by Grongnet *et al* (1981) especially during the first month of life: 0.77 at 4 weeks of age for skim milk and 0.83 for whey protein concentrate. In this trial, the apparent faecal digestibilities of the two ingredients increased only after six weeks of age to be similar at 0.90 and 0.88 respectively. The calves of this study were transported, after being sourced from different farms. Some scour problems were observed during the first three weeks, independent of the dietary treatments. This could explain why the calves had low digestibility values.

The apparent ileal digestibility of the protein N of whey protein concentrate, determined in calves of two months of age fed at a moderately high level (58 g/kg $BW^{0.75}$), was equal to 0.86 when the whey protein concentrate was pasteurized (74°C, 15s) and 0.88 when the whey protein concentrate was heated (72°C, 20 min.). The difference in digestibility between the two heat treatments was not significant (Caugant, Toullec, Guilloteau and Savoie, 1993b) and the apparent digestibility of whey protein concentrate was equal to 0.92 of that of skim milk powder. Corrected for the endogenous N loss and assuming that the true digestibility of skim milk is close to 1, the true ileal digestibility of the protein N of whey proteins was equal to 0.95 of that of skim milk (Caugant *et al*., 1993b).

In their trial, Caugant *et al* (1993b) detected a higher proportion of amino acids of dietary origin in the distal ileal content of calves receiving 0.50 of the CP from whey protein concentrate compared with calves receiving only skim milk powder. According to the authors, this higher proportion of amino acids of dietary origin could be explained by the presence of a peptide j (Roger 1979 in Caugant *et al*., 1993b), derived from the digestion of whey proteins by trypsin and chymotrypsin which would contribute to about 0.2 of the amino acids detected in the ileal digesta.

PERFORMANCE OF CALVES GIVEN SKIM MILK VERSUS WHEY PROTEIN CONCENTRATE

In Grongnet *et al* (1981), the calves receiving whey based milk replacers tended to gain less weight: 626 g/d for the native whey and 691 g/d for the partially denatured whey versus 726g/d for calves receiving skim milk based milk replacer. However, the differences were not significant and the number of animals involved was small (5 calves/treatment).

Terosky *et al* (1997) did not detect any difference in performance when skim milk powder was compared with its partial (0.33 or 0.67) or complete replacement by whey protein concentrate: 390, 360, 370 and 390 g of gain/d, respectively. They also detected no difference in biological value: 0.683, 0.712, 0.723 and 0.67 respectively.

Lammers, Heinrichs and Aydin (1998) used similar milk replacer formulas to Terosky *et al.* (1997). They compared four milk replacers of 200 g CP/kg in which the skim milk powder was substituted partially (0.33, 0.67) or completely by a whey protein concentrate. The milk replacers were fed at a similar feeding level (0.1 of birth weight for the first two weeks and 0.12 of birth weight thereafter). The number of days of observed scouring was less than for Terosky *et al* (1997): less than 1 d vs 3.3 d. The best average daily gain was observed when skim milk was replaced by whey protein concentrate at the 0.67 level and was significantly different from the average daily gain obtained with skim milk only: 260 g vs 199 g/d ($P<0.05$). For the milk replacers containing 0.33 of the protein from whey protein concentrate or from whey protein concentrate only, the average daily gains were intermediate: 231 g and 258 g/d, respectively.

However, these average daily gains were lower than that obtained by Terosky *et al* (1997), which was 375 g/d, despite similar feeding levels and a higher number of scouring days in Terosky *et al* (1997). In Terosky *et al* (1997), the calves were maintained in a room thermostatically regulated at around 20°C. In Lammers *et al* (1998), the calves were maintained indoors, but without any temperature regulation. The temperature during the trial was not reported, but it could have been below 20°C so that more energy was used by the calves to cover maintenance requirements.

Lammers *et al* (1998) conducted a second trial where calves received the same milk replacers, but with a commercial calf starter distributed ad libitum from the third day of life. The calves grew faster, due to intake of starter, but no significant difference was observed between treatments for average daily gain during the first six weeks of life: 452, 505, 470 and 447 g/d respectively for the treatment with skim milk only, 0.33, 0.67 of CP from whey protein concentrate and with whey protein concentrate only. In this type of trial, where the feeding level of milk replacer is maintained at around 600 g/d (about 0.12 of initial body weight at birth), growth rate is determined mainly by intake of starter.

In conclusion, it can be said that whey protein concentrate possesses an ileal digestibility equal to 0.95 of that of skim milk powder in calves greater than one month old. In younger animals, only faecal digestibility values are available and whey protein concentrate has a lower digestibility: 0.92 of that of skim milk powder, but the same value could almost certainly be applied to ileal digestibility. However, in growth performance trials, calves receiving a milk replacer based on skim milk powder do not seem to perform better than those receiving a milk replacer based on whey protein concentrate. Lammers *et al.* (1998) even concluded that whey protein concentrate has a higher biological value than skim milk powder. In conventional rearing systems, where the feeding rate of milk replacer is less than 600 g/d, growth rate is mainly determined by intake of starter (Lammers *et al.*, 1998). In this situation, whey proteins can be substituted successfully for skim milk powder.

ALTERNATIVE PROTEIN SOURCES

The number of alternative sources of milk proteins is quite limited as ingredients must fulfil certain physical criteria to ensure suitability for liquid application, such as colour and solubility. Nutritionally, ingredients should not contain any antinutritional factors or allergenic compounds and should have a high digestibility.

The two most commonly used CP sources in the EU market are soluble wheat gluten and soya protein concentrate. Their composition is given Table 6.1. Both ingredients have a high CP content : 800 and 665 g/kg respectively.

Soyabean flour contains many antinutritional factors for the young calf and digestibility is very low during the first month of life. Dawson, Morrill, Reddy,

Minocha, Ramsey (1988) measured an apparent faecal digestibility of only 0.418 for a milk replacer in which heated soya flour represented 0.75 of the CP. The antinutritional factors and the antigenic properties of soyabean flour are now well characterised and several reviews on the subject are available (Lallès 1993, Davis and Drackley, 1998b, Lallès and Toullec, 1998). Soyabean bean flour contains trypsin inhibitors, lectins, antigenic proteins and undigestible carbohydrates that can be removed if appropriate treatments are applied. Much research has been conducted to improve the nutritional value of soyabean meal because of potential interest by the veal industry. New soyabean products exist on the market that are very good sources of proteins suitable for calf milk replacers. However, it is important to know the processes applied to the soyabean meal to understand their possible effects on the antinutritional factors and thus, on digestibility (Lallès and Toullec, 1998).

Soya protein concentrate is obtained from soyabean meal by a number of different steps. The soyabean meal is first treated by hot water-alcohol to extract the undigestible carbohydrates and deactivate immunoreactive proteins such as glycinin and ß-conglycinin. The alcohol is then removed using a solvent and the treated soya is toasted and dried.

Apparent intestinal digestibility of soya protein concentrate is variable (Lallès and Toullec, 1998). A soya protein concentrate obtained by the process described above had an apparent ileal digestibility of CP of 0.84 (Caugant *et al.*, 1993a) or 0.91 (Montagne, Toullec and Lallès, 2001), corresponding to 0.93 of the value observed for skim milk powder in the same trial. This latter value is lower than that obtained by Tukur *et al.* (1995): 0.92, where the protein concentrate was obtained by precipitation at pH 4.5 with hydrochloric acid from water suspension (Table 6.3).

When apparent ileal digestibility was corrected for endogenous losses, the authors found true digestibilities of 0.92 (Caugant *et al.*, 1993a), 0.93 (Montagne *et al.*, 2001) and 0.98 (Tukur *et al.*, 1995). Analysis of the amino acid profile of the ileal content showed that a sub unit of the glycinin protein may still be present and could have accounted for 0.29 of the amino acids present in the distal ileum (Montagne *et al.*, 2001). However this value is lower than that estimated by Tukur *et al.* (1995): 0.07, showing that soya protein concentrate has variable digestibility. A more detailed study of the origin of the endogenous protein showed that digestion of soya protein was associated with a secretion of 'specific' endogenous CP, 0.5 of which came from the host and 0.5 from bacteria. The total flow of specific endogenous N was 9 g/kg of DMI in the diet where soya protein concentrate represented 0.7 of dietary CP, increasing the total endogenous flow of N in the ileum by 1.53 compared with a skim milk based milk replacer (Montagne *et al.*, 2001). According to the authors, the presence of some remaining 'components' in the protein concentrate may have enhanced mucus abrasion and cell desquamation. When total endogenous CP loss was considered (non specific and specific), it was possible to calculate a 'real' ileal digestibility, which was 0.95.

Soya protein isolate, obtained from alkali and acid treatment of soyabean flour (Davis and Drackley, 1998b) contains a higher level of CP (900 g/kg DM) than soya protein concentrate. It has a much lower antitrypsic activity than soyabeans, the glycinin and ß-conglycinin do not have any immunoreactive properties, and the level of carbohydrates is very low; less than 15 g/kg DM (Nunes do Prado, Toullec, Lallès, Guéguen, Hingand, Guilloteau, 1989a). The ingestion of soya isolate does not induce production of antibodies by the calf (Lallès, Toullec, Branco-Pardal, Sissons, 1995). In calves older than one month, soya isolate had a high N faecal digestibility: 0.93 (Nunes do Prado, Toullec, Guilloteau, Guéguen, 1989b), 0.90 (Lallès *et al*., 1995) and a high apparent ileal digestibility: 0.91. Corrected for total endogenous CP loss, the true ileal digestibility was 0.97; as good as dairy proteins (Nunes do Prado *et al*., 1989b). However, the authors determined that the supply of endogenous N and of bacterial N at the end of the ileum was increased by 1.9 when soya isolate was present in the milk replacer (soya isolate representing 0.73 of the CP), compared with a skim milk based milk replacer.

The processing used to produce soya isolate increases the price and makes this source uneconomical in a milk replacer formulation. Hence, this ingredient is not often used.

Wheat gluten is a co-product of the wheat starch industry and is obtained after a wet separation of the starch from wheat. The product, native gluten, possesses an unusual viscoelasticity and is not soluble. It needs to be processed further, enzymatically or chemically, to be suitable for milk replacer application. The process may depend on the suppliers, as commercial soluble wheat gluten in EU has pH values ranging from 4.5 to 6.

Wheat gluten does not contain any antinutrional factors. Toullec and Grongnet (1990) detected an increase in antibodies in calves after 5 weeks of ingestion of wheat gluten included at 0.05 in a milk replacer. However, the animals did not develop any symptoms of allergy.

Native wheat gluten has the same faecal digestibility as wheat gluten denatured by heat: 0.94 and 0.93 repectively Toullec and Grongnet (1990). Soluble wheat gluten has an apparent ileal digestibility for N of 0.87 (Toullec and Formal, 1998), which is slightly higher than that of native gluten: 0.84 (Branco-Pardal *et al*., 1995). The dietary fraction detected at the end of the ileum seemed to correspond to some subunit of the dietary protein (Toullec and Formal, 1998), especially rich in glutamic acid and proline. When corrected for endogenous CP, the true ileal digestibility of all amino acids was higher than 0.92. As with soya protein concentrate, the CP endogenous fraction seems to increase with inclusion of wheat gluten (Toullec and Formal, 1998), but to a lesser extent (Table 6.3).

PERFORMANCE OF CALVES GIVEN ALTERNATIVE PROTEIN SOURCES

There have been few evaluations of soya protein concentrate and hydrolysed wheat gluten in calf milk replacer for rearing calves. Trials conducted in the

US (review Davis and Drackley, 1998b) showed that growth rate of calves fed soya protein concentrate or hydrolysed wheat gluten decreased by 1.33 to 1.17, respectively during the first two weeks, compared with calves receiving a milk replacer based on whey protein concentrate. Over the entire period of the trials (42 d) growth rate was decreased by only 1.06 for both sources. The ingredients seemed to be better utilised by calves after one month of age. However, when milk replacer was fed at 0.08 of initial body weight, no difference in performance was observed between milk replacers containing only whey protein concentrate or variable amounts of hydrolysed wheat gluten during the first two weeks, or during the entire rearing period (Terui, Morrill and Higgins, 1996). In our commercial research facilities in the US, we found that performance was decreased by 1.25 over the entire period of the trial (42 d) when calves received a milk replacer containing 0.45 of protein from soya protein concentrate, compared with calves receiving a milk replacer based on whey protein concentrate (Hill, unpublished). This variation in the effect of soya protein concentrate on performance, compared with whey based milk replacer, could be due to variable quality of the product available on the market.

In the EU, published trials have focused on veal calves. Calves fed a milk replacer containing 0.05 of wheat gluten (0.2 of total dietary CP) for three months from 5 weeks of age performed as well as calves receiving a milk replacer containing only skim milk powder: 1204 and 1251 g/d respectively (Toullec and Grongnet, 1990). Under similar experimental conditions, live weight gain of calves receiving three different soya protein concentrates (contributing 0.65 to 0.72 of CP intake) for 100 d from 5 weeks of age were 1.135 to 1.18 lower than the live weight gain of calves receiving a milk replacer containing only skim milk powder: 1016 to 1047 g/d versus 1254 g/d (Toullec, Lallès and Bouchez, 1994). In this trial, the soya protein concentrates had low digestibility values: 0.77 to 0.87, which could lead to reduced performance.

CHARACTERISATION OF THE NITROGEN FRACTION

Milk replacers are formulated on their CP content, which varies between 200 and 260 g/kg. A fraction of dietary CP is not true protein (urea for example); any present in whole milk remains in the whey fraction during cheese manufacture and has no nutritional value.

In order to quantify the importance and possible variation in non-protein nitrogen (NPN) fractions of ingredients used in milk replacers, we collected 21 dairy ingredients within our group (Europe, the USA, South America). All the ingredients were available on the market: 18 were whey based products (6 sweet whey, 2 acid whey, 6 whey protein concentrate, 4 delactosed whey); 3 were skim milk powder. Two vegetable sources (1 soluble wheat gluten and 1 soya protein concentrate available in the EU market) were also analysed. The NPN fraction was evaluated using two methods. The first was the standard procedure, which defines NPN as the N fraction remaining soluble after precipitation with a solution of trichloracetic acid (TCA) at 12% (NPNstd). However, some very small peptides (di petides) can remain in the soluble fraction. To quantity these, the NPN was also determined using the tungstic acid method (Licitra, Hernandez and van Soest 1996), which precipitates all peptides. Only free amino acids and the other 'true' NPN components remain soluble (NPNtru). Total N was determined in all the samples and analysed by the Kjeldahl method. The results are shown in Table 6.4.

Table 6.4 CRUDE PROTEIN (CP; g/kg AIR-DRY MATTER) AND NON-PROTEIN NITROGEN (NPN; %N) OF DAIRY AND VEGETABLE INGREDIENTS. NPN DETERMINED BY PRECIPITATION WITH 12% TCA OR TUNGSTIC ACID (LICITRA ET AL., 1996)

		CP (a)			*NPN by TCA12%*			*NPN by tungstic acid*		
Ingredients	*n*	*Av.*	*Min.*	*Max.*	*Av.*	*Min.*	*Max*	*Av.*	*Min.*	*Max*
Skim milk powder	3	**338**	*319*	*351*	**6.3**	*6*	*6.7*	**5.9**	*5.5*	*6.2*
Sweet whey powder	6	**128**	*115*	*134*	**31**	*22.8*	*37.4*	**18.3**	*9.8*	*24.6*
Acid whey powder	2	**179**	*102*	*249*	**24.4**	*22.4*	*26.4*	**19.5**	*15.5*	*23.5*
Delactose whey powder	4	**236**	*211*	*255*	**24.2**	*22.8*	*25.3*	**22.1**	*21.3*	*22.6*
Whey protein concentrate powder	6	**332**	*316*	*345*	**10.5**	*8*	*12.3*	**5.1**	*4.4*	*5.7*
Soya protein concentrate	1	**638**			**2.4**			**1.3**		
Soluble wheat gluten	1	**768**			**33.1**			**26.4**		

a: N*6.38 for dairy ingredients, N*6.25 for vegetable ingredients

On average, the NPNstd was 0.195 and varied from 0.024 to 0.374. It was slightly higher than the average value for NPNtru: 0.136 (minimum: 0.04, maximum: 0.246), which indicates the presence of very small peptides in the ingredients. In skim milk powder NPNstd represented 0.063 and the NPNtru 0.059 of the N, which is in agreement with the generally accepted value of 0.05 for whole milk. In whey based products, the value for NPNstd varied from 0.08 to 0.37; the highest value being obtained with sweet whey: 0.31, then acid whey: 0.244, delactosed whey: 0.242, and whey protein concentrate 0.105. The difference in NPNstd between sweet whey and acid whey disappears when the NPNtru value is considered: 0.183 for sweet whey and 0.195 for acid whey. The presence of small peptides in sweet whey indicated that small proteic fractions were released during the manufacture of cheese. The NPNstd and the NPNtru values are very similar for the delactosed whey:

0.221 and 0.242 respectively, and similar to the NPNtru value of whey. In this product, lactose is taken out by a process of crystallization and thus, does not change the quality of N in the whey. The low value observed for whey protein concentrate is normal: 0.105 for NPNstd and 0.051 for NPNtru, as part of the lactose is taken out by an ultrafiltration membrane, which at the same time 'filters' molecules of low molecular weight such as urea (De Wit, 2001). Soluble wheat gluten had high proportions of NPNstd: 0.33 and NPNtru:0.264. This is in agreement with results obtained with another commercial soluble wheat protein where N from amino acids represented 0.747 of total N (Toullec and Formal, 1998), although wheat gluten contains a high proportion of glutamine, which is not considered in the amino acid analysis and could account for the high NPN value. The higher value for NPNstd may be explained by a release of peptides during the solubilisation treatment (Toullec and Formal, 1998). The NPNstd was 0.024 of that for soya protein concentrate.

In conclusion, it can be said that dairy ingredients vary in terms of the NPN fraction. The proportion is quite high for some ingredients and should be considered when formulating liquid feeds. The determination of NPN by tungstic acid seems to be more appropriate for evaluating the proportion of CP with nutritional value. The NPN value determined by TCA12% tends to overestimate the 'true' NPN fraction as it contains some peptides. Whey protein concentrate has the highest nutritional value in this respect, with more than 0.95 of CP composed of peptides and proteins; a value similar to that determined in skim milk powder.

AMINO ACID PROFILE

The profiles of amino acids in skim milk powder, whey protein concentrate, soya protein concentrate and soluble wheat gluten are given in Table 6.5 (values taken from Toullec 1989). A comparison of these profiles with ones provided by more recent publications and cited in this article showed very little variation and thus, are not reported.

The amino acid profiles required by calves growing at a low rate (250g/d), or more intensively (1100 g/d), are different (Table 6.5) as the amino acid requirements for maintenance are different from those for gain (Davis and Drackley, 1998b). Compared to the amino acid profile of these two requirements, skim milk powder appears slightly deficient in threonine, sulphur amino acids, lysine and isoleucine, especially for calves growing at a faster rate (1100 g/d), and has an excess of phenylalanine. Whey protein concentrate appears well balanced in lysine and sulphur amino acids compared with skim milk powder. It is slightly deficient in histidine and arginine, however, and contains more essential amino acids than skim milk powder: 0.511 versus 0.494. Its amino acid profile is more closely related to the requirements for high growth rates (Table 6.5) .

Table 6.5 AMINO ACID COMPOSITION OF DAIRY AND VEGETABLE INGREDIENTS COMMONLY USED IN MILK REPLACERS AND AMINO ACID PROFILE REQUIREMENTS OF THE YOUNG CALF (%INDICATED AA)

	Skim milk powder (1)	*Whey protein concentrate (1)*	*Soya protein concentrate (1)*	*Soluble wheat gluten (2)*	*Requirements*	
					(3)	*(4)*
Thr	8.7	12.6	7.9	6.3	9.9	9.3
Val	11.7	10.6	10.1	10.4	11.7	9.1
Met + cys	6.3	8.8	5.9	10.5	8.4	6.9
Ile	9.6	11.0	9.4	10.7	11.0	6.5
Leu	17.4	18.0	14.7	18.8	16.8	15.9
Phe + Tyr	18.3	12.0	18.0	22.7	11.7	14.0
Lys	14.6	16.2	12.3	3.7	16.5	14.7
His	4.9	3.6	5.3	5.5	5.1	5.6
Arg	6.5	4.3	13.8	9.0	7.3	16.2
Tryptophan	2.1	2.9	2.5	2.2	1.5	1.9
AA-N (%totN) (5)	49.4	51.1	45.3	34.8		

(1): Toullec 1989, (2): Toullec and Formal (1998), (3): Requirements of 50 kg live weight calf gaining 1100 g/d
Toullec 1989, (4): requirements of 50 kg live weight calf gaining 250 g/d (Williams and Hewitt, 1979 in Davis and Drackley 1998b)
(5): AA-N of indicated amino acids in total N of ingredients

Soya protein concentrate has a profile deficient in threonine, sulphur amino acids, leucine and lysine compared to the requirements. Soluble wheat gluten is especially deficient in lysine and threonine, but is a good source of valine, isoleucine and leucine. The two vegetable protein sources have a lower concentration of essential amino acids, 0.453 and 0.348 respectively, than skim milk powder and whey protein concentrate powder.

Synthetic amino acids can be used in formulation to improve the amino acid profile. Wheat gluten, for example, when included at 0.05 in liquid feed, can be supplemented with synthetic lysine and threonine to support high growth rates. A milk replacer formulated in this way gave performances in veal calves that were similar to a milk replacer based on skim milk (Toullec and Grongnet, 1990).

Carbohydrates sources

LACTOSE

The lactose content of milk is approximately 50 g/l or 400 g/kg DM (Mahaut *et al.*, 2000a, De Wit 2001). In milk replacers, its concentration can vary, depending on the fat and crude protein contents. However, lactose concentration is generally slightly higher than in whole milk: between 400 to 450 g/kg of powder. It contributes 0.25 of the gross energy in whole milk and 0.4 in milk replacer (NRC, 2001).

Lactose is a disaccharide composed of one unit of glucose and one unit of galactose. It is hydrolysed by lactase located in the brush border of the small intestine. Lactase activity is present in large amounts just after birth (Le Huerou, Guilloteau, Wicker, Mouats, Chayvialle, Bernard, Burton, Toullec and Puigserver, 1992). It decreases during the first week of life, remains stable until 8 weeks of age and slightly decreases thereafter. Lactase activity is expressed mainly at the proximal jejunum and to a lesser extent in the duodenum and medium jejunum (Le Huerou *et al.*, 1992). Lactose is considered to be completely digestible: 0.99 for the apparent faecal digestibility (Toullec, 1989). Ileal digestibility was found to be approximately 0.93 in milk fed calves older than one month of age (Hof, 1980). This suggests that some ingested lactose is fermented in the large intestine when the level of ingestion of lactose is high; 8 g of hexose equivalent/kg body weight (Hof, 1980), equivalent to the ingestion of 0.9 kg of dry matter of milk replacer containing 450g/kg of lactose. At an intake level of 10 g of hexose equivalent/kg of body weight, a drop in the dry matter content of faeces was observed: 0.103 versus 0.136. A drop in pH (7.1 vs 7.7) was also observed, indicating that a risk of scouring exists at this level. This confirms results obtained in the review of Davis and Drackley (1998b). For a calf of 45 kg live weight, 8,10 and 14 g of hexose equivalent correspond to a daily intake of lactose of 400, 500 and 700 g/d. Scouring problems with lactose may occur when milk replacer is fed at approximately 1.1 kg of dry matter per day, which is quite a high level of intake.

Intake of the first colostrum plays a major role in the development of the digestive tract, as demonstrated by Hammon and Blum (1997). In their experiment, intake of the first colostrum had a significant effect on the absorptive capacity of xylose at 5 days of age. As this sugar is absorbed proportionately to the surface of the intestine, the authors concluded that the first colostrum stimulated development of the digestive tract.

If management of the first colostrum is poor, one can expect some digestive problems, especially if the calf is given a large amount of liquid feed soon after birth. A high feeding rate of milk replacer, at least 1 kg per day for a calf of 45 kg live weight, is possible only if associated with good management of the first colostrum.

ALTERNATIVE CARBOHYDRATES SOURCES

It is difficult to replace lactose, as the calf has a very limited digestive capacity for other sources of carbohydrates.

Amylase, responsible for the hydrolysis of starch and secreted by the pancreas, is present in only small amounts in the digestive secretions after birth. The activity increases by 4 fold between day 7 and day 28 of age, and doubles again when the calf receives solid feed, compared to when the calf stays on milk (Le Huerou-Luron, 1992). Only a small amount of processed starch can be included in a milk replacer (Toullec, 1989).

Maltase and iso-maltase, two disaccharidases present in the brush border of the intestine, can be detected at birth and their activity increases with age. The increase is more important in ruminating calves (Le Huerou-Luron, 1992a). The activity of maltase is maximal at the distal segment of the small intestine.

When necessary, lactose can be replaced by glucose, maltose and processed starch, but their total amount should not exceed 80 to 100 g/kg dry matter (Toullec, 1989).

Fat sources

Fat represents around 0.5 of the gross energy in whole milk and 0.35 in a milk replacer. In whole milk, fat contains 0.95 triacylglycerols, emulsified in the form of very small globules of less than 1 μm to about 10 μm diameter, with those of 4 μm accounting for most of the mass (Jensen, 2002). They have a large surface area that facilitates hydrolysis by lipases and have a high digestibility, which increases up to 0.97 during the first month of age (Toullec, 1989). After ingestion, triacylglycerols are entrapped in the coagulum formed by caseins. Peak absorption of lipids is delayed by 5 to 7 hours compared with monogastrics (Bauchart and Aurrouseau, 1993). In a milk replacer without clotting proteins, absorption of the lipids is faster (Hocquette and Bauchart, 1999).

Short chain fatty acids are partially liberated in the abomasum by salivary lipase. The rest of the tryacylglycerols are digested further in the small intestine by pancreatic esterase (Bauchart, Gruffat and Durand, 1996).

The digestibility of fatty acids is determined by their number of carbons, by their degree of saturation and by their position in the triacylglycerols. Digestibility of short chain fatty acids (less than 10 C) is 1; digestibility of medium chain fatty acids (from 10 to 14 C) is 0.95; digestibility of long chain saturated fatty acids is between 0.8 and 0.9; and digestibility of long chain unsaturated fatty acids is greater than 0.95 (Toullec and Matthieu, 1969). The fatty acid located on position 2 remains with the glycerol. The fatty acids in position 1 and 3 are released as free fatty acids. If they are unsaturated, they are more sensitive to reacting with calcium to form a soap and thus, become less available for absorption.

Tallow inclusion is no longer permitted under EU legislation. It has been replaced by vegetable sources, such as coconut oil rich in C12 (0.45 of total fatty acids; Givens, Cottrill, Davies, Lee, Mansbridge and Moss, 2000), and palm oil rich in C18:1 (0.35 of total fatty acids; Givens *et al.*, 2000). These two sources have digestibilities greater than 0.95 (Toullec and Matthieu, 1969).

Few data are available on digestibility of fat in young calves. In Grognet *et al* (1981), a mixture of 0.5 tallow and 0.5 coconut oil, incorporated into skim milk and spray-dried, had a faecal digestibility of 0.91 during the first three weeks of age and 0.95 thereafter. Large individual variations (+/- 0.05 units of digestibility) were observed during the first three weeks of life.

Feeding a source of fat containing only coconut oil to milk-fed calves from two weeks of age led to fat infiltration of the liver in 19 days (Graulet, Gruffat-Mouty, Durand and Bauchart, 2000). The same effect was observed with hydrogenated coconut oil fed to young calves from the first week of life for six weeks (Jenkins and Kramer, 1986). The authors concluded that 0.01 of the energy should come from linoleic (C18:2) and linolenic acid (C18:3). However, when soya oil rich in C18:2 (0.5 of total fatty acids; Givens *et al.*, 2000) was fed to calves from one week of age for 17 days, fat infiltration of the liver occurred (Leplaix-Charlat, Durand and Bauchart, 1996).

Consequently, a good compromise, with good digestibility and good metabolic utilisation of fat sources, seems to be a mix of coconut oil, palm oil and soya oil. However, inclusion of polyunsatured fatty acids decreases the melting point and may impair the physical properties of fat sources. Most fat sources contain a mixture of 0.2 of coconut oil and 0.8 of palm oil, which contains 0.1 of the fat as C18:2 and C18:3.

A major factor that influences the digestibility of fat sources is particle size of the globules and their stability in the liquid feed after reconstitution. Fat sources obtained by homogenization with dairy ingredients such as sweet whey, followed by spray-drying, give very good emulsification and stability of the fat (Toullec, 1989). It should be emphasized here that the technologies used to homogenize and incorporate fat influence digestibility.

Minerals, vitamins and trace elements

Calcium (Ca) and phosphorus (P) concentrations of dairy and vegetable ingredients are shown in Table 6.1. Whey based products and vegetable ingredients have low Ca contents and additional sources must be used in liquid feeds. Dicalcium phosphate is a highly digestible source for the young calf and its apparent retention is as good as Ca and P from milk (Grognet *et al.*, 1981). Sources of Ca which do not contain P can be used also, such as Ca formiate. However, attention should be paid to the risk of an interaction with digestibility of fat if the final level of calcium is too high. Calcium may combine with bile acids, making them less available for intestinal absorption of fatty acids (Xu, Wensing and Beynen, 2000).

MACRO MINERALS AND RISK OF SCOURING

It is generally accepted that high ash (Ullerich and Kampheus, 1998) favours scouring problems due to an overload of minerals in the large intestine. This increases the osmotic pressure, lowers the resorption of water in this part of the digestive tract and increases the water content of faeces. However, the severity of the scours may depend on the mineral profile of the ash. Too high a level of calcium supplementation, in the form of Ca formiate, interferes with fat digestion and causes diarrhoea (Xu *et al.*, 2000).

Sulphate, in particular, increases the risk of watery faeces. Acid whey, obtained by precipitation of caseins with sulphuric acid, contains a high level of sulphate. A recent field survey by Kampheus, Stolte and Rust (2002) showed that liquid feeding with a sulphate concentration higher than 0.6 g/l resulted in markedly reduced dry matter content in calves faeces. In the 29 whey products they collected from the market, they found that sulphate concentration varied from 0.3 to 43 g/kg of dry matter. At that high level of concentration, a 0.1 inclusion level in a milk replacer powder is enough to obtain a concentration of 0.6 g/l of sulphate in the liquid feed when reconstituted at 125 g of powder /kg. The digestibility of sulphate is quite low, 0.516 in young calves (Kampheus *et al.*, 2002). The molecule may contribute to maintaining a high osmotic pressure along the digestive tract, increasing the risk of a malabsorption of water in the large intestine and contributing in this way to an increase in the water content of faeces.

Na, K and Cl contribute to the osmotic pressure of liquid feeds. Osmotic pressure can be measured by calculating the osmolality of the liquid feed, which corresponds to the number of solute particles (Na, K, Cl) per litre of water expressed in mOsm/l. Whole milk has a low osmolality: 103 mOsm/l of water (Toullec, 1989). For a milk replacer, osmolality will depend not only on the concentration of Na, K and Cl in the powder, but also on the dilution rate used to reconstitute the milk. If simple sugars such as glucose are added, they should be considered in the calculation of the osmolality.

Ash, Na, K, Cl and sulphate contents were determined for the 21 ingredients used for the survey on NPN (see earlier). The results are given in Table 6.6. All the products, except one whey based product, had ash, Na, K and Cl contents greater than whole milk. The three milk based products (skim milk powder) had levels closest to whole milk. Ash content of the 18 whey based products varied from 47 to 243 g/kg. The increase in ash in the delactosed whey is due partly to the partial removal of lactose. Within the 2 acids and the 4 delactosed wheys, two had a high level of sulphate: 63.6 and 59.2 g/kg. These levels were higher than those observed by Kampheus *et al.* (2002). The whey protein concentrates had the lowest values for ash. Among the two vegetable sources, the soya protein concentrate possessed a high level of K, while the hydrolysed wheat gluten was very low in ash.

Table 6.6 ASH, Na, K, Cl AND SO_4 CONTENT OF WHOLE MILK (g/kg DM) , DAIRY AND VEGETABLE INGREDIENTS (g/kg AIR DRY MATTER) EVALUATION OF OSMOLALITY (g/kg WATER) WHEN INGREDIENTS ARE DILUTED AT 125 g/kg OF LIQUID FEED.

Ingredients	*n*		*Ash*	*Na*	*K*	*Cl*	SO_4	*mOsml/l water*
Skim milk powder	3	**Av.**	**76**	**6**	**24**	**12**	**1**	**171**
		Max.	72	5	17	11	1	142
		Min.	78	6	29	12	1	187
Sweet whey powder	6	**Av.**	**75**	**9**	**32**	**14**	**2**	**228**
		Max.	47	7	21	5	2	140
		Min.	88	10	38	19	3	278
Acid whey powder	2	**Av.**	**155**	**15**	**29**	**18**	**33**	**319**
		Max.	66	3	8	4	2	60
		Min.	243	28	50	32	64	579
Delactose whey powder	4	**Av.**	**198**	**198**	**16**	**46**	**43**	**475**
		Max.	159	159	8	5	31	382
		Min.	231	231	22	49	69	557
Whey protein concentrate powder	6	**Av.**	**66**	**66**	**8**	**26**	**11**	**193**
		Max.	59	59	8	19	10	172
		Min.	75	75	9	29	12	207
Soya protein concentrate	1		**67**	**67**	**8.1**	**37.2**	**0.5**	**190**
Soluble wheat gluten	1		**19**	**6.4**	**1.6**	**1.2**	**0.4**	**51**

For each ingredient, it is possible to calculate an osmolality assuming they are diluted in water at a concentration similar to the DM content of whole milk: 125 g in 875 g of water. The calculated values ranged from 51 to 579 mOsm/l. No requirement for Osm/l exists, but a theoretical level can be calculated from Na, K and Cl requirements. Those values are 94 mOsm/l for Toullec (1989) and 59 mOsm/l for NRC (2001), slightly below the osmolality of whole milk (103 mOsm/l). Isotonic solutions contain 300 mOsm/l (Constable, 2002). It could be considered that liquid feeding should maintain a hypotonic or an isotonic level. Ullerich and Kampheus (1998) did not detect any drop in dry matter content of faeces when the osmolality of a liquid feed was increased from 125 to 250 mOsm/l by adding Na, K and Cl sources, even though the ash content of the milk replacer was quite high: 135 g/Kg DM.

VITAMINS AND TRACE ELEMENTS

Whole milk is deficient in trace elements (Fe, Mn, Zn, Cu, I, Co, Se) and in soluble vitamins, compared to the requirements of young calves fed on milk

(Toullec, 1989, NRC 2001). Vitamin A is present in adequate amounts: 11 500 IU/kg DM (Toullec, 1989) compared to requirements, 9 000 IU/Kg DM. However most liquid feeds are fortified in vitamin A, to between 20 000 and 40 000 IU/kg DM. Vitamins D and E are present in lower concentrations than requirements. Thus, when whole milk is fed to young dairy calves, vitamin and trace element supplementation is necessary.

Only traces of fat soluble vitamins are detected in skim milk and whey powders (Mahaud *et al.*, 2000b). Concentrations of water soluble vitamins are higher in liquid whey (De Wit, 2001) than in whole milk (Table 6.7). However, vitamins such as thiamin (vitamin B1), cobalamin (vitamin B12) and vitamin C are thermosensitive (Mahaut *et al.*, 2000a); consequently, whey based powders can contain variable amounts of water soluble vitamins. Milk replacer containing skim milk or whey based powders should be supplemented with fat soluble and with water soluble vitamins to cover calf requirements.

Liquid whey contains a higher level of trace minerals than whole milk (De Wit 2001, Toullec 1989): 9 mg of Fe /kg of DM instead of less than 4 for whole milk and 3 mg of Cu/kg versus less than 1 for whole milk, for example. However, concentrations remain below requirements (Table 6.7) and milk replacers should be supplemented with trace minerals.

Table 6.7 VITAMIN, MAJOR AND TRACE ELEMENT CONTENTS OF WHOLE MILK (DM BASIS), SKIM MILK POWDER (AIR DRY BASIS) AND REQUIREMENTS OF VITAMINS AND TRACE ELEMENTS OF THE YOUNG MILK-FED CALF (mg/kg DM, EXCEPT VIT A AND D, IU AND MAJOR MINERALS, g)

	Whole milk	*Requirements*			*Whole milk*	*Requirements*	
	(1)	*(1)*	*(2)*		*(1)*	*(1)*	*(2)*
Vitamin A	11500	48000	9000	Ca	10	10 - 13	10
Vitamin D3	307	2800	600	P	7.5	7	7
Vitamin E	7.8	10-30	50	Mg	1	1.5	0.7
Vitamin K	0.64		—	K	12	10	6.5
				Na	4	4	4
Ascorbic acid	120	100	—	Cl	8.5	8	2.5
Thiamin (B1)	3.3	1.0 - 5.3	6.5				
Riboflavin (B2)	12.2	0.8 - 2.5	6.5	Fe	1 - 4	40	100
Niacin	9.5	10 - 14	10	Cu	0.1 - 1.1	5 -10	10
Pyridoxine (B6)	4.4	2.0	6.5	Zn	15 - 38	50	40
Pantothenic a.(B5)	25.9	5 - 11	13	I	0.1 - 0.2	0.5	0.5
Folic a.	0.56	0.8	0.5	Se	0.02 - 0.15	0.1	0.3
Vit B12	0.05	0.02 - 0.06	0.07	Co	0.004 - 0.008	0.1	0.11
Biotin	0.3	0.11	0.1				
Choline	1080	1440	1000				

(1): Toullec 1989, Vitamins A, D3, E and Fe requirements are for veal calves, (2): NRC (2001) for 45 kg live weight calf eating 530 g of DM from liquid feed.

DISCUSSION

It is essential to consider the digestibility of nutrients when formulating liquid feeds. A high digestibility is required to limit digestive problems because digestibility affects the quantity of digestible nutrients the animals will be able to use for maintenance and growth. NRC (2001) developed an equation to evaluate the metabolisable energy content of liquid feed. The equation relates each nutrient (protein, fat and lactose) to its respective heat of combustion to evaluate gross energy. Nutrients coming from milk or milk derived ingredients are assumed to have a digestibility of 0.97 and their digestible energy is assumed to be converted to metabolisable energy with an efficiency of 0.96.

A digestibility coefficient of 0.97 for the first month of life may overestimate the digestible energy from whey based ingredients for which CP is 0.92 as digestible as skim milk powder. Furthermore, some CP is only NPN (around 0.05), which has no nutritional value, and should not be included in the equation.

No data are available for the digestibility of vegetable fats in very young calves. The fatty acid profile and technology applied to incorporate the fat in the raw materials influence largely the digestibility. However, it can be assumed that a fat mixture of 0.2 coconut oil and 0.8 palm oil (containing 0.15 of C:12 and C:14, 0.39 of C:16, 0.05 of C18:0, 0.33 of C18:1, 0.10 of C18:2 and C18:3) has a digestibility of about 0.92.

Consequently, when the metabolisable energy content of a liquid feed has to be evaluated for a young calf of less than one month old, some readjustments should be made to the NRC equation, otherwise overestimation may occur.

However, the evaluation of metabolisable energy content of liquid feed is not the only criterion to consider when predicting performance of very young calves. Issues such as calving, housing, environment and feeding management can influence largely the health and performance of the calf by influencing its digestive capacities and its metabolism.

Whey based ingredients, especially whey protein concentrate, represent very good alternatives to skim milk powder. Their CP fraction is highly digestible, even in the first month of life; equal to 0.92 in that period and 0.95 thereafter. In conventional rearing systems, where about 500 g/d of milk replacer is given to a calf, whey based milk replacers can perform as well as milk replacers based on skim milk.

Vegetable proteins, such as hydrolysed wheat gluten and soya protein concentrate, also have a high ileal digestibility. However, the ileal digestibility values of those ingredients were determined in calves of more than one month old. In younger animals, digestive capacity to digest non-dairy products is quite limited, so conclusions drawn on older animals may not apply to younger ones. Furthermore, digestion of vegetable proteins is associated with a higher endogenous loss of nitrogen, which indirectly increases requirements for energy and protein. Consequently, when vegetable proteins are used, they should be included at a level that the digestive tract can cope with, and that limits the increase in endogenous losses.

In modern dairy rearing systems, the young Holstein heifer calf should gain at least 800 g/d during the first months of life in order to achieve a correct body weight at 6 months of age and at breeding (Tanan and Newbold, 2002). Such a growth rate can be achieved in calfhood only if a certain quality of liquid feed is given in the first weeks of life. With 500 g/d of milk replacer powder, calves can grow only a limited amount (around 200-300 g/d according to NRC 2001 recommendations) and in such a system growth rate is mainly dependent on the amount of solid feed intake. Higher growth rates can be achieved at an early stage by feeding a higher quantity of liquid feed at about 750 to 1000 g/d. Liquid feed should contain a good level of CP, 240-260 g/kg DM and a good fat ME/lactose ME ratio of around 1 (Tanan and Newbold, 2002) to optimize muscle development and limit fat deposition. Whole milk, with its high fat ME /lactose ME ratio of 2, may not be the best source of nutrients for such a rearing system. When a milk replacer is used, the substitution of skim milk powder by non-clotting ingredients, such as whey protein concentrate, increases the rate of gastric emptying and thus increases the risk of overwhelming the digestive capacity of the small intestine. At this level of feeding, the digestive tract of the young calf must be well prepared, and that can be done only if good colostrum management is practiced.

Whey based milk replacers have about 0.7 of their digestible energy and protein available for metabolism within 3-4 hours post feeding, contrary to skim milk based milk replacer where only 0.5 of the digestible energy and 0.2 of the digestible protein may be available in the first 3-4 hours. This difference in supply of nutrients over time might have an impact on voluntary intake of solid feed, which should always be supplied ad libitum. Non-clotting milk replacers provide digestible nutrients faster and may 'starve' the young calf more quickly than milk replacers based on skim milk. In consequence, whey based products seem to be more appropriate than skim milk based products when fed at a high level for early weaning systems.

Milk replacers for young heifer calves are quite costly due to their high level of CP. Special attention should be paid to the quality of CP, especially the NPN content and amino acid profile.

Several aspects related to digestion and nutrient sources have not been discussed in this chapter because they could be the subject of a complete article on their own. These aspects include the possible manipulation of digestive functions of the young calf and the possible bioactive properties of dairy ingredients. Digestive capacities of the young calf are limited in the first month of life. There is an interest in stimulating them in order to support better utilization of nutrients. Investigation of additives to stimulate them, such as specific essential oils or organic acids, could be of interest in a feeding strategy to replace antibiotics as growth promoters.

Whole milk contains some bioactive substances, such as immunoglobulins, lactoferrin and lactoperoxidase (De Wit, 2001). These molecules, as well as some peptides derived from digestion of milk proteins, are important in the defence against micro-organisms (Shah, 2000). Whey based products, such as whey protein concentrate, might still contain some bioactive compounds; however, most of the bioactive proteins are sensitive to heat treatment, especially immunoglobulins (Mahaut *et al.*, 2000a). Consequently, bioactivities of whey protein concentrate not specifically designed for that purpose might be questionable. Sweet whey contains 0.15 to 0.20 of the protein as glycomacropeptide derived from the hydrolysis of κ-casein by chymosin (Saito, *et al.*, 1991). This peptide has interesting bioactive properties against E-coli and viruses (Brody, 2000). However, applications related to calf health require further investigation.

Conclusion

In early weaning systems where liquid feed is provided in a limited amount, the quality of the ingredients, especially their digestibility, is crucial to limit health problems and support a good start in life. In systems where a higher feeding rate of liquid feed is used to promote early growth rate, the current approach of formulating liquid feed on CP and fat content has to be re-evaluated. It has to go beyond the criteria of digestibility of ingredients. It has to be more nutritional, taking into consideration the content and the quality of the energy (fat, lactose, fatty acids composition), the content and quality of the CP (NPN, amino acid profile), and the kinetics of supply of the digestible nutrients. A better knowledge of ingredient composition, and of the processes they have gone through, is essential to better understand and control their nutritional value and to optimize their use in liquid feed formulation.

References

Bauchart, D., Gruffat, D., Durand, D., (1996) Lipid absorption and hepatic metabolism in ruminants. *Proc. Nutr. Soc*, **55**, 39-47.

Bauchart, D., Aurrouseau, B (1993) Digestion et métabolisme des lipides chez le veau de boucherie; conséquences sur la composition en lipides des tissus. *Viande Prod. Carnés*, **14**, 172-182.

Blum, J.W., Hammon, H. (2000) Colostrum effects on the gastrointestinal tract, and on nutritional, endocrine and metabolic parameters in neonatal calves. *Livestock Production Science*, **66**, 151-159.

Branco-Pardal, P., Lallès, J-P., Formal, M., Guilloteau, P., Toullec, R. (1995) Digestion of wheat gluten and potato protein by the preruminant calf: digestibility, amino acid composition and immureactive proteins in ileal digesta. *Reprod. Nutr. Dev.*, **35**, 639-654.

Brody, E. P. (2000) Biological activities of bovine glycomacropeptide. *British Journal of Nutrition*, **84**, Suppl. 1, S39-S46.

Caugant, I., Petit, H.V., Charbonneau, R., Savoie, L., Toullec, R., Thirouin, S., Yvon, M. (1992) In vivo and in vitro gastric emptying of protein fractions of milk replacers containing whey proteins. *J. Dairy Sci.*, **75**, 847-856.

Caugant, I., Toullec R., Formal M., Guilloteau P., Savoie, L (1993a) Digestibility and amino acid composition of digesta at the end of the ileum in preruminant calves fed soyabean protein. *Reprod. Nutr. Dev.*, **33**, 335-347.

Caugant, I., Toullec, R., Guilloteau, P., Savoie, L (1993b) whey protein digestion in the distal ileum of the preruminant calf. *Animal Feed Science and Technology*. 41, 223-236.

Cruywagen, C. W., Horn-Quass, J. G. (1991) Effect of curd suppression of a calf milk replacer fed at increasing levels on nutrient digestibility and body mass-gain. *S. Afr. J. Anim. Sci.*, **21**(3), 153-156.

Constable, P. D (2002) The treatment of the diarrheic calf: an update. In *Recent developments and perspectives in bovine medecine. Keynote lectures of the XXII World Buiatric Congress, August 19-23rd 2002, Hannovre, Germany*. Ed by Kaske, Scholz, Höltershinken, Hannover, Germany: Klinik für Rinderkrankheiten.

Davis, C.L. and Drackley, J.K. (1998a) Gross anatomy, development, and function of the digestive system. In *The Development, Nutrition and Management of the Young Calf*, pp. 13-37. Edited by Iowa University press. Ames, Iowa, the USA.

Davis, C.L. and Drackley, J.K. (1998b) Milk replacers: formulation and use. In *The Development, Nutrition and Management of the Young Calf*, pp. 207-257. Edited by Iowa University press, Ames, Iowa, the USA.

Dawson, D. P., Morrill, J. L., Reddy, P. G., Minocha, H. C., Ramsey, H. A. (1988) Soya protein concentrate and heated soy flours as protein sources in milk replacer for preruminant calves. *J. Dairy Sci.*, **71**, 1301-1309.

De Wit, J.N. (2001) *Lecturer's handbook on whey,* first edition. Edited by The European Whey Products Association, Brussels, Belgium.

Emmons, D.B., Lister E.E. (1976) quality of protein in milk replacers for young calves. I. Factors affecting in vitro curd formation by rennet (chymosin, rennin) from reconstituted skim milk powder. *Can. J. Anim. Sci.*, **56**,317-325.

FEFAC – Federation Européenne des fabricants d'aliments composés, European federation of feed manufacturers. http://www.fefac.org.

Givens, D. I., Cottrill, B.R., Davies, M., Lee P. A., Mansbridge, J., Moss, A. R. (2000) corrigendum to 'sources of N-3 polyunsaturated fatty acids additional to fish for livestock diets – A review. *Nutrition Abstracts and Reviews,* Series B, **70** -(8),1-13.

Graulet, B., Gruffat-Mouty, D., Durand, D., Bauchart, D. (2000) Effects of milk diets containing beef tallow or coconut oil on the fatty acid metabolism of liver slices from preruminant calves. *British Journal of Nutrition*, **84**, 309-318.

Grongnet, J-F., Patureau-Mirand, P., Toullec R., Prugnaud J (1981) Utilisation des protéines du lait et du lactosérum par le jeune veau préruminant. Influence de l'âge et de la dénaturation des protéines de lactosérum. *Annales de Zootechnie*, **30**, 443-464.

Guilloteau, P., Toullec, R., Grongnet, J-F., Patureau-Mirand, P., Prugnaud, J. (1986) Digstion of milk, fish and soyabean protein in the preruminant calf : flow of digesta, apparent digestibility at the end of the ileum and amino acid composition of ileal digesta. *Br. J. Nutr.*, **55**, 571-592.

Hammon, H., Blum, J. W. (1997) Prolonged colostrum feeding enhances xylose absorption in neonatal calves. *J. Anim. Sci.*, **75**, 2915-2919.

Hocquette, J-F, Bauchart, D (1999) Intestinal absorption, blood transport and hepatic and muscle metabolism of fatty acids in preruminant and ruminant animals. *Reprod. Nutr. Develop.*, **39**, 27-48.

Hof, G. (1980) An investigation into the extent to which various dietary components, particularly lactose, are related to the incidence of diarrhoea in milk-fed calves. *PhD. Thesis*, Wageningen Agricultural University, Wageningen, the Netherlands.

INRA-AFZ (2002) *Tables de composition et de valeur nutritive des matières premières destinées aux animaux d'élevage*. Ed. by D. Sauvant, J-M. Perez and G. Tran. INRA Editions, Paris.

Jenkins, K.J., Kramer, J.K.G. (1986) Influence of low linoleic and linolenic acids in milk replacers on calf performances and lipids in blood plasma, heart and liver. *J. Dairy Sci.*, **69**, 1374-1386.

Jensen, R. G. (2002) Invited review: the composition of bovine milk lipids: January 1995 to december 2000. *J. Dairy Sci.*, **85**, 295-350.

Kampheus, J., Stolte M., Rust, P. (2002) High sulfate content in whey products and milk replacers – a potential reason for changed faeces composition/ watery faeces in calves. Abstract In *Proceedings of the XXII World Buiatrics Congress, August 19-23rd 2002*, Hannovre, Germany. Ed. By World Association of Buiatrics.

Lallès, J-P. (1993) Nutritional and antinutritional aspects of soyabean and field pea proteins used in veal calf production : A review. *Livestock Production Science*, **34**, 181-202.

Lallès, J.P., Toullec, R., Branco Pardal, P., Sissons, J.W. (1995) Hydrolyzed soya protein isolate sustains high nutritional performance in veal calves. *J. Dairy Science*, **78**, 194-204.

Lallès, J-P., Toullec, R. (1998) Soyabean products in milk repalcers for farm animals : processing, digestion and adverse reactions. *Recent Res. Devel. in Agricultural and Food Chem.*, **2**, 565-576.

Lammers, B. P., Heinrichs, A. J., Aydin, A. (1998) The effect of whey protein concentrate or dried skim milk in milk replacer on calf performance and blood metabolites. *J. Dairy Sci.*, **81**, 1940-1945.

Leplaix-Charlat, L., Durand, D., Bauchart, D (1996) Effects of diets containing tallow and soyabean oil with and without cholesterol on hepatic metabolism of lipids and lipoproteins in the preruminat calf. *J. Dairy Sci.*, **79**, 1826-1835.

Le Huerou-Luron, I., Guilloteau, P., Wicker-Planquart, C., Chayvialle, J-A., Burton, J., Mouats, A., Toullec, R. and Puigserver, A. (1992) Gastric and pancreatic enzyme activities and their relationship with some gut regulatory peptides during postnatal development and weaning in calves. *J. Nutr.*, **122**, 1434-1445.

Le Huerou, Guilloteau, P., Wicker, C., Mouats, A., Chayvialle, J-A., Bernard C., Burton, J., Toullec, R., Puigserver, A (1992) Activity distribution of seven digestive enzymes along small intestine in calves during development and weaning. *Digestive Diseases and Sciences*, **37**, 40-46.

Licitra, G., Hernandez, H.M., Van Soest, P.J. (1996) Standardization of procedures for nitrogen fractionation of ruminant feeds. *Animal Feed Science and Technology*, **57**, 347-358.

Longenbach, J. I., Heinrichs, A. J. (1998) A review of the importance and physiological role of curd formation in the abomasum of young calves. *Animal Feed Science and Technology*, 73, 85-97.

Mahaut, M., Jeantet, R., Brulé, G., Schuck, P. (2000a). Lait de consommation In *Les Produits Industriels Laitiers*,pp1-23. Edited by Tec & Doc, Paris.

Mahaut, M., Jeantet, R., Schuck, P., Brulé, G. (2000b). Produits déshydratés In *Les Produits Industriels Laitiers*, pp 49-90. Edited by Tec & Doc, Paris.

Montagne, L., Toullec, R. Lallès, J-P (2001) Intestinal digestion and endogenous proteins along the small intestine of calves fed soyabean or potato. *J. Animal Sci.*, **79**, 2719-2730.

National Research Council (2001) Nutrient requirements of the young calf In *Nutrient Requirements of Dairy Cattle*, 7th revised edition, p. 214-233. Edited by National Academy Press, Washington DC, the USA.

Niro Atomizer (1978) Determination of undenatured whey protein nitrogen in non-fat dry milk (WPNI) In *Analytical Methods for Dry Milk Powders, 4th edition*, pp 76-79. Ed. A/S Niro Atomizer, Copenhagen, Denmark.

Nunes Do Prado, I., Toullec, R., Lallès J.P., Guéguen, J., Hingand, J., Guilloteau, P. (1989a) Digestion des protéines de pois et de soja chez le veau préruminant. I. Taux circulants de nutriments, formation d'anticorps et perméabilité intestinale aux macromolécules. *Reprod. Nutr. Dev.*, **29**, 413-424.

Nunes Do Prado, I., Toullec, R., Guilloteau, P., Guéguen, J (1989b) Digestion des protéines de pois et de soja chez le veau préruminant. II. Digestibilité apparente à la fin de l'iléon et du tube digestif. *Reprod. Nutr. Dev.*, **29**, 425-439.

Petit, H. V., Ivan, M., Brisson, G. J. (1988) Digestibility and blood parameters in the preruminant calf fed a clotting or a nonclotting milk replacer. *J. Anim. Sci.*, **66**, 986-991.

Petit, H.V., Ivan M., Brisson, G.J. (1989) Digestibility measured by faecal and ileal collection in preruminant calves fed a clotting or a nonclotting milk replacer. *J. Dairy Sci.*, **72**,123-128.

Roy, J. H. B. (1970) Protein in milk replacers for calves. *J Sci. Food Agric.*, **21**, 346-351.

Saito, T., Yanaji, A., Itoh, T. (1991) A new isolation method of caseinglycopeptide from sweet cheese whey. *J. Dairy Science*, **74**, 2831-2837.

Scanff, P., Savalle, B., Miranda, G., Pelissier, J-P., Guilloteau, P., Toullec, R. (1990) In vivo gastric digestion of milk proteins. Effect of technological treatments. *Journal of Agricultural and Food Chemistry*, **38**(8), 1623-1629.

Scanff, P., Yvon, M., Pélissier, J-P., Guilloteau, P., Toullec, R. (1992) Affect of some technological treatments of milk on in vivo gastric emptying of immunoreactive whey proteins. *Lait*, **72**, 43-51.

Shah, N. P. (2000) Effect of milk-derived bioactives: an overview. *British Journal of Nutrition*, **84**, Suppl 1, S3-S10.

Strudsholm, F., Lykkeaa, J. (1988) Observations at slaughter in the gastrointestinal tract from preruminant calves fed different milks and milk replacers. *Acta Agric. Scand.*, **38**, 329-336.

Tanan, K.G., Newbold J.R. (2002) The Nutrition of the dairy heifer calf In *Recent Advances in Animal Nutrition – 2002*, pp 119-149. Edited by P. C. Garnsworthy and J. Wiseman. Nottingham University Press, Nottingham, UK.

Terosky, T. L., Heinrichs, A. J., Wilson, L. L. (1997) A comparison of milk protein sources in diets of calves up to eight weeks of age. *J. Dairy Sci.*, **80**, 2977-2983.

Terui, H., Morrill, J.L., Higgind, J.J. (1996). Evaluation of wheat gluten in milk replacers and calf starters. *J. Dairy Sci.*, **79**, 1261-1266.

Toullec R., Matthieu, C. M. (1969) Utilisation digestive des matières grasses et de leur principaux acides gras par le veau pré-ruminant à l'engrais. Influence sur la composition corporelle. *Ann. Biol. Anim. Bioch. Biophys.*, **9**, 136-160.

Toullec, R. (1989) Veal Calves. In *Ruminant Nutrition. Recommended Allowances and Feed Tables*, pp. 109-120. Edited by Jarrige R. John Libbey Eurotext, London- Paris.

Toullec, R., Grongnet, J-F. (1990) Remplacement partiel des protéines du lait par celles du blé ou du maïs dans les aliments d'allaitement : influence sur l'utilisation digestive chez le veau de boucherie. *INRA Prod. Anim.*, **3**(3), 201-206.

Toullec, R., Lallès, J.P., Bouchez, P. (1994) Replacement of skim milk with soya bean protein concentrates and whey in milk replacers for veal calves. *Animal Feed Science and Technology*, **50**, 101-112.

Toullec,R., Formal, M. (1998) Digestion of wheat protein in the preruminant calf: ileal digestibility and blood concentrations of nutrients. *Animal Feed Science and Technology*, **73**, 115-130.

Tukur, H.M., Branco-Pardal, P., Formal M., Toullec R., Lallès J-P., Guilloteau, P. (1995) Digestibility, blood levels of nutrients and skin responses of calves fed soyabean and lupin proteins. *Reprod. Nutr. Dev.*, **35**, 27-44.

Ullerich, A., Kampheus, J. (1998) Effects of high concentrations of sodium and potassium in milk replacers on intestinal processes and faeces composition in young calves. *J. Anim. Physiol. and Anim. Nutr.*, **80**, 194-200.

Xu, C., Wensing, T., Beynen, A.C., (2000) high intake of calcium formiate depresses macronutrient digestibility in veal calves fed milk replacers containing either dairy proteins or whey protein plus soya protein concentrate. *J. Anim. Physiol. and Anim. Nutr.*, **83**, 49-54.

7

NUTRIENT SOURCES FOR SOLID FEEDS AND FACTORS AFFECTING THEIR INTAKE BY CALVES

T. MARK HILL, JIM M. ALDRICH, AND RICK L. SCHLOTTERBECK
Akey, Lewisburg, Ohio, USA

Introduction

Various calf researchers and advisors have recommended that the first solid feed that a calf is offered should be highly digestible and palatable, and should contain a limited but defined degree of bulk. Recommendations regarding high digestibility are related to the young calf having limited rumen fermentation and responding to volatile fatty acids produced by rumen fermentation. Solid food should be palatable in order to encourage consumption, which leads to rumen development and facilitates the transition from a liquid to a solid diet. Defined bulkiness has been suggested to build rumen and omasal musculature, to facilitate digestion of forage, as the calf matures.

This paper will review selected published research and trials from our unit that relate to how solid feeds for calves less than three months of age are formulated. It will also review factors that affect intake of solid feed by calves. Nutrient concentrations will be expressed on an as-fed basis since most solid feeds have a moisture content of 110 to 130 g/kg (DM content 870 to 890 g/kg).

Trends and philosophy of diet composition

Grains used to be the predominant source of energy for solid feeds in the US. They still are in many feeds, but more feeding programs are using fibrous co-products like wheat midds and soyabean hulls. Additionally, the use of steam-rolled or flaked corn rather than dry, rolled or other dry, processed corn is becoming more popular. Wheat midds and soyabean hulls are typically lower cost ingredients that at times can significantly lower the cost of a calf feed. Additionally, some farms are using one solid feed from birth until approximately four months of age, and using significant amounts of fibrous ingredients in the diet reduces the appearance of scours and acidosis in the weaned calf. The moisture-processed grains in calf feeds are marketed to be more digestible; however, some of these grains are not highly processed (thick flakes with little gelatinized starch).

There appears to be more usage of moisture-processed grains with extensive gelatinized starch and high sugar ingredients in Europe than in the US; high fibre feeds appear to be used more frequently in other areas of the world.

Feeds (like grains) that easily yield volatile fatty acids (VFA) upon digestion promote the development of rumen epithelial tissue. Forages, which are high in fibre that is slowly digested, yield less VFA than grains when digested. Forages may provide excess bulk and limit DM intake in young calves, so it might be best to delay feeding them until after weaning (Davis and Drackley, 1998). Also, feed intake can be variable with long hay (Thomas and Hinks, 1968 from Beharka, Nagaraja, and Morrill, 1991). However, Dr. Jim Morrill's group (at Kansas State University) have suggested that a degree of abrasive fibre of adequate particle size might be required to prevent keratinization and abnormal development of rumen papillae, and to strengthen the rumen and omasal lining and musculature; and calves may consume bedding if forage is not provided to achieve this stimulation (Greenwood, Morrill, Titgemeyer, and Kennedy, 1997b; Beharka, Nagaraja, Morrill, Kennedy, and Klemm, 1998; using finely ground control diets).

Dry forage and form of solid feed presentation

Davis and Drackley (1998) recommended that feeding hay should be delayed until after weaning, and that the production of volatile fatty acids from concentrate fermentation is the key to development of the rumen epithelium. When looking at several published trials where hay was fed, its consumption was typically less than 150 g/kg of the total dry matter intake from solid feeds (Jasper and Weary, 2002; Chua, Coenan, van Delen, and Weary, 2002). Intake of long hay is hard to measure accurately due to its waste by calves. We measured intake and waste of straw and a high quality grass hay by 22- to 42-day-old calves fed milk replacer and solid feed free choice while housed in crates without bedding. Straw intake was 20 g/kg total solid feed intake and hay intake was 50 g/kg (706 and 741 g/day), and the calves wasted 3 times more forage than they consumed. This might suggest that straw bedding could meet the calf's need for abrasive, large particles, and that straw consumption is too small to alter performance.

Consumption of hay 2 to 4 weeks post-weaning often does not increase, whilst concentrate intake increases rapidly (Funaba, Kagiyama, Inki, and Abe, 1994; Jasper and Weary, 2002; Chua *et al.*, 2002), meaning that hay might represent less than 50 g/kg total solid feed intake. Quigley, Steen, and Boehms (1992) fed restricted concentrates (4.5 kg DM) and free choice hay (160 g CP /kg, 620 g NDF /kg) to calves initially 16 weeks old, and observed a mean hay intake of 0.8 kg DM/day, or approximately 18 g/kg solid feed intake. If approximately 250 g of hay was limit-fed and consumed by calves during their first 2 to 4 weeks post-weaning, this would represent 50 to 100 g/kg total intake.

Suggestions have been made to grind hay and incorporate 100 to 150 g/kg in complete solid feed mixtures for calves (Morrill, personal communication). This could pose problems for manufacturing and increase the heterogeneity of feeds.

No firm conclusions can be made to state that a pellet or textured diet is better or worse for a calf. Many people recommend not to feed finely ground diets, stating that calves select against fine particles. The limited published trials yield mixed results. We have fed diets with similar nutrient profiles and compositions, either as a complete feed pellet or textured (combination of coarsely processed grains and a protein supplement pellet coated with liquid molasses) and observed no differences in calf performance (calves bedded with straw). Franklin, Amaral-Phillips, Jackson, and Campell (2003) observed solid feed intake and gains to be greatest for calves fed a textured feed, lowest for calves fed a pelleted feed, and intermediate for calves fed a ground feed; however, the ingredient composition and nutrient profiles of the diets differed. Calves were housed in hutches with no mention of bedding. Warner (1991) cited a thesis of Porter, who observed greater calf solid feed intake and gains with a meal feed having 850 g/kg of the particles retained on an 1190 micron sieve vs. a pellet having 200 g/kg of the particles retained on an 1190 micron sieve (no calf bedding material). Our textured feed mentioned above had approximately 500 g/kg remaining on a 1,190 micron sieve, not counting the pellets in the diet. Warner (1991) also cited a 1960 research trial where it was found that the rumens of calves fed only milk were packed with wood shavings (bedding) and had no papillary development. This suggested that the calves could not ferment the lignified shavings (the milk by-passed the rumen) and that fermentation-inert ingredients, even though they provide abrasive fibre, are of little value. It could be speculated that if a calf is bedded with straw, the particle size of the diet is less critical than if the calf is raised without bedding (crates, elevated pens, etc.). When no bedding is used, a solid feed with significant particle size might be required. However, the size cannot be defined from gleaning the published literature.

Conclusion: If calves are housed with bedding materials, especially straw, there is no evidence that feeding hay pre-weaning will be beneficial. If calves are housed with no bedding, feeding hay, straw, or coarsely textured grains does have merit.

Protein

The Dairy NRC (2001, 1989) lists 180 g CP /kg (as-fed basis; 200 g/kg dry matter basis) as the requirement for calf solid feeds. The calf model in the

Dairy NRC (2001) suggests that energy, not protein, limits gain in calves weighing 60 to 90 kg and receiving only a dry feed. Trials by Akayezu, Linn, Otterby, Hansen, and Johnson (1994), Luchini, Lane, and Combs (1991), and Hill, Aldrich, Proeschel, and Schlotterbeck (1991) substantiate that 180 g CP / kg diets are adequate. However, Drackley, Bartlett, and Bloom (2003) found that calves fed solid feed containing 220 g CP /kg were more efficient than calves fed solid feeds with 180 g CP /kg. Our trials evaluated the CP content of solid feeds fed with milk-replacer (MR) powders containing 200 g CP/kg (as-fed) fed at 454 g/day, or containing 260 g CP/kg (as-fed) fed at 680 g/day (Hill *et al.* 2001; Figure 7.1). Exceeding 180 g CP/kg in the solid feed did not affect live-weight gain, solid feed intake, hip width, body condition score, or health measurements when calves gained 620 g/day (200 g CP /kg MR) or 767 g/day (260 g CP /kg MR) from 0 to 56 days on-test. Akayezu *et al.* (1994) was the only trial using solid feeds less than and exceeding 180 g CP /kg; they observed a plateau in live-weight gain at ~180 g CP /kg (Figure 7.1).

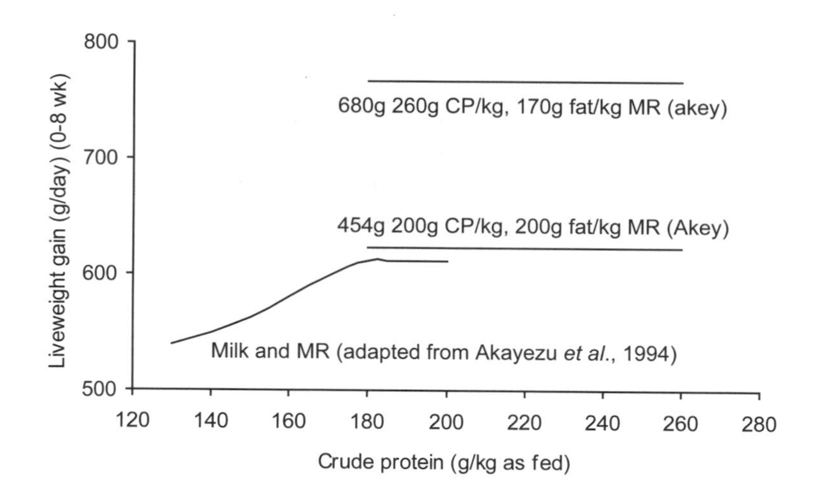

Figure 7.1 Effect of crude protein concentration of solid feed on live-weight gain.

Swartz, Heinrichs, Varga, and Muller (1991), Holtshausen and Cruywagen (2000), and Abdelgadir Morrill, and Higgins (1996a) observed no differences in live-weight gains of calves fed diets with different rumen undegradable protein (RUP) values. Abdelgadir, Morrill, and Higgins (1996b) observed improved performance when RUP sources were fed. Maiga, Schingoethe, Ludens, Tucker, and Casper (1994) observed better live-weight gains when calves were fed extruded soyabean meal vs. soyabean meal. Abdelgadir *et al.* (1996b) observed improved performance when calves were fed protein and corn sources with similar rates of ruminal degradation; however, Abdelgardir *et al.* (1996a) saw similar live-weight gains regardless of corn and protein degradation rates. Bunting, Fernandez, Fornea, White, Froetschel, Stone, and

Ingawa (1996) found improved live-weight gains from RUP sources during hot weather, but not during cold weather. McCoy, Ruppert, and Hutjens (2003) observed trends for slower live-weight gains when RUP sources were fed. Fiems, Bouchque, Cottyn, and Bysse (1987) observed no difference in calf performance when formaldehyde-treated soyabean meal replaced soyabean meal. Warner (1984) reviewed research prior to 1984 and reported no advantage to using RUP sources in calf starters. These trials show mixed results from adding RUP sources, with more trials reporting no benefit than trials reporting benefits.

Vazquez-Anon, Heinrichs, Aldrich, and Varga (1993) and Holtshausen and Cruywagen (2000) observed that rumen degradability of CP sources increased with age post-weaning, suggesting that this trend probably extends back into the pre-weaning phase of calf development. The lack of complete development of the rumen and its microbial population in pre-weaned calves might explain why there are no conclusive benefits from formulating solid feeds with higher RUP contents.

Few studies have compared sources of protein used in solid feeds. Fiems, Bouchque, Cottyn, and Bysse (1985) observed numerically lower live-weight gains when rapeseed (canola) meal replaced soyabean meal. Intake of the rapeseed meal diets was less than that of the soyabean meal diets, and poor intake (poor palatability) of rapeseed meal diets might have more influence than digestibility on live-weight gain, especially if the glucosinilate level is high. Sharma, White, and Ingalls (1986) observed poorer live-weight gains and poorer digestibility of diets when rapeseed meal or extruded or pelleted whole cottonseed replaced soyabean meal. However, they observed similar performance in calves fed unprocessed whole cottonseed seed, unprocessed whole sunflower seed, and soyabean meal-based diets. Fiems *et al.* (1986) observed poorer digestibility, live-weight gain, and efficiency of gain when cottonseed meal replaced soyabean meal in the solid feed. Replacing soyabean meal with corn gluten feed resulted in poor efficiency of gains, greater intakes, and similar live-weight gains as diets with soyabean meal. Replacing soyabean meal with urea (approximately half of the protein in the diets from urea) supported slower live-weight gains and poorer efficiency (Fiems *et al.*, 1987). As a group, these studies indicate that soyabean meal is as good, or better, than most protein sources for calf feeds.

A combination of corn and soyabean meal offers a good balance of amino acids, is free of many anti-nutritional factors, is low in fibre, and is consistently digestible (as reviewed by Chiba, 2001, for pigs). Soyabean meal was the base protein in the calf trials discussed above, and calves fed the soyabean meal-based diets performed as well or better than calves fed diets containing other sources of protein.

Conclusion: Solid feeds formulated to 180 g CP /kg on an as-fed basis using soyabean meal appear to be optimum.

Grains, processing and other carbohydrate sources

Some consultants suggest than heat processing improves palatability of grains, but we are not aware of studies that show this. Abdelgadir *et al.* (1996b) demonstrated that roasting corn increased starch gelatinization three-fold and tended to improve calf performance when combined with soyabean meal but not roasted soyabeans (suggesting an effect of synchronising rumen carbohydrate and protein). Abdelgadir *et al.* (1996a) observed conglomerating corn (grind corn, add water, pellet, and then roast) increased starch gelatinization five-fold but, when mixed with combinations of urea, soyabean meal, and roasted soyabeans, did not alter calf performance. Abdelgadir and Morrill (1995) observed calves fed roasted sorghum grain to have 5% better feed efficiency than calves fed raw sorghum. Also, they observed a 20% lower intake of dry feed and 11% better feed efficiency, and a trend for reduced live-weight gain in calves fed conglomerated vs. raw sorghum. It seems logical to think that processing to make the starch more digestible would improve its utilization by calves and thus calf performance; however, the available research does not show a clear benefit from gelatinizing grains.

Maiga *et al.* (1994) compared diets containing corn, barley, or whey and observed the best live-weight gains with corn. Williams, Fallon, Innes, and Garthwaite (1987) observed similar live-weight gains when a beet/citrus pulp combination replaced barley, but intake was greater, and efficiency was poorer.

Work from the lab of Dr. Jud Heinrichs at The Penn State University compared whole, dry rolled, roasted-rolled, and steam flaked corn (330 g/kg inclusion) in textured solid feeds fed to calves weaned at 28 days of age, but maintained on the solid feeds until 42 days of age (Lesmeister, 2003). Live-weight gains did not differ pre-weaning, but favoured calves fed whole or dry-rolled corn post-weaning (29 to 42 days of age). Similarly, dry feed intake was greater, and feed efficiency tended to be better, for calves fed whole and dry-rolled corn post-weaning, but did not differ pre-weaning. Lesmeister (2003) observed a tendency for calves fed roasted, rolled and steam-flaked corn to have greater total VFA concentrations than calves fed whole or dry, rolled corn.

Hancock and Behnke (2001) suggest a 1.2 to 1.4% improvement of gain:feed in growing pigs (most of the data from pigs over 20 kg) for each 100-micron reduction in mean particle size of corn. Similar observations were reported for barely, sorghum, and wheat. Data of this nature are not available from calf trials. However, the mean particle size of dry processed corn in textured calf feeds is typically very large (> 1,500 microns) compared with its mean size when included in pelleted feeds (< 900 microns).

Interestingly, when we fed a novel, highly refined carbohydrate similar to starch but with less cross-linkages, we observed consistent improvements in live-weight gain of about 4%, as well as improvements in intake of solid feed by calves (Figure 7.2). This ingredient was also used quite well in a piglet trial. When this ingredient was combined with butyrate and fed to calves, the response more than doubled (Figure 7.2).

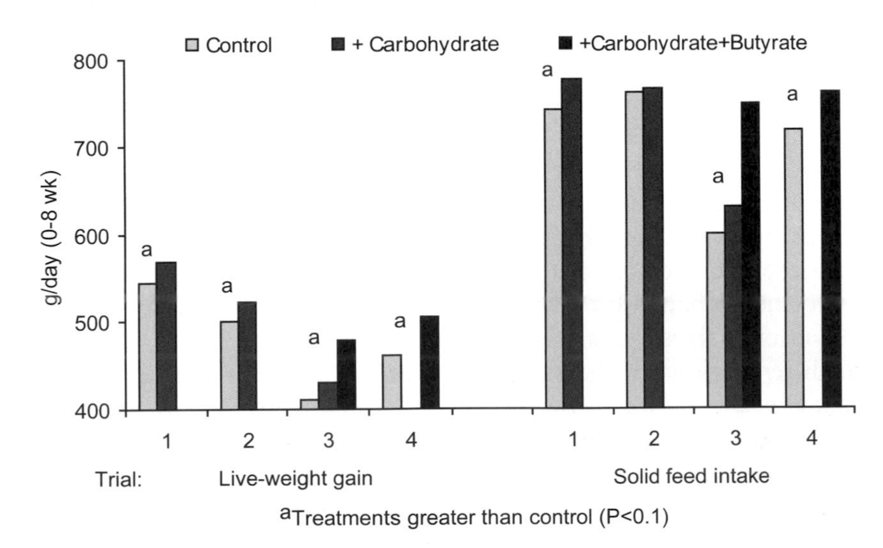

Figure 7.2 Effect of a novel carbohydrate and butyrate on live-weight gain and solid feed intake.

When we fed textured dry feeds with different levels of total sugar, we observed a trend for the higher (80 g/kg added) sugar diets (from cane molasses or molasses plus sucrose) to support 5 to 7% less live-weight gain and solid fed intake, and have less firm faeces (but not an enteric scour) post-weaning vs. calves fed the lower (40 g/kg added) sugar diets. Low intakes post-weaning typically yield a firm faecal output and the observed faecal scores may indicate too much total sugar in the diet creating an osmotic scour. Lesmeister (2003) fed textured diets with 50 and 120 g/kg cane molasses (DM basis) and observed greater live-weight gains, greater dry feed intake, and less scouring in calves fed the 50 g/kg molasses treatment. However, calves fed the 120 g/kg molasses diet had greater blood concentrates of total VFA, plus longer and wider rumen papillae, than calves fed the 50 g/kg molasses diet.

We saw no difference in calves fed a control diet or one with 50 g/kg sweet whey; the control diet and the diet with whey both contained 50 g/kg molasses.

Conclusion: Dry-processed grain with no more than 50 g/kg added sugar appears optimum.

Fibre sources

Low inclusion (120 to 200 g/kg of the diet) of beet pulp, cottonseed hulls, and whole fuzzy cottonseed have resulted a loss of feed efficiency when replacing corn, indicating they are not digested as well as corn, although live-weight gains and feed intake were not statistically different (Figure 7.3). However, a diet containing 500 g/kg soyabean hulls supported 6% slower live-weight gains, and had a 4% poorer feed efficiency than a control diet based on 620 g/kg corn. This depressed performance is surprisingly good and suggests than there is a great deal of rumen fermentation in calves less than 8 weeks of age, as reported in the literature (Quigley, Caldwell, Sinks, and Heitmann, 1991; Vazquez-Anon *et al.*, 1993; Holtshausen and Cruywagen, 2000). Williams *et al.* (1987) observed similar live-weight gains when a beet/citrus pulp combination replaced barley, but intake was greater, and efficiency was poorer. Hill, Hopkins, Davidson, Bolt, Brownie, Brown, Huntington, and Whitlow (2003a) observed greater solid feed intake and live-weight gains, but worse feed efficiency when solid feeds containing 150 g/kg cottonseed hulls were fed to Holstein calves vs. solid feeds without cottonseed hulls. However, they saw no differences when the same solid feeds were fed to Jersey calves (Hill, Hopkins, Davidson, Bolt, Brownie, Brown, Huntington, and Whitlow, 2003b).

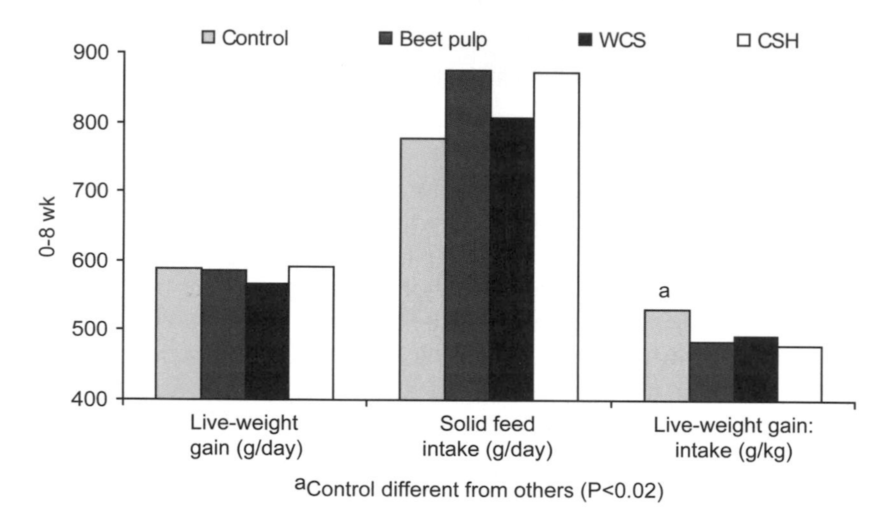

Figure 7.3 Effect of replacing a portion of corn in solid feeds with beet pulp, whole cotton seed (WCS) or cotton seed hulls (CSH).

Conclusion: If the solid feed is going to be fed from birth to approximately two months of age, use limited amounts of fibre sources in the diet to maximize live-weight gain and efficiency, unless grains are cost-prohibitive or unavailable. If the solid feed is going to be fed considerably longer than the first two months, fibre sources will have more merit.

Fat

Fallon, Williams, and Innes (1986) fed 0, 50, 100, and 200 g/kg fat from Ca-soaps and observed depressed intakes and live-weight gains with added fat. Caffrey, Mill, Brophy, and Kelleher (1988) observed 35 g/kg fat addition to barley-based diets to depress intake. Doppenberg and Palmquist (1991) observed 100 g/kg added fat to depress intake and live-weight gain. Kuehn, Otterby, Linn, Olson, Chester-Jones, Marx, and Barmore (1994) replaced soyabean meal with roasted soyabeans and observed solid feed intake to be 10% lower post-weaning and live-weight gains to be 8% slower pre-weaning and 12% slower post-weaning (fat and protein source were confounded). Kuehn *et al.* (1994) cited a study from 1959 using 100 and 200 g/kg fat addition that observed reductions in intake and live-weight gain. Bunting *et al.* (1996) observed no response to adding 50 g/kg hydrolyzed tallow. McCoy *et al.* (2003) replaced soyabean meal with a high fat, extruded cottonseed-soyabean product and observed solid feed intake to decrease by 34% and live-weight gain to decrease by 4% (fat and protein source were confounded).

Conclusion: Limit the amount of fat in a solid feed to avoid depressing solid feed intake and live-weight gain.

Coccidiostats

Coccidiostats added to solid feeds appear to increase live-weight gain by calves, even when there is no diagnosis of clinical coccidiosis. Most studies show that when added to the solid feed coccidiostats reduce the shedding of coccia (Hoblet, Charles, and Howard, 1989; Heinrichs, Swartz, Drake, and Travis, 1990; Heinrichs and Bush, 1991; Eicher-Pruiett, Morrill, Nagaraja, Higgins, Anderson, and Reddy, 1992; Quigley, Drewry, Murray, and Ivey, 1997). Our studies (non-challenge trials with no clinical coccidiosis observed), using approved levels of decoquinate, lasalocid, and monensin in solid feeds, have shown consistent improvements in live-weight gain and solid feed intake compared with no coccidiostat. Adequate intake of the solid feed containing a cocciostat is needed to achieve an effective dose, and infection with coccidia often occurs before sufficient solid feed intake is achieved. Thus, including a coccidiostat in the liquid diet will provide more complete control.

Conclusion: Inclusion of an approved level of coccidostat in the solid feed is merited.

Microbial additives, prebiotics, and probiotics

The use of bacterial, yeast, and fungal products has shown benefits in some trials, but the results have been highly variable and inconclusive (Jenny, Vandijk, and Collins, 1991; Quigley *et al.*, 1992; Higginbotham and Bath, 1993; Morrill, Feyerherm, and Laster, 1995; Abe *et al.*, 1995; Cruywagen *et al.*, 1996; Morrill, Dayton, and Mickelson, 1997; Lesmeister, 2003; Heinrichs, Jones, and

Heinrichs, 2003). Our trials evaluating several yeast, bacterial, and oligosaccharide products have shown no positive effects. Some bacterial products do not have good viability in normal storage conditions due to the presence of moisture and oxygen, and some cannot withstand the temperatures of pelleting. For some products, trials have not been conducted to establish a dose response or effective dose, and for others dose response trials yield confusing results.

Conclusion: The use of most microbial additives and prebiotics might be best reserved for use as an adjunct to other treatments when calves are scouring.

Botanicals, plant extracts, essential oils, herbs

There has been little research with botanicals, plant extracts, essential oils, herbs, and similar compounds in calves. Donovan, Franklin, Chase, and, Hippen (2002) compared an allicin-based product to a combination of oxytetracycline and neomycin and found no difference in calf performance. Olsen, Epperson, Zeman, Fayer, and Hildreth (1998) observed that a similar product did not reduce the effects of *Cryptosporidium parvum* in calves. We have evaluated four commercial plant extract-based products and observed no response with three products. We have evaluated one commercial blend of several botanicals and observed consistent improvements in live-weight gain, some reduction in scouring, and improvements in solid feed intake, in six comparisons (Figure 7.4). This blend improved live-weight gain when added via the liquid feed, via the solid feed, and when included in both the liquid and solid feeds (Hill, Aldrich, and Schlotterbeck, 2003c). The botanical blend also improved live-weight gain when added to an all milk protein MR, to a 500 g/kg soya protein (from soya protein concentrate) MR (Hill, Aldrich, and Schlotterbeck, 2003d), to a 600 g/kg corn solid feed, and to a solid feed with 500 g/kg soyabean hulls.

Conclusion: We are enthused by one particular blend of botanicals but, as a general group of ingredients, the functions and use of botanticals are not understood.

Flavours and aromas

Morrill and Dayton (1998) evaluated a blend of ethyl butyrate, butterscotch, maple, and saccharin in solid feed and saw increased intake and live-weight gain. They observed an even greater response when the flavours were added to both the milk and solid feed. Thomsen and Rindsig (1980) observed that a flavour added to solid feed improved solid feed intake and live-weight gain. Adding a maple flavour and milk aroma yielded greater intake of solid feed and greater live-weight gain than the butter flavour. However, they observed that adding the same flavour to both the liquid and solid diet only increased

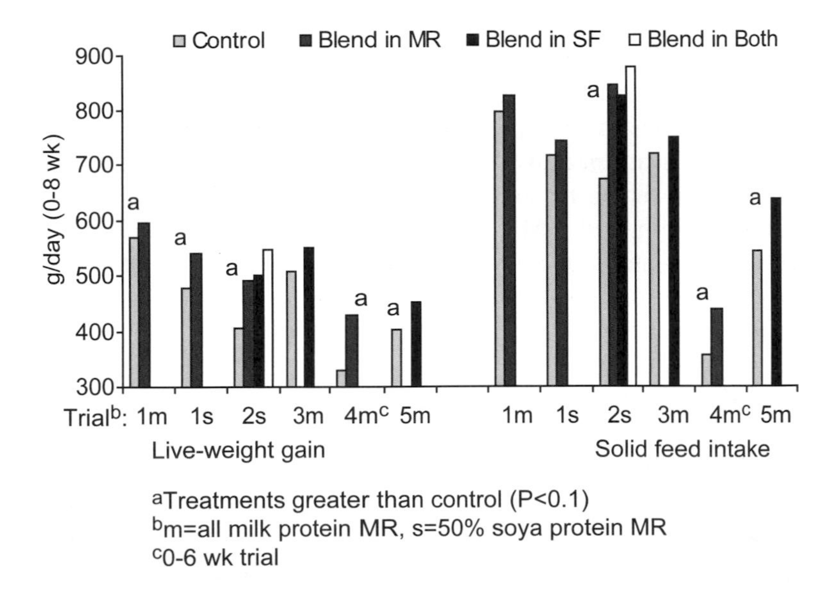

Figure 7.4 Effect of Apex™ botanicals on live-weight gain and solid feed intake when fed via the milk replacer (MR), solid feed (SF) or both.

solid feed intake and live-weight gain when using the milk aroma and comparing it with the control that had no flavour added to either feed. Eastridge and Roseler (2003) added a maple flavour to the solid feed only and observed no response in solid feed intake or live-weight gain. We have attempted some preference trials and trials where we have switched calves from one diet to another. Our results suggest that the calf prefers the diet that it has been previously fed, regardless of flavour type or the absence of flavour. There will also be individual calves that respond completely the opposite to the majority of calves.

Conclusion: Flavours and aromas might add to the consistency of a solid feed, and calves appear to like consistent diets (possibly justifying their use), but there is no repeatable research that justifies their use.

Amount of liquid fed in "conventional" programs and solid feed intake

Calves are often raised on a restricted-fed liquid diet to encourage solid feed intake at an early age. In the US, it is most common for calves to be fed a fixed amount of MR powder (typically between 450 to 570 g/day) or 3.8 L of whole milk daily from birth to 30 - 60 days of age ("conventional program"). Solid feed is offered free choice from birth (or less than one week of age). Some, but very few, calves are fed an increasing amount of liquid feed as a percentage of body weight or some incremental level ("step-up" method), as the calf grows or ages. Davis and Drackley (1998) cite data from the Milk Specialties Co.

showing that a recommended "step-up" method, which increased live-weight gain between 0 and 42 days by 9%, required 27% more MR powder and depressed solid feed intake, compared with calves fed 454 g/day of MR powder. Catherman (2000) observed depressed solid feed intakes when a 200 g CP/kg, 200 g fat/kg MR was increased from 454 g/day to 568 g/day. He did not observe changes in intake of solid feed when he fed the same two amounts of a 220 g CP/kg, 200 g fat/kg MR and live-weight gains were similar among all 4 treatments. Davis and Drackley (1998) noted low solid feed intake and greater reductions in growth rates at the time of weaning for calves fed greater amounts of milk relative to calves fed restricted amounts of milk in several referenced trials.

Conclusion: If conventional MR are limited to 450 to 550 g of powder daily, and whole milk is limited to approximately 4 L daily, solid feed intake will not be depressed.

"Enhanced" liquid feeding programs and solid feed intake

Quigley, Wolf, and Elsasser (2003) observed a 19% reduction in solid feed intake, increased days scouring (62%), and increased mortality (22.1 vs. 2.8%), but 27% faster live-weight gains when they fed 908 g/day of a 280 g CP/kg, 150 g fat/kg MR (an "enhanced" program), compared with feeding 454 g/day of a 200 g CP/kg, 200 g fat/kg MR (a "conventional" program). When we fed a "step-up, enhanced" program (280 g CP/kg, 200 g fat/kg MR, target maximum intake of powder per calf either 1,130 g/day or 1,360 g/day) to calves that had be transported for 10 hours, we achieved an average MR intake of 770 g/day and 860 g/day and depressed solid feed intake by 48 and 49%, respectively, compared with feeding 454 g/day of a 200 g CP/kg, 200 g fat/kg MR powder. When we fed a "step-up, enhanced" program (280 g CP/kg, 200 g fat/kg MR, target maximum intake of 1,135 g powder per calf per day) to non-transported calves, we achieved an average MR intake of 980 g/day, depressed solid feed intake by 25%, and saw no improvement in live-weight gain. Compared with feeding 454 g/day of a 200 g CP/kg, 200 g fat/kg MR, feeding 680 g/day of a 280 g CP/kg, 200 g fat/kg MR depressed solid feed intake by 10% and feeding 680 g/day of a 260 g CP/kg, 170 g fat/kg MR did not depress solid feed intake, with both groups of calves receiving 680 g/day of MR gaining faster than calves receiving 454 g/day of MR. Pollard, Dann, and Drackley (2003) compared feeding a conventional MR (220 g CP/kg, 200 g fat/kg; 16 kg fed) at 12.5 g/kg birth weight to feeding an "enhanced" MR (280 g CP/kg, 200 g fat/kg; 39 kg fed) at 20 to 25 g/kg body weight in two trials. They observed 54 to 55% reductions in solid feed intake but 13 to 23% improved growth rates over eight weeks with the "enhanced" program. Quantity of MR fed and MR fat content affect the consumption of solid feed by calves.

A detriment of feeding too much liquid feed is also its depression of solid feed intake and utilization (feed efficiency) post-weaning. Jasper and Weary (2002) fed whole Holstein milk at 4.9 kg/day (restricted amount by bucket) and 8.8 kg/day (ad libitum; nursed by the cow) for 36 days and observed extremely low solid feed intakes of 0.2 and 0.1 kg/day, respectively. During the 5-day weaning period, solid feed intakes were 1.0 and 0.75 kg/day, and 1.94 and 2.01 kg/day during the 20-day post-weaning period, respectively. Total live-weight gain was 17.3 and 28.1 kg pre-weaning, 2.7 and 1.8 kg during weaning, and 17.0 and 13.6 kg post-weaning for the low and high milk-fed groups, respectively. Strzetelski *et al.* (2001) fed either 1.0 or 1.6 kg/day of a 220 g CP/kg, 170 g fat/kg MR and observed solid feed intakes of 0.41 and 0.34 kg/day in the 49-day pre-weaning period and 2.52 and 2.65 kg/day dry-feed intake in the 63-day post-weaning period, respectively. Total live-weight gains were 20.0 kg pre-weaning and 81.5 kg post-weaning for the low MR group and 37.7 kg pre-weaning and 81.2 kg post-weaning for the high MR group. Calves fed the higher level of liquid diet in both trials (Jasper and Weary, 2002; Strzetleski, Niwinska, Kowalezyk, and Jurkiewiez, 2001) were slightly more efficient pre-weaning and slightly less efficient post-weaning. These results are consistent with our observations in trials where we weaned at 4, 5, 6, or 7 weeks of age with several "enhanced" MR feeding programs. A moderate feeding rate of 680 g/day using a 260 or 280 g CP/kg MR, 170 g fat/kg MR, did not reduce solid feed intake or its efficiency of utilization pre- or post-weaning in our trials; greater rates of feeding MR, or greater concentrations of fat in the MR, reduced solid feed intake and utilization.

Conclusion: High feeding rates and high fat (200 g/kg or more) MR depress solid feed intake and result in slow live-weight gains during the weaning and post-weaning periods. When calf management is above average, a moderate feeding rate of 680 g/day of a 260 to 280 g CP/kg MR, 170 g fat/kg MR did not reduce solid feed intake and resulted in approximately 40% greater live-weight gains than conventional programs.

Fatty acid profile of liquid diet and solid feed intake

The inclusion of a specific combination of medium chain and 18:2, 18:3 fatty acids, combined with animal fats in MR ranging from 150 to 200 g fat/kg have consistently resulted in improved live-weight gains (Figure 7.5) and reduced scouring (Figure 7.6; Hill, Aldrich, and Schlotterbeck, 2003e). In most, but not all trials, there was an increase in solid feed intake (Figure 7.5). Whole milk supplemented with this fatty acid combination supported greater live-weight gains than milk alone. The addition of medium chain fatty acids alone or 18:2, 18:3 fatty acids alone supported only marginally greater performance than MR containing all animal fat.

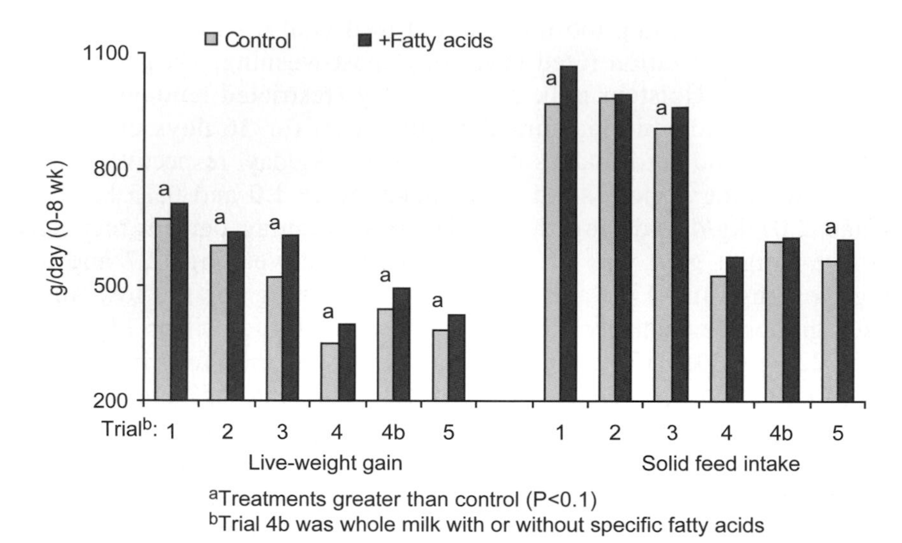

Figure 7.5 Effect of the addition of specific fatty acids to the milk replacer on live-weight gain and solid feed intake.

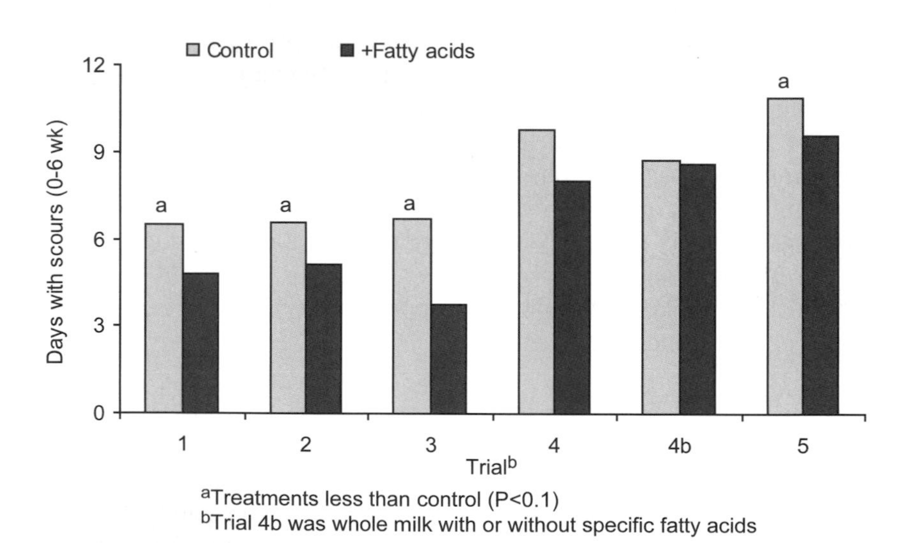

Figure 7.6 Effect of the addition of specific fatty acids to the milk replacer on days with scours.

Conclusion: Feeding an animal-fat-based MR fortified with medium chain and 18:2, 18:3 fatty acids resulted in a 13% improvement in live-weight gain, 6% improvement in solid feed intake, and 18% reduction in scouring, compared with an unfortified MR.

Weaning age and solid feed intake

It is typical to wean calves at about six weeks of age. When calves are weaned at less than six weeks of age, the total solid feed consumed prior to eight weeks of age is typically greatly increased, compared with weaning at six weeks of age. Additionally, overall live-weight gain for the eight-week period is typically equal or greater for calves weaned at less than six weeks of age, compared with calves weaned at six weeks of age (Quigley *et al.*, 1991; Greenwood, Morrill, and Titgemeyer, 1997a; unpublished Akey data).

Greenwood *et al.* (1997a) weaned calves based on solid feed intakes of 10, 15, or 20 g/kg initial body weight and observed similar live-weight gains from 0 to 8 weeks and also up to 20 weeks of age among all 3 groups. Calves weaned at 10 g solid feed intake per kg initial body weight consumed the most solid feed from 0 to 8 weeks and tended to grow faster than calves weaned at greater solid feed intakes. The average age at weaning was 32, 43 and 45 days for calves weaned at solid feed intakes of 10, 15, or 20 g/kg initial body weight, respectively.

Quigley *et al.* (1991) weaned calves from milk (1.8 kg per head daily) at either 28 or 56 days of age. They saw no differences in rate of gain because of weaning age, and calves weaned at 28 days of age consumed more solid feed but similar total amounts of dry matter over the 98-day trial. Live-weight gains were 0.26 kg/day, 0.53 kg/day, and 0.66 kg/day from 0 to 28 days, 29 to 56 days, and 57 to 98 days, respectively.

We have evaluated feeding high levels (817 g of powder) of a 300 g CP/kg, 200 g fat/kg MR for two or three weeks, followed by a conventional MR and feeding level (454 g powder) for three weeks (thus 5 or 6 week weaning), compared with a convention level (454 g of a 200 g CP/kg, 200 g fat/kg MR) for five or six weeks. Our idea was to see if we could feed calves to grow fast during their first two or three weeks with liquid nutrition and then reduce the liquid nutrition to encourage solid feed intake in an attempt to avoid a slump in growth rate at weaning. We saw no difference in rate of gain from 0 to 56 days of age, even though the calves fed the high rate, high CP MR grew faster initially. Those calves consumed less solid feed and were less efficient than calves fed conventionally. Additionally, the calves weaned at five weeks of age consumed more solid feed and were equally as efficient as those weaned at six weeks of age.

Conclusion: With excellent management, calves can be successfully weaned at four to five weeks of age.

Summary

If calves are housed with bedding materials, especially straw, there little evidence that feeding hay pre-weaning will benefit them. If calves are housed with no bedding, feeding hay, straw, or a coarse solid diet could be needed. There is limited information on the effects of grains and grain processing and its effects on the calf, but there are no consistent data that show benefits from processing to increase gelatinization. Solid feeds formulated to be 180 g CP/kg (as-fed basis) are adequate, based on published calf trials. There are no consistent data to merit the use of rumen undegraded protein (RUP) sources of protein. Replacing corn with 100 g/kg or more fibrous feeds has resulted in a worse (approximately 10%) feed efficiency, but not always a lower rate of live-weight gain (but up to approximately 10% lower live-weight gain). Adding fat to solid feed has frequently depressed solid feed intake and live-weight gain. The inclusion of approved coccidiostats in solid feeds has reduced coccidia shedding, and often improved live-weight gain and intake of solid feed. However, infection with coccidia could occur before there is sufficient intake of a dry feed containing a coccidiostat. Results from using various microbial products and oligosaccharides have been inconsistent. Some microbials are very unstable in the presence of moisture and oxygen, and some live organisms cannot withstand the temperatures of pelleting. One botanical or plant-extract type product has shown consistent improvements in live-weight gains, and its effect on live-weight gains was additive when added to both liquid and solid diets. Flavours have yielded inconsistent responses in a limited number of trials. Feeding high levels of liquid feeds (greater than approximately 525 g dry matter daily) has depressed solid feed intake. An exception has been when calves were fed an "enhanced" MR program, where the powder contained 260 to 280 g CP/kg, ~ 170 g fat/kg, and fed at 680 g/day (maximum). Supplementation with specific medium and 18:2, 18:3 fatty acids via the liquid diet has consistently stimulated live-weight gain and solid feed intake. Under good management, weaning calves at four or five weeks of age has resulted in greater intake of solid feed, and similar to greater live-weight gain from 0 to 8 weeks of age, than weaning calves at six or more weeks of age.

References

Abdelgadir, I. E. O., and J. L. Morrill. (1995) Effect of processing sorghum grain on dairy calf performance. *Journal of Dairy Science*, **78**, 2040-2046.

Abdelgadir, I. E. O., J. L. Morrill, and J. J. Higgins. (1996a) Ruminal availabilities of protein and starch: effects on growth and ruminal and plasma metabolites of dairy calves. *Journal of Dairy Science,* **79**, 283-290.

Abdelgadir, I. E. O., J. L. Morrill, and J. J. Higgins. (1996b) Effect of roasted soyabeans and corn on performance and ruminal and blood metabolites of dairy calves. *Journal of Dairy Science,* **79**, 465-474.

Abe, F., N. Ishibashi, and S. Shimamura. (1995) Effect of administration of bifidobacteria and lactic acid bacteria to newborn calves and piglets. *Journal of Dairy Science*, **78**, 2838-2846.

Akayezu, J. M., J. G. Linn, D. E. Otterby, W. P. Hansen, and D. G. Johnson. (1994) Evaluation of calf starters containing different amounts of crude protein for growth of Holstein calves. *Journal of Dairy Science,* **77**, 1882-1889.

Beharka, A. A., T. G. Nagaraja, and J. L. Morrill. (1991) Performance and ruminal function development of young calves fed diets with Asperigillus oryzae fermentation extract. *Journal of Dairy Science,* **74**, 4326-4336.

Beharka, A. A., T. G. Nagaraja, J. L. Morrill, G. A. Kennedy, and R. D. Klemm. (1998) Effects of form of the diet on anatomical, microbial, and fermentative development of the rumen of neonatal calves. *Journal of Dairy Science,* **81**, 1946-1955.

Bunting, L. D., J. M. Fernandez, R. J. Fornea, T. W. White, M. A. Froetschel, J. D. Stone, and K. Ingawa. (1996) Seasonal effects of supplemental fat or undegradable protein on growth and metabolism of Holstein calves. *Journal of Dairy Science,* **79**, 1611-1620.

Caffrey, P. J., C. Mill, P. O. Brophy, and D. L. Kelleher. (1988) The effects of method of processing of starters, tallow inclusion and roughage supplementation on the performance of early-weaned calves. *Animal Feed Science and Technology,* **19**, 231-246.

Catherman, D. R. 2000. Effect of increased milk replacer consumption on performance and economics in Holstein heifer calves. *Journal of Dairy Science,* **83** (Suppl. 1), 235. (Abstr.).

Chiba, L. I. (2001) *Swine Nutrition*, 2nd Edition. pp 803-837. Edited by Lewis, A. J., and L. L. Southern. CRC Press, Boca Roton, FL.

Chua, B., E. Coenen, J. van Delen, and D. M. Weary. (2002) Effects of pair versus individual housing on the behavior and performance of dairy calves. *Journal of Dairy Science,* **85**, 360-364.

Cruywagen, C. W., I. Jordaan, and L. Venter. (1996) Effect of lactobacillus acidophilus supplementation of milk replacer on pre-weaning performance of calves. *Journal of Dairy Science,* **79**, 483-486.

Davis, C. L. and J. K. Drackley. (1998) *The Development, Nutrition, and Management of the Young Calf.* Iowa State University Press, Ames, Iowa.

Donovan, D. C., S. T. Franklin, C. C. L. Chase, and A. R. Hippen. (2002) Growth and health of Holstein calves fed milk replacers supplemented with antibiotics or Enteroguard. *Journal of Dairy Science,* **85**, 947-950.

Doppenbery, J. and D. L. Palmquist. (1991) Effect of dietary fat level on feed intake, growth, plasma metabolites and hormones of calves fed dry or liquid diets. *Livestock Production Science*, **29**, 151-166.

Drackley, J. K., K. S. Bartlett, and R. M. Blome. (2003) Protein content of milk replacers and calf starters for replacement calves. http://traill.outreach.uiuc.edu/dairynet/paperDisplay.cfm?ContentID=269

Eastridge, M. L., and D. K. Roseler. (2003) Grain intake and growth of pre-weaned dairy calves. http://ohioline.osu.edu/sc163/sc163_9.html

Eicher-Pruiett, S. D., J. L. Morrill, T. G. Nagaraja, J. J. Higgins, N. V. Anderson, and P. G. Reddy. (1992) Response of young dairy calves with lasalocid delivery varied in feed sources. *Journal of Dairy Science,* **75**, 857-862.

Fallon, R. J., P. E. V. Williams, and G. M. Innes. (1986) The effects on feed intake, growth and digestibility of nutrients of including calcium soaps of fat in diets for young calves. *Animal and Feed Science Technology,* **12**, 103-115.

Fiems, L. O., C. V. Bouchque, B. G. Cottyn, and F. X. Bysse. (1985) Evaluation of rapeseed meal with low and high glucosinolates as a protein source in calf starters. *Livestock Production Science,* **12**, 131-143.

Fiems, L. O., C. V. Bouchque, B. G. Cottyn, and F. X. Bysse. (1986) Cottonseed meal and maize gluten feed versus soyabean meal as protein supplements in calf starters. *Archives of Animal Nutrition,* **36**, 731-740.

Fiems, L. O., C. V. Bouchque, B. G. Cottyn, and F. X. Bysse. (1987) Effect of formaldehyde-treated soya bean meal and urea in starters on nitrogen quality, degradability in sacco, sheep digestibility and calf performance. *Animal and Feed Science Technology* **16**, 287-295.

Franklin, S. T., D. M. Amaral-Phillips, J. A. Jackson, and A. A. Campell. (2003) Health and performance of Holstein calves that suckled or were hand-fed colostrums and were fed one of three physical forms of starter. *Journal of Dairy Science,* **86**, 2145-2153.

Funaba, M, K. Kagiyama, T. Iriki, and M. Abe. (1994) Changes in nitrogen balance with age in calves weaned at 5 or 6 weeks of age. *Journal of Animal Science,* **72**, 732-738.

Greenwood, R. H., J. L. Morrill, and E. C. Titgermeyer. (1997a) Using dry feed intake as a percentage of initial body weight as a weaning criterion. *Journal of Dairy Science,* **80**, 2542-2546.

Greenwood, R. H., J. L. Morrill, E. C. Titgemeyer, and G. A. Kennedy. (1997b) A new method of measuring diet abrasion and its effect on the development of the forestomach. *Journal of Dairy Science,* **80**, 2534-2541.

Hancock, J. D., and K. C. Behnke. (2001) Use of ingredient and diet processing technologies to produce quality feeds for pigs. In *Swine Nutrition*, 2nd Edition. pp 469-497. Edited by Lewis, A. J., and L. L. Southern. CRC Press, Boca Roton, FL.

Heinrichs, A. J., C. M. Jones, and B. S. Heinrichs. (2003) Effects of mannan oligosaccharide or antibiotics in neonatal diets on health and growth of dairy calves. *Journal of Dairy Science,* **86**, 4064-4069.

Heinrichs, A. J., and G. J. Bush. (1991) Evaluation of decoquinate or lasalocid against coccidiosis from natural exposure in neonatal dairy calves. *Journal of Dairy Science,* **74**, 3223-3227.

Heinrichs, A. J., L. A. Swartz, T. R. Drake, and P. A. Travis. (1990) Influence of decoquinate fed to neonatal dairy calves on early and conventional weaning systems. *Journal of Dairy Science,* **73**, 1851-1856.

Higginbotham, G. E., and D. L. Bath. (1993) Evaluation of Lactobacillus fermentation cultures in calf feeding systems. *Journal of Dairy Science,* **76**, 615-620.

Hill, S. R., B. A. Hopkins, S. Davidson, S. M. Bolt, C. Brownie, T. Brown, G. B. Huntington, and L. W. Whitlow. (2003a) The effects of conttonseed hulls added to diets with and without live yeast or mannoligosaccharide in Holstein calves. *Journal of Dairy Science,* **86** (Suppl. 1), 31 (abstract).

Hill, S. R., B. A. Hopkins, S. Davidson, S. M. Bolt, C. Brownie, T. Brown, G. B. Huntington, and L. W. Whitlow. (2003b) The effects of conttonseed hulls added to diets with and without live yeast or mannoligosaccharide in Jersey calves. *Journal of Dairy Science,* **86** (Suppl. 1), 33 (abstract).

Hill, T. M., J. M. Aldrich, A. J. Proeschel, and R. L. Schlotterbeck. (2001) Protein levels for neonatal calf starters. *Journal of Dairy Science,* **84** (Suppl. 1), 265 (abstract).

Hill, T. M., J. M. Aldrich, and R. L. Schlotterbeck. (2003c) Responses to feeding Apex plant extracts to neonatal calves via the milk replacer and starter. *Journal of Dairy Science,* **86** (Suppl. 1), 21 (abstract).

Hill, T. M., J. M. Aldrich, and R. L. Schlotterbeck. (2003d) Evaluation of a plant extract, Apex, included in calf milk replacers. *Journal of Animal Science,* **81** (Suppl. 2), 71 (abstract).

Hill, T. M., J. M. Aldrich, and R. L. Schlotterbeck. (2003e) Effect of feeding neonatal calves milk replacers containing a blend of vegetable and animal fats. *Journal of Dairy Science,* **86** (Suppl. 1), 21 (abstract).

Hoblet, K. H., T. P. Charles, and R. P. Howard. (1989) Evaluation of lasalocid and decoquinate against coccidiosis resulting from natural exposure in weaned dairy calves. *American Journal of Veterinary Research*, **50**, 1060-1064.

Holtshausen, L., and C. W. Crywagen. (2000) The effect of age on in sacco estimates of rumen dry matter and crude protein degradability in real calves. *South African Journal of Animal Science*, **30**, 212-219.

Jasper, J. and D. M. Weary. (2002) Effects of ad libitum milk intake on dairy calves. *Journal of Dairy Science,* **85**, 3054-3058.

Jenny, B. F., H.J . Vandijk, and J. A. Collins. (1991) Performance and fecal flora of calves fed a Bacillus subtilis concentrate. *Journal of Dairy Science,* **74**, 1968-1973.

Kuehn, C. S., D. E. Otterby, J.G. Linn, W. G. Olson, H. Chester-Jones, G.D. Marx, and J.A. Barmore. (1994) The effect of dietary energy concentration on calf performance. *Journal of Dairy Science,* **77**, 2621-2629.

Lesmeister, Keith Eric. (2003) *Dietary alterations and their influence on rumen development in neonatal dairy calves.* Ph.D. Thesis. The Penn State University.

Luchini, N. D., S. F. Lane, and D. K. Combs. (1991) Evaluation of starter crude protein level and feeding regime for calves weaned at 26 days of age. *Journal of Dairy Science,* **74**, 3949-3955.

Luchini, N.D., S. F. Lane, and D. K. Combs. (1993) Pre-weaning intake and post-weaning dietary energy effects on intake and metabolism of calves weaned at 26 days of age. *Journal of Dairy Science,* **76**, 255-266.

Maiga, H. A., D. J. Schingoethe, F. C. Ludens, W. L. Tucker, and D. P. Casper. (1994) Response of calves to diets that varied in amounts of ruminally degradable carbohydrate and protein. *Journal of Dairy Science,* **77**, 278-283.

McCoy, G. C., L. D. Ruppert, and M. F. Hutjens. (2003) Feeding on extruded cottonseed-soyabean-based calf starter. http://trail.outreach.uiuc.edu/ dairynet/paperdisplay.cfm? Content ID = 269

Morrill, J. L., J. M. Morrill, A. M. Feyerherm, and J. F. Laster. (1995) Plasma proteins and a probiotic as ingredients in milk replacer. *Journal of Dairy Science,* **78**:902-907.

Morrill, J. L., A. D. Dayton, and R. Mickelsen. (1977) Cultured milk and antibiotics for young calves. *Journal of Dairy Science,* 60, 1105-1109.

Morrill, J. L., and A. D. Dayton. (1998) Effect of feed flavor in milk and calf starter on feed consumption and growth. *Journal of Dairy Science,* **61**, 229-232.

National Research Council, (1989) *Nutrient Requirements of Dairy Cattle*, 6th rev. ed. Washington, D.C.: National Academy Press.

National Research Council, (2001) *Nutrient Requirements of Dairy Cattle*, 7th rev. ed. Washington, D.C.: National Academy Press.

Olson, E. J., W. B. Epperson, D. H. Zeman, R. Fayer, and M. B. Hildreth. (1998) Effects of an allicin-based product on cryptospiridosis in neonatal calves. *Journal of the American Veterinary Medical Association*, **212**, 987-989.

Pollard, B. C., H. M. Dann, and J. K. Drackley. (2003) Evaluation of intensified liquid feeding programs for dairy calves. *Journal of Dairy Science,* **86** (Suppl. 1). 174 (Abstr.).

Quigley, III, J. D., L. A. Caldwell, G. D. Sinks, and R. N. Heitmann. (1991) Changes in blood glucose, nenesterified fatty acids, and ketones in response to weaning and feed intake in young calves. *Journal of Dairy Science,* **74**, 250-257.

Quigley III, J. D., T. M. Steen, and S. I. Boehms. (1992) Postprandial changes of selected blood and ruminal metabolites in ruminating calves fed diets with and without hay. *Journal of Dairy Science,* **75**, 228-235.

Quigley, III, J. D., J. D. Drewry, L. M. Murray, and S. J. Ivey. (1997) Effects of lasalocid in milk replacer or calf starter on health and performance of calves challenged with eimeria species. *Journal of Dairy Science,***80**, 2972-2976.

Quigley III, J. D., T. A. Wolfe, and T. H. Elsasser. (2003) Effect of plasma IgG concentration and milk replacer feeding on hormone and growth responses in stressed calves. *Journal of Dairy Science,* **86** (Suppl. 1), 135 (Abstr.).

Scharma, H. R., B. White, and J. R. Ingalls. (1986) Utilization of whole rape (canola) seed and sunflower seeds as sources of energy and protein in calf starter diets. *Animal Feed Science and Technology* **15**, 101-112.

Strzetelski, J., B. Niwinska, J. Kowalezyk, and A. Jurkiewiez. (2001) Effect of milk replacer feeding frequency and level of concentrate intake and rearing performance of calves. *Journal of Animal Feed Science,* **10**, 413-420.

Swartz, L. A., A. J. Heinrichs, G. A. Varga, and L. D. Muller. (1991) Effects of varying dietary undegradable protein on dry matter intake, growth, and carcass composition of Holstein calves. *Journal of Dairy Science,* **74**, 3884-3890.

Thomsen, N. K., and R. B. Rindsig. (1980) Influence of similarly flavored milk replacers and starters on calf starter consumption and growth. *Journal of Dairy Science,* **63**, 1864-1868.

Vazquez-Anon, M., A. J. Heinrichs, J. M. Aldrich, and G. A. Varga. (1993) Effect of post-weaning age on rate of in site protein disappearance in calves weaned at 5 weeks of age. *Journal of Dairy Science,* **76**, 2749-2757.

Warner, R. G. (1984) The impact of protein solubility in dairy calf starters. In *Proceedings of the Cornell Nutrition Conference for Feed Manufacturers.* Rochester, NY. pp 42-45.

Warner, R. G. (1991) Nutritional factors affecting the development of a functional rumen – a historical perspective. In *Proceedings of the Cornell Nutrition Conference for Feed Manufacturers.* Rochester, NY. pp 1-12.

Williams, P. E. V., R. J. Fallon, G. M. Innes, P. Garthwaite. (1987) Effect on food intake, rumen development and live weight of calves of replacing barley with sugar beet-citrus pulp in a starter diet. *Animal Production,* **44**, 65-73.

8

PASSIVE IMMUNITY IN NEWBORN CALVES

J. D. QUIGLEY, C. J. HAMMER, L. E. RUSSELL, J. POLO[1]
APC, Inc., Ankeny, IA, USA; [1]APC Europe, 08400 Granollers, Spain

Introduction

In the United States, the sale of cattle and calves and dairy products generated approximately $38 and $21 billion, respectively, in 2002 (USDA ERS, 2003). Dairy and beef replacement enterprises contribute significantly to the total economic output of farming operations. Indeed, in both dairy and beef operations, production of calves for sale or for replacement animals is an essential part of the operation. Recently, changes in the dairy sector have resulted in development of a unique industry, wherein farms are operated to specifically raise dairy calves as a stand-alone enterprise. These "calf farms" or "calf ranches" have developed rapidly and now contribute to the continued specialization of the dairy industry.

Calves are born with a predetermined genetic potential, which may be permanently affected by management decisions implemented throughout the rearing period and by environmental factors. Studies have shown that level of management has a profound effect on calf morbidity and mortality (Jenny *et al.*, 1981; James *et al.*, 1984; Curtis *et al.*, 1985; Waltner-Toews *et al.*, 1986a, 1986b). Proper management of young stock, particularly during the neonatal period, can markedly reduce morbidity and mortality, whereas improper management will lead to economic losses from increased cost of veterinary intervention, deaths, reduced growth, and suboptimal reproductive performance. In addition, poor management of young stock can reduce the lifetime productivity of the individual cow and the herd as a whole.

The most critical time in the life of a dairy replacement is during the first few days, when morbidity and mortality are greatest. The National Dairy Heifer Evaluation Project (NDHEP; NAHMS, 1992, 1996) indicated clearly the inadequacy of current colostrum feeding and management practices in the U.S. The NDHEP reported that nearly 41% of dairy calves had inadequate circulating concentrations of IgG at 24 hours of age. Failure of passive transfer of immunity (FPT) is often greater in bull calves; analysis of plasma IgG in over 1,100 calves from 3 to 8 days of age indicated that over 57% had circulating plasma IgG concentrations less than 10 g/L (Quigley, unpublished data). This is similar to data of McVicker *et al.* (2002), who reported that 56% of calves from 2 to 10 days of age had FPT. McDonough *et al.* (1994) also reported that only 22% of 460 special-fed veal calves reared in California had adequate transfer of passive immunity.

Mortality of dairy calves born alive from birth to weaning is greater than 10% in the U.S. (NAHMS, 1996). Mortality after weaning is about 2%. Costs associated with high mortality and reduced productivity are significant and exceed $200 million annually, excluding costs associated with use of therapeutic and sub-therapeutic antibiotics, lost feed and labour inputs, and overhead costs. The most important factor associated with pre-weaning mortality is the consumption of colostrum and acquisition of passive immunity within the first 24 hours of birth. The USDA (NAHMS, 1992) estimated that 50% of mortality that occurred in pre-weaned calves was directly related to inadequate acquisition of passive immunity.

Absorption of immunoglobulins

Absorption of intact macromolecules across the intestinal epithelium into the neonatal circulation is possible for approximately 24 hours after the calf is born. Absorption of Ig occurs non-selectively by pinocytosis, which moves proteins into the epithelium. Proteins are transported through the cell into the lymphatic system and, subsequently, to the blood. Absorption of protein appears to increase along the length of the small intestine, with the lower small intestine (60 to 80%) the site of maximal IgG absorption (Fetcher *et al.*, 1983). After leaving the epithelium, Ig molecules move into the lymph and then to the circulation. Several authors have reviewed mechanisms of IgG absorption (Bush and Staley, 1980; Butler, 1999; Jochims *et al.*, 1994; Staley and Bush, 1985).

Maturation of the small intestine, including intestinal cell turnover, increasing abomasal acidity, development of intestinal secretions, and appearance of intra-epithelial digestive vacuoles, begins shortly after birth, and the ability of the intestine to absorb macromolecules without digestion is lost by about 24 hours after birth. The exact time of cessation of macromolecular transport (also known as closure) varies by immunoglobulin type, but ranges from 20 to 24 hours of age.

In addition to the maturation of intestinal cells, the secretion of digestive enzymes may also contribute to lower apparent efficiency of absorption (AEA) of IgG by degrading IgG prior to absorption. At birth, and for a limited period thereafter, the secretion of digestive enzymes remains limited to allow macromolecules such as IgG to escape digestion (Guilloteau *et al.*, 1983; Thivend *et al.*, 1980). By about 12 hours after birth, enzyme secretion becomes more marked, thereby reducing the ability of IgG to reach the peripheral circulation without being degraded. Supplementation of colostrum with soyabean trypsin inhibitor increased the absorption of IgG (Quigley *et al.*, 1995), indicating the deleterious effects of proteolytic enzymes on AEA.

The role of the neonatal Fc receptor (FcRn) in absorption of IgG across the intestinal epithelium of rodents and humans has been reported (Praetor *et al.*, 1999; Rodewald, 1976, 1980). Few data are available in ruminants, however, and the role of the receptor in IgG absorption across the gut epithelium is unclear. Laegreid *et al.* (2002) reported that allelic variation in genes coding for the alpha chain of FcRn may influence the odds of FPT in calves and suggested that certain genotypes may have higher odds of FPT. However, it is not clear whether FPT was caused by dam or calf differences and colostral IgG concentration was not determined in the study. Mayer *et al.* (2002) reported that the FcRn contributed significantly to movement of IgG_1 into colostrum of ewes, but they did not find FcRn in the duodenal enterocytes of lambs. However, the authors reported the presence of FcRn in the apical region of crypt cells and suggested that FcRn may play a role in re-secretion of IgG from the circulation back into the intestine to provide local immunity.

Failure of passive transfer. Traditionally, successful transfer of passive immunity has been determined by measuring the concentration of IgG in the serum of the calf at 24 to 48 hours after birth. If serum IgG concentration exceeds some critical level, then the calf is thought to be relatively well protected against pathogens. The critical level for determining FPT is usually <10 g/L (1,000 mg/dl), although some researchers have defined FPT as 8 to 10 g of IgG (or IgG_1)/L of serum or plasma (Blood and Radostits, 1989; Hancock, 1985; McGuirk, 1989; Odde, 1986; Wittum and Perino, 1995). Calves with <10 of IgG/L of serum are at greater risk of disease than calves with greater serum IgG concentrations. Of course, the concentration of serum IgG is a continuum of risk – that is, calves with <10.1 of IgG/L of serum are not at markedly greater risk than calves with 9.9 g of IgG/L. Generally, it is well accepted that the greater the concentration of IgG in the circulation of calves at 24 to 48 hours after birth, the greater the protection against the array of pathogens to which the calf might be exposed.

It is important to note that development of disease is affected not only by immune competence, but also by exposure. Good sanitation and management can lead to reduced chance for pathogenic exposure, thereby keeping calves healthy and growing.

Calculation of IgG needs

There are many factors that influence the concentration of IgG in the blood of the calf at 24 to 48 hours after birth. These include:

- Mass of IgG consumed
- Apparent efficiency of IgG absorption (AEA)
- Plasma or serum volume of the calf

These factors can be summarized as:

$$\text{Serum IgG (g/L)} = \text{IgG consumed (g)} \times \text{AEA (\%)} / \text{serum volume (L)} \quad [1]$$

Equation [1] can be used to calculate the efficiency with which IgG are absorbed:

$$\text{AEA (\%)} = \text{serum IgG (g/L)} \times \text{serum volume (L)} / \text{IgG consumed (g)} \quad [2]$$

Intake of IgG is a function of colostrum quality and amount ingested:

$$\text{IgG consumed (g)} = \text{Volume of colostrum consumed (L)} \times \text{IgG concentration (g/L)} \quad [3]$$

Apparent efficiency of IgG absorption is the proportion of ingested IgG recovered in the circulation shortly after cessation of macromolecular IgG transport (closure), usually at 24 to 48 hours after birth. Because absorbed IgG equilibrate with non-vascular liquid pools, AEA cannot equal 100%. Most research suggests approximately a 1:1.2 ratio of vascular to non-vascular liquid pools in the neonate (Kruse, 1970; Payne *et al.*, 1967; Wagstaff *et al.*, 1992); therefore, maximal AEA is approximately 45% in neonates. The concept of AEA is not well understood by many veterinarians or nutritionists, but encompasses many of the concepts universally accepted as important to successful passive transfer. For a complete review of factors that affect AEA, see Quigley and Drewry (1998).

Blood volume

The amount of IgG in the bloodstream is, necessarily, affected by the size of the plasma or serum pool. Intuitively, it is logical that calves with a larger blood volume will attain a lower IgG concentration than calves with smaller blood volume if they are fed the same mass of IgG. This, then requires prediction of plasma volume in some manner. Prediction of plasma, serum or whole blood volume is often carried out using dye dilution methods (McEwan *et al.*, 1968, 1970). McEwan *et al.*, (1970) reported a mean plasma volume of 8.3% of BW. Others have reported mean values of 8.7 to 9.3% (McEwan *et al.*, 1968; Quigley *et al.*, 1998a) and 6.5% (Möllerberg *et al.*, 1975). The value of 7% of BW has been used widely in other research trials in which AEA was estimated. However, Matte *et al.* (1982) reported that plasma volume changed from 14.5% of BW at 12 hours of age to 8 to 9% of BW at 36 to 48 hours of age. This reported variability suggests that there is considerable error in prediction of plasma volume as a percentage of BW, particularly if estimates of AEA are made repeatedly.

Colostrum and Ig intake

Colostrum, defined as the lacteal secretions from the mammary gland during the first 24 hours after birth, has long been identified as the primary source of passive immunity for species with epitheliochorial placentation (Smith and Little, 1922). The importance of acquisition of adequate passive immunity is well established – effects of passive transfer on neonatal morbidity, mortality, and performance (growth, efficiency) have been documented extensively. For various reasons, the acquisition of passive immunity might be inadequate, and the risk of morbidity and mortality are increased.

The concentration of Ig in colostrum varies according to the cow's disease history, age, volume of colostrum produced, season of the year, breed, and other factors (Foley and Otterby, 1978). Research from Washington (Pritchett *et al.*, 1991) indicated the average concentration of IgG_1 in colostrum from 919 Holstein cows was 48.2 g/L with a range of 20 to >100 g/L. Others reported colostrum concentrations ranging from 32.1 g/L (n = 25; Andrew, 2001) to 76.7 g/L (n = 77; Tyler *et al.*, 1999). Jersey (Quigley *et al.*, 1994b) and Guernsey (Tyler *et al.*, 1999) cows generally produce colostrum with greater concentrations of IgG than Holsteins.

Cows exposed to more pathogens produce colostrum with greater Ig than cows exposed to fewer pathogens. Prepartum milking or leaking of milk from the udder prior to calving can reduce the concentration of Ig in colostrum (Roy, 1991). Colostral quality may be affected by transport of IgG from the blood to the mammary gland as well as dilution. Guy *et al.* (1994) reported that premature lactogenesis caused cessation of IgG_1 transfer into colostral secretions. Possibly, as dairy cattle are bred to produce greater amounts of milk, this contributes to reduced colostral IgG by premature lactogenesis. The negative relationship between colostral volume and IgG was also reported by Pritchett *et al.* (1994).

Differences between colostral quality in beef and dairy cattle suggest a relationship between volume of colostrum produced and colostral quality. Guy *et al.* (1994) reported that reduced IgG_1 concentration in colostrum was associated with greater lactogenic activity in dairy cows, evidenced by fivefold higher alpha-lactalbumin concentration in sera. Dilution of IgG_1 in colostrum may be responsible for breed differences in colostral IgG_1 concentrations. Presumably, as milk production potential continues to increase, the potential for reduced colostral quality will increase also.

Variation in Ig content makes accurate colostrum management and feeding difficult. Colostral IgG can be measured accurately in the laboratory; unfortunately, assays involved are time-consuming and expensive. Measurement of colostrum specific gravity, using a device called a colostrometer, is one method to estimate Ig content of colostrum (Fleenor and Stott, 1980). This device is based on the relationship between Ig in colostrum and specific gravity. Unfortunately, components of colostrum other than Ig affect specific gravity, so the relationship is variable (Quigley *et al.*, 1994b). Also, the relationship between specific gravity and IgG is dependent on temperature (Pritchett *et al.*, 1994; Mechor *et al.*, 1991, 1992). However, the colostrometer may give a qualitative estimate of colostrum quality - particularly if the colostrum is of poor quality. More recent development of lateral flow immunoassay (Mcvicker *et al.*, 2002) has improved accuracy of prediction of colostral IgG content, but tests are expensive and not widely used.

Amount of colostrum consumed by the calf is the only factor in the equation of serum IgG (Equation [1]) that can be manipulated easily on the farm. Therefore, many veterinarians and dairy professionals have increased the recommended amount of colostrum in an attempt to reduce the incidence of FPT (NAHMS, 2002; Pritchett *et al.*, 1991). In some cases, up to 8 L of colostrum may be administered by oesophageal feeder within the first 24 hours of birth. While this approach serves a useful purpose, it does not address all factors that need to be considered in attempting to maximize successful passive transfer of immunity. Furthermore, administration of large volumes of colostrum may pose health risks through multiple intubation with oesophageal feeders and aspiration of regurgitated colostrum.

Colostrum in transmission of disease

Colostrum and transition milk have long been known as vectors for transmission of disease in calves. For example, Meylan *et al.* (1996) reported that a primary route of transmission of *Mycobacterium paratuberculosis* is by consumption of colostrum from infected cows, although others indicate that ingestion of faeces may be the primary route (Streeter *et al.*, 1995; Sweeney, 1996). Because colostrum is identified as a major vector for transmission of Johne's disease, most experts recommend that cows should be tested prior to calving and colostrum from positive or suspected positive cows should not be fed. Indeed, Johne's management programs focus on sanitation and management of calves as a means of reducing spread of Johne's (Wells, 2000; Groenendaal *et al.*, 2003). Instead, calves should be fed frozen colostrum obtained from cows that have tested negative for Johne's. In herds with many Johne's positive cows, this can often lead to inadequate supplies of colostrum to feed to calves.

Other infective agents identified in mammalian colostrum include bovine immunodeficiency virus (Moore *et al.*, 1996; Meas *et al.*, 2002), bovine leukaemia virus (Hopkins and DiGiacomo, 1997), caprine Mycoplasma mycoides (East *et al.*, 1983), bovine leucosis virus (Rusov, 1993), Neospora caninum (Uggla *et al.*, 1998), brucella, tuberculosis and many others.

Modern management practices can lead to significant bacterial contamination of maternal colostrum. Fecteau *et al.* (2002) reported that nearly 36% of colostrum samples collected from Canadian dairy herds contained >100,000 bacteria/ml. Andrew (2001) reported similar prevalence of contaminated from U.S. herds. Poulsen *et al.* (2002) reported that 82% of colostrum from dairies in Wisconsin contained >100,000 bacteria/ml and many contained >1,000,000 cfu/ml. Bacterial contamination of colostrum can negatively affect acquisition of passive immunity (James *et al.*, 1981; Poulsen *et al.*, 2002). In addition, colostrum is often stored at room temperature for extended periods on many dairy farms. According to the USDA (NAHMS, 2002), nearly 11% of dairy operations routinely store first milking colostrum at room temperature. Another 19% use refrigeration as a means of storing colostrum, which may be inadequate in many situations. Research conducted at the University of California, Davis (P. Jardon; unpublished data) showed that when colostrum was left at room temperature, growth of bacteria increased exponentially; indeed, bacterial counts can double approximately every 20 minutes. Within six hours, the number of bacteria in colostrum exceeded 10^6 cfu/ml. These bacteria can markedly affect the health of the calf.

Cow and maternity area are sources of infection for many dairy calves. Numerous researchers have reported increased risk of infection when calves are left in the calving environment for more than a few hours (Jenny *et al.*, 1981; James *et al.*, 1984; Quigley *et al.*, 1994a; Sweeney, 1996). Ingestion of maternal faeces has been cited as a source of infection for many different calfhood diseases, including Johne's disease, rotavirus, coronavirus, and *Cryptosporidium parvum*. Removing calves immediately from the calving environment has been shown to reduce the risk of transmission of several organisms (Quigley *et al.*, 1994a) and is generally recommended by veterinarians. Unfortunately, nearly 44% of calves were left with the dam up to 24 hour after birth (NAHMS, 2002), thereby increasing exposure to faeces.

Effects of pasteurization on colostrum

One approach to reducing the infectivity of colostrum is pasteurization. In theory, pasteurization should reduce the level of infectivity of pathogens below threshold levels, thereby reducing the risk of transmission of disease. Traditionally, pasteurization has been used to reduce bacterial counts in waste milk prior to feeding to calves as a source of nutrition (Jumaluddin *et al.*, 1996a, b). More recently, researchers have attempted to pasteurize colostrum prior to feeding within the first 24 hours of life. The primary consideration regarding pasteurizing colostrum is the destruction of pathogens and concomitant destruction of functional proteins, including IgG (Stabel, 2001). Most of the research that has been done to date has explored the effects of pasteurization on the amount of destruction of IgG.

Godden *et al.* (2003) reported that batch pasteurization (63 °C for 30 min) reduced the IgG content of colostrum by an average of 26.2% when compared to pre-pasteurized colostrum samples. There was an effect of batch size, with larger batches (95 L) producing a greater reduction in colostral IgG content than smaller batches (57 L).

Meylan *et al.* (1996) also tested effects of batch pasteurization on survival of colostral IgG, but in a laboratory setting. Their data indicated that IgG in pasteurized colostrum was reduced by more than 12% compared to unpasteurized samples.

Most of the research evaluating pasteurization has used IgG as the indicator molecule for determining the degree of damage caused by pasteurization. However, there are many other proteins in colostrum that may be damaged upon exposure to heat. A few researchers have looked at the effects of heating on other proteins. For example, German researchers (Steinbach *et al.*, 1981) reported that heating colostrum to 55 °C for 30 min had no effect on either IgG or IgM; however, heating to 60 °C for 10 min reduced IgM dramatically. Liebhaber *et al.* (1977) reported that pasteurization reduced IgA in human colostrum by 33% and viable immune cells were reduced by over 50%. On the other hand, Jansson *et al.* (1985) reported that activity of epidermal growth factor was unaffected by pasteurization. In light of the uncertainties related to success of pasteurization, and concomitant effects on efficacy of pasteurized colostrum, it is difficult to justify widespread recommendation to pasteurize all colostrum.

Colostral supplements and replacers

INTRODUCTION

Recognition of high rates of FPT in calves and difficulties in managing colostrum quality prompted researchers to find ways to provide additional IgG to improve the quality of maternal colostrum. Although every farmer knows the importance of colostrum management and early administration of high quality colostrum, the incidence of FPT on most dairy farms remains stubbornly high. High incidence of Johne's disease on many dairies in the U.S. also limits availability of high quality colostrum.

Colostrum supplements were introduced into the market in the mid to late 1980s and have become an important tool for producers. Most colostrum supplements approved by the U.S. Department of Agriculture have been approved as a means of preventing specific diseases (usually obtained from cows vaccinated with specific vaccines) rather than a means of increasing overall serum IgG concentration.

All products designed for administration to neonatal calves are intended to provide a source of IgG. Other components of colostrum that contribute to an animal's resistance to disease, including leukocytes, growth factors and hormones, are not formulated into products because they are difficult to obtain, process and preserve, and absolute requirements are unknown.

Although most products contain relatively little IgG, they are currently marketed as colostrum replacers, but they have neither a sufficient mass of IgG nor the nutritional composition required by the calf. Recent introduction of highly concentrated IgG preparations has allowed discrimination between two classes of products. Quigley *et al.* (2002a) attempted to define terms to provide a more consistent framework within which to regulate and utilize these products.

The term "colostrum supplement" should refer to those preparations intended to provide < 100 g of IgG/dose and are not formulated to completely replace colostrum. Supplements should be formulated to be fed in conjunction with colostrum, to increase IgG concentration, and to provide nutrients that are inherently variable in maternal colostrum (e.g., vitamin E). Colostral IgG supplements should be separately categorized as those intended to provide targeted IgG (from hyperimmunized animals) and those providing non-specific IgG.

In addition to an adequate mass of IgG (>100 g of IgG/dose), colostral replacers must provide nutrients required by the calf. Energy as carbohydrate and lipid is needed to allow the calf to thermoregulate and to establish homeostasis. Digestible protein sources are required as a source of amino acids for gluconeogenesis and protein synthesis, and vitamins and minerals are essential to successful colostral replacer formulation. Colostrum is a highly concentrated source of fat soluble vitamins, which are needed because placental transfer of these vitamins is limited.

There are three sources of IgG used in colostrum supplements and replacers - lacteal secretions (colostrum and milk), blood and eggs. Each IgG source has different characteristics, advantages and limitations. All are widely available, although the infrastructure for large-scale collection and processing of bovine colostrum is currently limited. The concentration of IgG in bovine milk is quite low and costs of processing to concentrate IgG are high. Collection of colostrum or milk from other species of animals (sows, mares) is not currently possible in large quantities. Collection and processing of egg IgY is directed primarily to production of antibodies against specific pathogens. Colostral and blood-derived IgG have broad specificity and are more appropriate to production of colostrum supplements.

Each source of animal protein must be evaluated for availability, cost, ease of manufacture, and safety related to potential contamination with infectious or noxious agents. In several countries, the feeding of blood fractions to ruminants is prohibited, making this source of IgG unavailable. Products derived from colostrum are widely promoted for human health and performance, thereby increasing the price of products for neonatal dairy calves.

Colostrum supplements

Colostrum supplements derived from lacteal secretions. Colostral supplements derived from whey and cow colostrum are generally produced by collection of colostrum from dairy farms. Colostrum may be directly dried (lyophilization or spray-drying) or processed to remove components (e.g., fat) prior to drying. Other products are produced by concentration of IgG in whey. Supplements are available as powders, pastes and boluses. Haines *et al.* (1990) reported that the IgG concentration of commercially available colostrum supplements provided between 0.1 and 13.5 g of IgG/dose. Most modern supplement products contain from 25 to 45 g of IgG/dose (a dose is usually 200 to 500 g of powder).

Collection and processing of colostrum for use in supplements and replacers is not without biosecurity risk. Although colostrum used in manufacturing supplements is obtained from cows on Grade A dairies, these secretions are often collected outside the normal milking parlour; thus there is a risk of contamination from faeces, mud and other contaminants from unwashed udders when cows are milked in the maternity area. Colostrum may be stored in different vessels at temperatures that may be undefined and unregulated. It may be transported to the processing facility by means other than those regulated by normal milk handling and processing regulations. It is neither evaluated for numbers of somatic cells nor for the presence of specific pathogens. Furthermore, the risk of infection with mastitis pathogens such as *Staphylococcus aureus* is significant, particularly in colostrum from heifers (Roberson *et al.*, 1998). The risk of antibiotic residues is also significant (Andrew, 2001).

Administration of 4 L of colostrum provides approximately 800 g of DM (assuming colostrum contains 200 g DM/L). Colostrum containing 30 g IgG/L that is subsequently dried would provide 150 g IgG/kg of dried product. Administration of 800 g of this product should provide sufficient IgG to protect calves against FPT. Therefore, manufacturing standards that allow collection of colostrum containing at least 30 g/L should provide products that could effectively replace colostrum. Unfortunately, most published literature indicates that supplements and replacers derived from lacteal secretions do not provide adequate passive immunity to calves.

Efficacy of colostrum supplements have been evaluated at several locations. Absorption of IgG from supplements derived from lacteal secretions have generally been reported to be poor (Abel and Quigley, 1993; Garry *et al.*, 1996; Grongnet *et al.*, 1986; Harman *et al.*, 1991; Hopkins and Quigley, 1997; Ikemori *et al.*, 1997; Mee *et al.*, 1996; Morin *et al.*, 1997; Zaremba *et al.*, 1993), although the reasons for poor IgG absorption are not defined clearly. Colostrum supplements are designed to be fed in conjunction with maternal colostrum; however, in some experiments, colostrum supplements have been fed as a substitute for maternal colostrum.

Abel and Quigley (1993) added a colostral supplement to maternal colostrum from 32 cows. No effect of colostral supplement was observed in serum IgG concentrations of calves at 24 or 48 hours after birth. When only the poor quality colostrum samples (<20 g IgG/L) were evaluated, results were similar. Mee *et al.* (1996) fed Colostrx® (Schering-Plough, Union, NJ) to four groups of 29 calves. Calves were fed 2 L of bovine colostrum or 500 g of Colostrx mixed in 1.2 L of warm water. In Experiment 2, calves were fed either 2 L of colostrum or 500 g of Colostrx in 1 L of water mixed with 1 L of colostrum. In Experiment 1, mortality of calves fed Colostrx was 27.6% compared with 3.5% for calves fed colostrum ($P < 0.05$). In Experiment 2, there was no difference between treatments, and mean mortality was 13.8% for calves fed colostrum and 3.5% for calves fed colostrum + Colostrx. Serum IgG concentrations at 24 to 36 h after birth were 17.8, 3.0, 18.4 and 9.5 g/L for calves fed colostrum, Colostrx, colostrum (Experiment 2), and colostrum + Colostrx, respectively.

Garry *et al.* (1996) fed normal bovine colostrum or three colostrum supplements (First Milk™ Formula, Procor Technologies; Colostrx, Immu-Start®, Imu-Tek Animal Health, Inc.) at the rate of 4 L of colostrum or 2 doses (packages) of product. The meals were fed within 2 and 12 hours of birth. Serum IgG concentrations at 24 hours were approximately 21, 6, 5, and 4 g/L for calves fed colostrum, First Milk, Colostrx, and Immu-Start, respectively. At no point did serum IgG in calves fed any colostrum supplement reach levels indicative of successful passive transfer. Further, AEA was < 10% for all supplements versus >20% for maternal colostrum. These data clearly indicate that none of these products are effective in alleviating the condition of FPT.

Hopkins and Quigley (1997) added a colostrum derived supplement (First Milk™ Formula) to maternal colostrum and fed 15 calves. Plasma IgG of these calves at 24 hours of age was lower ($P < 0.01$) than in calves fed only maternal colostrum and averaged 21.0 and 16.0 g/L for calves fed colostrum or colostrum plus supplement, respectively. The AEA of IgG was reduced from 40 to 30% when the supplement was added. Similarly, Morin *et al.* (1997) reported marked depression in AEA of IgG_1 when calves were fed 272 g of supplement (18%), compared with calves fed maternal colostrum alone (33%).

Hunt *et al.* (1988), in a review of colostrum supplement products at that time, stated that "*Colostrx should not be combined with colostrum since the resulting mixture is too thick and may be excessively hyperosmotic. Each Colostrx® bag contains a minimum of 24 grams (g) of bovine IgG. This product does not claim to be a total immunoglobulin substitute, and 24 g of bovine IgG represents only one-sixth to one-tenth of the oral immunoglobulin mass necessary to protect calves against septicemic colibacillosis. Electrophoretogram analysis of the mixed solution showed 87% of the total protein present was neither globulin nor albumin, and could not be identified with available laboratory standards... Therefore, much of the product may offer little systemic benefit to the calf, or may represent an immunoglobulin aggregate which we cannot presently identify.*

Colostrum deprived calves fed Colostrx® absorb such low levels of measurable bovine immunoglobulin that they must still be considered to have FPT[A]".

These data indicate that most current products derived from whey or processed colostrum do not provide significant IgG when fed according to normal recommendations. Clearly, additional research is necessary to further define processing methods responsible for impaired IgG absorption in supplements from lacteal secretions.

Supplements derived from chicken eggs. Preparations derived from chicken eggs have been evaluated in some studies (Erhard *et al.*, 1995, 1997). Typically, these preparations contain IgY obtained from hyperimmunization of chickens. However, absorption of the IgY into the circulation appears to be relatively low and, therefore, these preparations may be most useful in post-closure applications (Erhard *et al.*, 1997).

Supplements derived from bovine serum. A colostrum supplement based on serum proteins is currently used in the U.S. to provide absorbable IgG. Data indicated improved survival of neonatal calves when this product was fed alone or added to maternal colostrum (Arthington *et al.*, 2000a,b; Quigley *et al.*, 1998, 2000, 2001). Immunoglobulin preparations derived from bovine plasma have the advantage of ease of collection, high degree of governmental regulation, and available methods to fractionate the IgG. A summary of research conducted with bovine serum as a colostrum substitute is in Table 8.1.

Table 8.1 PLASMA IGG AND APPARENT EFFICIENCY OF ABSORPTION (AEA) IN CALVES FED COLOSTRUM SUPPLEMENT CONTAINING BOVINE SERUM

n	*Plasma IgG (g/L @ 24 h)*	*AEA*	*Author*
10	6.5	15	Quigley *et al.*, 1998b
12	8.3	...	Arthington *et al.*, 2000a
10	6.8	28	Arthington *et al.*, 2000b
8	5.7	30	Davenport *et al.*, 2000
48	6.5	20	Quigley *et al.*, 2000
14	8.4	25	Quigley *et al.*, 2001
20	7.4	33	Quigley *et al.*, 2001
16	7.4	33	Quigley *et al.*, 2002a
11	10.6*	...	Quigley *et al.*, 2002b
12	6.4	...	Holloway *et al.*, 2002
161	7.3	24	

*Included Jersey calves.

Other studies have evaluated the use of bovine serum colostrum supplement in conjunction with maternal colostrum. Arthington *et al.* (2000a) fed calves within 3 hours of birth colostrum of varying quality, and with different amounts of bovine serum supplement to provide equal IgG intake. Mean serum IgG concentrations 12 hours after feeding were 6.72 g/L for calves fed colostrum only and 10.5 g/L for calves fed colostrum + bovine serum supplement.

Holloway *et al.* (2002) reported that calves fed a bovine serum supplement had lower serum IgG concentrations at 2 days of age (6.4 g/L) compared with calves fed maternal colostrum (33.5 g/L). However, the authors noted that the supplement product had similar efficacy to routine colostrum administration practices.

Finally, there appears to be species specificity to absorption of serum-derived IgG. Immunoglobulin obtained from porcine serum is not absorbed as efficiently as IgG from bovine serum (Arthington *et al.*, 2000a; Drew and Owen, 1988).

Colostrum replacers

True colostrum replacers must provide high concentrations of IgG in addition to nutrients required for establishment of homeostasis and thermoregulation within the first 24 hours of birth. Colostrum and plasma are the two sources of IgG with sufficient concentration of Ig to allow economical and efficient manufacture of colostrum replacers.

Replacers derived from colostrum. Chelack *et al.* (1993) fed 9 calves highly concentrated spray-dried colostrum or frozen-thawed pooled colostrum. The spray-dried product provided 126 g of Ig and was reconstituted in 3 L of water. Serum IgG achieved at 48 hours of age were 11.6 g/L for calves fed colostrum and 10.6 g/L for calves fed spray-dried colostrum. The calculated AEA at 10% serum volume was 45 and 47%, respectively. These data suggest that products for treatment of FPT derived from whey and/or processed colostrum (i.e., lacteal secretions) can provide sufficient IgG if they are formulated, manufactured, fed and managed properly.

Replacers derived from blood. Fractionation of IgG from bovine plasma is widely reported in the literature and many different techniques exist for fractionation of IgG. A product derived from fractionated bovine plasma contains 125 g of IgG per dose. The product is also formulated to contain highly digestible protein, carbohydrates and lipids needed for neonatal thermoregulation and optimal IgG absorption. Several trials have been conducted to measure the absorption of IgG from calves fed only this colostrum replacer product (Table 7.2). In all cases, mean circulating IgG concentration exceeded average minimal IgG concentration for successful passive transfer. However, method of fractionation and IgG concentration of the final product is important, since intake of IgG in one feeding was more efficiently absorbed than a similar mass of IgG administered in two feedings at a 12-hour interval (Hammer *et al.*, 2004).

Table 8.2 PLASMA IGG AND APPARENT EFFICIENCY OF ABSORPTION (AEA) IN CALVES FED COLOSTRUM SUPPLEMENT CONTAINING BOVINE IG CONCENTRATE*.

n	*Plasma IgG (g/L @ 24 h)*	*AEA*	*Author*
40	10.7	31	Quigley *et al.*, 2001
12	13.6	20	Quigley *et al.*, 2001
39	14.0**	20	Jones *et al.*, 2004
16	13.3	30	Quigley *et al.*, 2002a
11	13.9	…	Quigley *et al.*, 2002b
29	10.8	30	Hammer *et al.*, 2004
148	12.2	26	

*Mean IgG intake = 186 g.
**Included Jersey calves.

Circulating IgG measurements can be affected by IgG source administered to neonatal calves. Quigley *et al.* (2002a) fed calves either maternal colostrum or a colostrum replacer containing IgG from fractionated bovine plasma and measured plasma IgG at birth and at 24 hours of age. The relationship between plasma IgG and total protein differed by IgG source (Figure 8.1), indicating that indirect measurements of plasma IgG (e.g., total protein) should be used with care when non-colostrum sources of IgG are used.

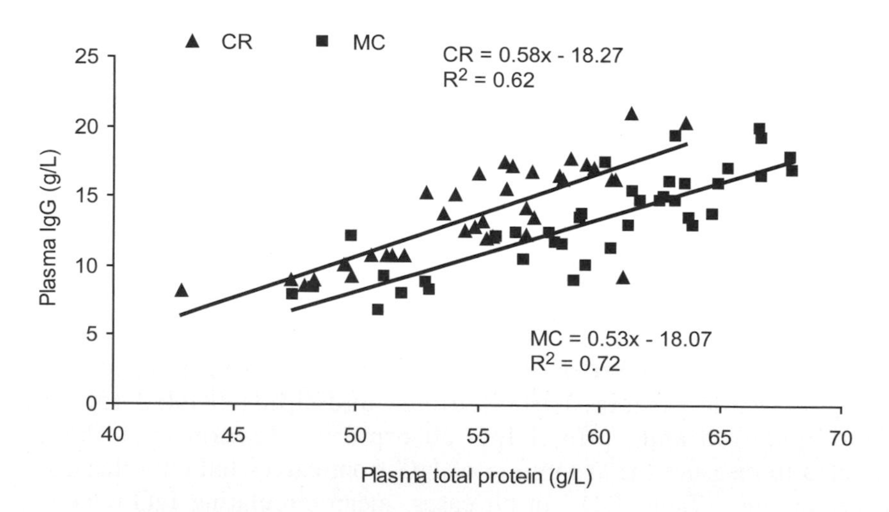

Figure 8.1 Relationship between plasma IgG and total protein concentrations at 24 hours of age in calves fed maternal colostrum (MC) or colostral replacer (CR). From Quigley *et al.*, 2002a.

Two trials have evaluated calf survival and health of calves fed only colostrum replacer product derived from bovine plasma. Poulsen *et al.* (2003) fed colostrum or the replacer product to calves on eight dairy farms in Wisconsin and reported that acquisition of passive immunity (measured by the proportion of calves with FPT) did not differ between calves fed maternal colostrum (n = 142) or those fed the colostrum replacer (n = 147). In addition, there was no difference between groups in mortality or number of veterinary interventions required to 14 days of age. Jones *et al.* (2004) fed calves (n = 79) either pooled maternal colostrum or colostrum replacer at equal IgG intakes. Concentrations of plasma IgG at 24 hours of age were similar between groups, and mortality, morbidity and growth to 29 days of age were unaffected by treatment. Concentration of IgG measured on days 8, 15, 22 and 29 indicated that plasma IgG was higher in calves fed colostrum replacer on days 8 and 15 (Figure 8.2). This might be due to the different profile of IgG isotypes in plasma-derived products (approximately 50% IgG_1) compared to colostrum (approximately 95% IgG_1). IgG_1 may be cleared from circulation more rapidly than IgG_2 (Mayer *et al.*, 2002).

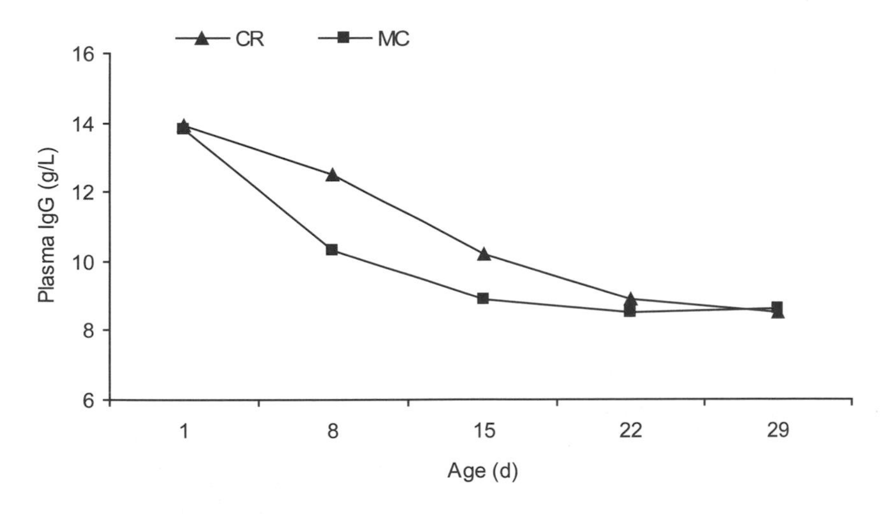

Figure 8.2 Concentration of plasma IgG in calves fed maternal colostrum (CM) or a colostrum replacer (CR) at birth. From Jones *et al.*, 2004. *Weekly means differ ($P < 0.05$).

Quigley *et al.* (2001) reported similar morbidity and mortality to 60 days of age in calves (n = 160) fed either maternal colostrum or colostrum replacer. These data suggest that the potential exists to replace colostrum with highly concentrated preparations of bovine plasma.

Colostrum replacers derived from bovine colostrum should be reasonable in cost. The primary concerns regarding the use of bovine colostrum are cost, infrastructure for collection and methods of processing. Methods of fractionating IgG from colostrum have been reported (Su and Chiang, 2003; Semotan and Kalab, 1997). To date, however, no published study reports complete replacement of maternal colostrum with commercially prepared products.

Summary

Management of colostrum is essential to maintain health of calves, and to maintain productivity and profitability of the dairy enterprise. Unfortunately, it is clear that colostrum is often deficient in IgG concentration, and/or might be contaminated with important disease-causing organisms. Thus, the need exists for supplementing or replacing colostrum to enhance the survival and health of young dairy and beef calves. Colostrum supplement and replacement products have been introduced to provide IgG to these animals. Products available in the market are derived from lacteal secretions or bovine plasma. Published literature suggests that supplements and replacements derived from bovine plasma are highly effective in improving the survival and health of young calves, thereby improving the economics of dairy and beef operations. These products fill an important niche on most farms, and their continued availability is essential to continued profitability of the dairy sector.

References

Abel Francisco, S.F. and J. D. Quigley, III. (1993) Serum immunoglobulin concentration in response to maternal colostrum and colostral supplementation in dairy calves. *Am. J. Vet. Res.* **54**, 1051-1054.

Andrew, S. M. (2001) Effect of composition of colostrum and transition milk from Holstein heifers on specificity rates of antibiotic residue tests. *J. Dairy Sci.* **84**, 100–106

Arthington, J. D. M. B. Cattell, J. D. Quigley, III, G. C. McCoy, and W. L. Hurley. (2000a) Passive immunoglobulin transfer in newborn calves fed colostrum or spray-dried serum protein alone or as a supplement to colostrum of varying quality. *J. Dairy Sci.* **83**, 2834–2838.

Arthington, J. D., M. B. Cattell and J. D. Quigley, III. (2000b) Effect of dietary IgG source (colostrum, serum, or milk-derived supplement) on the efficiency of IgG absorption in newborn Holstein calves. *J. Dairy Sci.* **83**, 1463-1467.

Blood, D. C. and O. M. Radostits. (1989) In *Veterinary Medicine,* 7th ed, pp 619-637. London: Bailleire Tindall.

Bush, L, J. and T. E. Staley. (1980) Absorption of colostral immunoglobulins in newborn calves. *J. Dairy Sci.* **63**, 672-680.

Butler, J. E. (1999) Immunoglobulins and immunocytes in animal milks. In *Mucosal Immunology, 2nd ed*, pp 1531-1534. Edited by P. L. Ogra. New York: Academic Press.

Chelack, B. J., P. S. Morley, and D. B. Haines. (1993) Evaluation of methods for dehydration of bovine colostrum for total replacement of normal colostrum in calves. *Can. Vet. J.* **34**, 407-412.

Curtis, C. R., H. N. Erb, and M. E. White. (1985) Risk factors for calfhood morbidity and mortality on New York dairy farms. *Proceedings of Cornell Nutrition Conference*, pp. 90-99.

Davenport, D. F., J. D. Quigley, III, J. E. Martin, J. A. Holt, and J. D. Arthington. (2000) Addition of casein or whey protein to colostrum or a colostrum supplement product on absorption of IgG in neonatal calves. *J. Dairy Sci.* **83**, 2813-2819.

Drew, M. D. and B. D. Owen. (1988) The provision of passive immunity to colostrum-deprived piglets by bovine or porcine serum immunoglobulins. *Can. J. Anim.* Sci. **68**, 1277-1284.

East, N. E., A. J. DaMassa, L. L. Logan, D. L. Brooks, and B. M. McGowan. (1983) Milkborne outbreak of Mycoplasma mycoides subspecies mycoides infection in a commercial goat dairy. *J. Amer. Vet. Med. Assoc.* **182**, 1338-1341.

Erhard, M. H., E. Gobel, B. Lewan, U. Losch, and M. Stangassinger. (1997) Systemic availability of bovine immunoglobulin G and chicken immunoglobulin Y after feeding colostrum and whole egg powder to newborn calves. *Arch. Tierernahr.* **50**, 369-380.

Erhard, M. H., U. Losch, and M. Stangassinger. (1995) Intestinal absorption of homologous and heterologous immunoglobulin G in newborn calves. *Z. Ernahrungswiss.* **34**, 160-163.

Fecteau, G., P. Baillargeon, R. Higgins, J. pare, and M. Fortin. (2002) Bacterial contamination of colostrum fed to newborn calves in Quebec dairy herds. *Can. Vet J.* **43**, 523-527.

Fetcher, A. C. C. Gay, T. C. McGuire, D. D. Barbee, and S. M. Parish. (1983) Regional distribution and variation of gamma-globulin absorption from the small intestine of the neonatal calf. *Am. J. Vet Res.* **44**, 2149-2154.

Fleenor, W. A., and G. H. Stott. (1980) Hydrometer test for estimation of immunoglobulin concentration in bovine colostrum. *J. Dairy Sci.* **63**, 973-977.

Foley, J. A. and D. E. Otterby. (1978) Availability, storage, treatment, composition, and feeding value of surplus colostrum: a review. *J. Dairy Sci.* **61**, 1033-1041.

Garry, F., R. Adams, M. B. Cattell, and R. P. Dinsmore. (1996) Comparison of passive immunoglobulin transfer to dairy calves fed colostrum or commercially available colostral-supplement products. *J. Amer. Vet. Med. Assoc.* **208**, 107-110.

Godden, S. M., S. Smith, J. M. Feirtag, L. R. Green, S. J. Wells, and J. P. Fetrow. (2003) Effect of on-farm commercial batch pasteurization of colostrum on colostrum and serum immunoglobulin concentrations in dairy calves *J. Dairy Sci.* **86**, 1503–1512.

Groenendaal, H., M. Nielen, and J. W. Hesselink. (2003) Development of the Dutch Johne's disease control program supported by a simulation model. *Prev. Vet. Med.* **60**, 69-90.

Grongnet, J. F., E. Grongnet-Pinchon, D. Levieux, M. Piot and J. Lareynie. (1986) Newborn calf intestinal absorption of immunoglobulins extracted from colostrum. *Reprod. Nutr. Dev.* **26**, 731-743.

Guilloteau, P., T. Corring, P. Garnot, P. Martin, R. Toullec, and D. Durand. (1983) Effects of age and weaning on enzyme activities of abomasum and pancreas of the lamb. *J. Dairy Sci.* **66**, 2373.

Guy M. A., T. B. McFadden, D. C. Cockrell, and T. E. Besser. (1994) Effects of unilateral prepartum milking on concentrations of immunoglobulin G1 and prolactin in colostrum. *J. Dairy Sci.* **77**, 3584-3591.

Haines, D. M., B. J. Chelack, and J. M. Naylor. (1990) Immunoglobulin concentrations in commercially available colostrum supplements for calves. *Can. Vet. J.* **31**, 36-37.

Hammer, C. J., J. D. Quigley, L. Ribeiro, and H. D. Tyler. (2004) Characterization of a colostrum replacer containing IgG concentrate and growth factors. *J. Dairy Sci.* **87**, 106–111.

Hancock, D. D., (1985) Production symposium: Immunological development of the calf; assessing efficiency of passive transfer in dairy herds. *J. Dairy Sci.* **68**, 163-183.

Harman, R. J., T. V. Doherty, and M. J. McCloskey. (1991) Colostrx® supplement protects against challenge with enterotoxigenic *E. coli* in neonatal dairy calves. *Agri-Pract.* **12**, 12-14.

Holloway, N. M., J. W. Tyler, J. Lakritz, S. L. Carlson, R. K. Tessman and J. Holle. (2002) Serum immunoglobulin G concentrations in calves fed fresh colostrum or a colostrum supplement. *J. Vet. Intern. Med.* **16**, 187-191.

Hopkins, S. G. and R. F. DiGiacomo. (1997) Natural transmission of bovine leukemia virus in dairy and beef cattle. *Vet. Clin. North Am. Food Anim. Pract.* **13**, 107-128.

Hopkins, B. A., and J. D. Quigley, III. (1997) Effects of method of colostrum feeding and colostral supplementation on serum IgG concentrations in neonatal calves. *J. Dairy Sci.* **80**, 979-983.

Hunt, E., S. D. Van Camp and S. Fleming. (1988) Developing technologies for prevention of bovine failure of passive transfer of antibody (FPTA). *Bovine Practitioner.* **23**, 131-134.

Ikemori Y, M. Ohta, K. Umeda, F. C. Icatlo, M. Kuroki, H. Yokoyama, and Y. Kodama. (1997) Passive protection of neonatal calves against bovine coronavirus-induced diarrhea by administration of egg yolk or colostrum antibody powder. *Vet Microbiol.* **58**, 105-111.

James, R. E., M. L. McGilliard, and D. A. Hartman. (1984) Calf mortality in Virginia Dairy Herd Improvement herds. *J. Dairy Sci.* **67**, 908-911.

James, R. E., C. E. Polan, and K. A. Cummins. (1981) Influence of administered indigenous microorganisms on uptake of I^{125}-g-globulin in vivo by intestinal segments of neonatal calves. *J. Dairy Sci.* **64**, 52-61.

Jansson, L., F. A. Karlson, and B. Westermark. (1985) Mitogenic activity and epidermal growth factor content in human milk. *Acta Paed. Scand.* **74**, 250-253.

Jenny, B. F., G. E. Gramling, and T. M. Glaze. (1981) Management factors associated with calf mortality in South Carolina dairy herds. *J. Dairy Sci.* **64**, 2284-2289.

Jochims, K., F. J. Kaup, W. Dromer and M. Pickel. (1994) An immunoelectron microscopic investigation of colostral IgG absorption across the intestine of newborn calves. *Res. Vet. Sci.* **57**, 75-80.

Jones, C. M., R. E. James, J. D. Quigley, III, and M. L. McGilliard. (2004) Influence of pooled colostrum or colostrum replacement on IgG and evaluation of animal plasma in milk replacer. *J. Dairy Sci.* **87**, 1806-1814.

Jumaluddin, A. A., D. W. Hird, M. C. Thurmond, and T. E. Carpenter. (1996a) Effect of preweaning feeding of pasteurized and nonpasteurized milk on postweaning weight gain of heifers on a California dairy. *Prev. Vet. Med.* **28**, 91-99.

Jumaluddin, A. A., T. E. Carpenter, D. W. Hird, and M. C. Thurmond. (1996b) Economics of feeding pasteurized colostrum and pasteurized waste milk to dairy calves. *J. Amer. Vet. Med. Assoc.* **209**, 751-756.

Kruse, V. (1970) Absorption of immunoglobulin from colostrum in newborn calves. *Anim. Prod.* 12, 627-630.

Laegreid, W. W., M. P. Heaton, J. E. Keen, W. M. Grosse, C. G. Chitko-McKown, T.P.L. Smith, J. W. Keele, G. L. Bennett and T. E. Besser. (2002) Association of bovine neonatal Fc receptor a-chain gene (FCGRT) haplotypes with serum IgG concentration in newborn calves. *Mammalian Genome* **13**, 704–710.

Liebhaber, M., N. J. Lewiston, M. T. Asquith, L. Olds-Arroyo and P. Sunshine. (1977) Alterations of lymphocytes and of antibody content of human milk after processing. *J. Pediatrics.* **91**, 897-900.

Matte, J. J., C. L. Girard, J. R. Seoane, and G. J. Brisson. (1982) Absorption of colostral immunoglobulin G in the newborn dairy calf. *J. Dairy Sci.* **65**, 1765-1770.

Mayer, B., A. Zolnai, L. V. Frenyó , V. Jancsik, Z. Szentirmay, L. Hammarström and I. Kacskovics. (2002) Redistribution of the sheep neonatal Fc receptor in the mammary gland around the time of parturition in ewes and its localization in the small intestine of neonatal lambs. *Immunology.* **107**, 288–296.

McDonough, S. P., C. L. Stull, and B. I. Osburn. (1994) Enteric pathogens in intensively reared veal calves. *Am. J. Vet. Res.* **55**, 1516-1520.

McEwan, A. D., E. W. Fisher, and I. E. Selman. (1968) The effect of colostrum on the volume and composition of the plasma of calves. *Res. Vet. Sci.* **9**, 284-286.

McEwan, A. D., E. W. Fisher, and I. E. Selman. (1970) An estimation of the efficiency of the absorption of immune globulins from colostrum by newborn calves. *Res. Vet. Sci.* **11**, 239-242.

McGuirk, S. M. (1989) Practical colostrum evaluation. *Proc. 21st Annual Meeting of Bovine Practitioners.* **21**, 79-82.

Mcvicker, J. K., G. C. Rouse, M. A. Fowler, B. H. Perry, B. L. Miller, and T. E. Johnson. (2002) Evaluation of a lateral-flow immunoassay for use in monitoring passive transfer of immunoglobulins in calves. *Am. J. Vet. Res.* **63,** 247-250.

Meas S., T. Usui, K. Ohashi, C. Sugimoto, and M. Onuma. (2002) Vertical transmission of bovine leukemia virus and bovine immunodeficiency virus in dairy cattle herds. *Vet. Microbiol.* **84**, 275-282.

Mechor, G. D., Y. T. Gröhn, and R. J. Van Saun. (1991) Effect of temperature on colostrometer readings for estimation of immunoglobulin concentration in bovine colostrum. *J. Dairy Sci.* **74**, 3940-3943.

Mechor, G. D., Y. T. Gröhn, L. R. McDowell, and R. J. Van Saun. (1992) Specific gravity of bovine colostrum immunoglobulins as affected by temperature and colostrum components. *J. Dairy Sci.* **75**, 3131-3135.

Mee, J. F., K. J. O'Farrell, P. Reitsma, and R. Mehra. (1996) Effect of a whey protein concentrate used as a colostrum substitute or supplement on calf immunity, weight gain, and health. *J. Dairy Sci.* **79**, 886-894.

Meylan, M., D. M. Rings, W. P. Shulaw, J. J. Kowalski, S. Bech and G. F. Hoffiss. (1996) Survival of Mycobacterium paratuberculosis and preservation of immunoglobulin G in bovine colostrum under experimental conditions simulating pasteurization. *Am. J. Vet Res.* **57**, 1580-1585.

Möllerberg, L., L. Ekman, and S. O. Jacobsson. (1975) Plasma and blood volume in the calf from birth till 90 days of age. *Acta Vet. Scand.* **16**, 178-181.

Moore, E. C., D. Keil, K. St. Cyr. (1996) Thermal inactivation of bovine immunodeficiency virus. *Appl. Env. Microbiol.* **62**, 4280-4283.

Morin, D. E., G. C. McCoy and W. L. Hurley. (1997) Effects of quality, quantity, and timing of colostrum feeding and addition of a dried colostrum supplement on immunoglobulin G1 absorption in Holstein bull calves. *J. Dairy Sci.* **80**, 747-753.

National Animal Health Monitoring System. (1992) *Dairy herd management practices focusing on preweaned heifers.* USDA, Animal and Plant Health Inspection Service, Veterinary Services, Fort Collins, CO.

National Animal Health Monitoring System. (1996) *Dairy herd management practices focusing on preweaned heifers.* USDA, Animal and Plant Health Inspection Service, Veterinary Services, Fort Collins, CO.

National Animal Health Monitoring System. (2002) *Colostrum Feeding.* USDA, Animal and Plant Health Inspection Service, Veterinary Services, Fort Collins, CO.

Odde, K. G. (1986) Neonatal calf survival. *Vet. Clin. North Am. Food Anim. Pract.* **4**, 501-508.

Oliver S. P., R. T. Duby, R. W. Prange, and J. P. Tritschler. (1984) Residues in colostrum following antibiotic dry cow therapy. *J. Dairy Sci.* **67**, 3081-3084.

Payne, E., J. W. Ryley, and R.J.W. Gartner. (1967) Plasma, blood, and extracellular fluid volume in grazing Hereford cattle. *Res. Vet. Sci.* **8**, 20-26.

Poulsen, K. P., F. A. Hartmann, S. M. McGuirk. (2002) Bacteria in colostrum: impact on calf health. *J. Vet. Intern. Med. (Abstr.).*

Poulsen, K.P., V. Egglestone, M. Collins, and S.M. McGuirk. (2003) The efficacy of a colostrum replacement product used in dairy calves. *J. Vet. Int. Med.* **17**, 391 (Abstr).

Praetor, A., I. Ellinger and W. Hunziker. (1999) Intracellular traffic of the MHC class I-like IgG Fc receptor, FcRn, expressed in epithelial MDCK cells. *J. Cell Sci.* **112**, 2291-2299.

Pritchett, L. C., C. C. Gay, T. E. Besser, and D. D. Hancock. (1991) Management and production factors influencing immunoglobulin G1 concentration in colostrum from Holstein cows. *J. Dairy Sci.* **74**, 2336-2341.

Pritchett L. C., C. C. Gay, D. D. Hancock, and T. E. Besser. (1994) Evaluation of the hydrometer for testing immunoglobulin G1 concentrations in Holstein colostrum. *J. Dairy Sci.* **77**, 1761-1767.

Quigley, J. D., III and J. J. Drewry. (1998) Nutrient and immunity transfer from cow to calf pre- and post-calving. *J. Dairy Sci.* **81**, 2779-2790.

Quigley, J. D., III, J. J. Drewry, and K. R. Martin. (1998a) Estimation of plasma volume in Holstein and Jersey calves. *J. Dairy Sci.* **81**, 1308-1312.

Quigley, J. D., III, D. L. Fike, M. N. Egerton, J. J. Drewry, and J. D. Arthington. (1998b) Effects of a colostrum replacement product derived from serum on immunoglobulin G absorption by calves. *J. Dairy Sci.* **81**, 1936-1939.

Quigley, J. D., III, P. French, and R. E. James. (2000) Effect of pH on absorption of immunoglobulin G in neonatal calves. *J. Dairy Sci.* **83**, 1853-1855.

Quigley, III, J. D., C. J. Kost and T. M. Wolfe. (2002a) Absorption of protein and IgG in calves fed a colostrum supplement or replacer. *J. Dairy Sci.* **85**, 1243–1248.

Quigley, J. D. and T. A. Wolfe. (2002b) Absorption of IgG from maternal colostrum or fractions of bovine or porcine plasma proteins. *J. Anim. Sci.* **85**(Suppl. 1), 336.

Quigley, J. D., III, K. R. Martin, D. A. Bemis, L.N.D. Potgieter, C. R. Reinemeyer, B. W. Rohrbach, H. H. Dowlen, and K. C. Lamar. (1994a) Effects of housing and colostrum feeding on the prevalence of selected organisms in feces of Jersey calves. *J. Dairy Sci.* **77**, 3124-3131.

Quigley, J. D., III, K. R. Martin, H. H. Dowlen, and K. C. Lamar. (1995) Addition of soybean trypsin inhibitor to bovine colostrum: Effects on serum immunoglobulin concentrations in Jersey calves. *J. Dairy Sci.* **78**, 886-892.

Quigley, J. D., K .R. Martin, H. H. Dowlen, L. B. Wallis, and K. Lamar. (1994b) Immunoglobulin concentration, specific gravity, and nitrogen fractions of colostrum from Jersey cattle. *J. Dairy Sci.* **77**, 264-269.

Quigley, J. D., III, R. E. Strohbehn, C. J. Kost, and M. M. O'Brien. (2001) Formulation of colostrum supplements, colostrum replacers and acquisition of passive immunity in neonatal calves. *J. Dairy Sci.* **84**, 2059–2065.

Roberson, J. R., L. K. Fox, D. D. Hancock, J. M. Gay, and T. E. Besser. (1998) Sources of intramammary infections from Staphylococcus aureus in dairy heifers at first parturition. *J. Dairy Sci.* **81**, 687-693.

Rodewald R. (1976) pH-dependent binding of immunoglobulins to intestinal cells of the neonatal rat. *J. Cell Biol.* **71**, 666–669.

Rodewald, R. (1980) Distribution of immunoglobulin G receptors in the small intestine of the young rat. *J. Cell Biol.* **8**5, 18–32.

Roy, J.H.B. (1991) *The Calf.* Volume 1. Management of Health. Butterworths, Boston, MA.

Rusov, C. (1993) Milk from leukotic cows as a potential danger to the health of man and animals. *Veterinarski-Glasnik.* **47**, 123-127.

Semotan, K. and D. Kalab. (1997) A new method of preparation of bovine colostral immunoglobulins for parenteral administration in calves. *Vet. Med.* (Praha). **42**, 249-252.

Smith, R. and R. Little. (1922) The significance of colostrum to the newborn calf. *J. Exp. Med.* **36**, 181–198.

Stabel, J. R. (2001) On-farm batch pasteurization destroys Mycobacterium paratuberculosis in waste milk. *J. Dairy Sci.* **84**, 524-527.

Staley, T. E. and L. J. Bush. (1985) Receptor mechanisms of the neonatal intestine and their relationship to immunoglobulin absorption and disease. *J. Dairy Sci.* **68**, 184-205.

Steinbach, G., B. Kreutzer, and H. Meyer. (1981) Effect of heating on the immunobiological value of bovine colostrum. *Monat. Fur Veter.* **36**, 29-31.

Streeter, R. N., G. F. Hoffsis, S. Bech-Nielsen, W. P. Shulaw, and D. M. Rings. (1995) Isolation of Mycobacterium paratuberculosis from colostrum and milk of subclinically infected cows. *Am. J. Vet. Res.* **56**, 1322-1324.

Su, C. K. and B. H. Chiang. (2003) Extraction of immunoglobulin-G from colostral whey by reverse micelles. *J. Dairy Sci.* **86**, 1639-1645.

Sweeney, R. W. (1996) Transmission of paratuberculosis. Vet. Clin. North Am. *Food Anim. Pract.* **12**, 305-312.

Thivend, P., R. Toulec, and P. Guilloteau. (1980) Digestive adaptation in the preruminant. Page 561 In *Digestive Physiology and Metabolism in Ruminants,* pp 561-568. Edited by Y. Ruckebusch and P. Thivend. AVI Publ. Co., Westport, CT.

Tyler, J. W., B. J. Steevens, D. E. Hostetler, J. M. Holle, and J. L. Denbigh, Jr. (1999) Colostrum immunoglobulin concentrations in Holstein and

Guernsey cows. *Am. J. Vet. Res.* **60**, 1136-1139.

Uggla, A., S. Stenlund, O. J. Holmdahl, E. B. Jakubek, P. Thebo, H. Kindahl, and C. Bjorkman. (1998) Oral Neospora caninum inoculation of neonatal calves. *Int. J. Parasitol.* **28**, 1467-1472.

USDA Economic Research Service. (2003) United States State Fact Sheet. http://www.ers.usda.gov/StateFacts/US.HTM.

Wagstaff, A. J., I. Maclean, A. R. Mitchell, and P. H. Holmes. (1992) Plasma and extracellular volume in calves: comparison between isotopic and 'cold' techniques. *Res. Vet. Sci.* **53**, 271–273.

Waltner-Toews, D., S. W. Martin, A.H. Meek, and I. McMillan. (1986a) Dairy calf management, morbidity and mortality in Ontario Holstein herds. *Prev. Vet. Med.* **4**, 103-117.

Waltner-Toews, D., S. W. Martin, and A.H. Meek. (1986b) Dairy calf management, morbidity and mortality in Ontario Holstein herds. IV. Association of management with mortality. *Prev. Vet. Med.* **4**, 159-171.

Wells, S. J. (2000) Biosecurity on dairy operations: hazards and risks. *J. Dairy Sci.* **83**, 2380-2386.

Wittum, T. E. and L. J. Perino. (1995) Passive immune status at postpartum hour 24 and long-term health and performance of calves. *Am. J. Vet. Res.* **56**, 1149-1154.

Zaremba, W., W. M. Guterbock and C. A. Holmberg. (1993) Efficacy of a dried colostrum powder in the prevention of disease in neonatal Holstein calves. *J. Dairy Sci.* **76**, 831-836.

9

DIGESTIVE SECRETIONS IN PRERUMINANT AND RUMINANT CALVES AND SOME ASPECTS OF THEIR REGULATION

PAUL GUILLOTEAU[1], ROMUALD ZABIELSKI[2]
[1]*Institut National de la Recherche Agronomique, Unité Mixte de Recherche chez le Veau et le Porc, Domaine de la Prise, 35590 Saint-Gilles, France;* [2]*Department of Physiological Sciences, Faculty of Veterinary Medicine, Warsaw Agricultural University, 02-766 Warsaw, and The Kielanowski Institute of Animal Physiology and Nutrition, 05-110 Jablonna, Poland*

Abbreviations

Ach : Acetylcholine; BW : Body weight; CNS : Central nervous system; CCK : Cholecystokinin; CCK_1 and CCK_2 : Receptors for CCK and/or gastrin of type 1 (or A) and type 2 (or B), respectively; DM : Dry matter; DVD : Dorsal vagal complex; GI : Gastrointestinal; GRP : Gastrin releasing peptide; MMC : Migrating motor complex; PACAP : Pituitary adenylate cyclase activating polypeptide; PP : Pancreatic polypeptide; PPS : Periodic pancreatic secretion; VFA : Volatile fatty acid; VIP : Vasoactive intestinal polypeptide

Introduction

The main role of the gastrointestinal (GI) tract is to transform the ingested diet into nutrients that can be absorbed across the GI wall to the blood circulation in order to supply body tissues with essential energy and structural components. The work of the GI tract involves transport (thanks to GI motility), degradation (provided by digestive secretions and microorganisms) and absorption (transcellular and paracellular). The GI tract is a complex system with continuous cross-talk among epithelial cells (producing digestive secretions), the local immune system and microflora. In this review, only the digestive secretions produced by the epithelial cells and the digestive glands are taken into account, even if they are not independent of the presence of gut microflora. Thus, in the ruminant calf, major absorption of carbohydrates and lipids (by way of volatile fatty acids, VFA) occurs in the rumen mucosa, and that of proteins (in the form of short peptides and single amino acids) in the small intestinal mucosa, and rumen microorganisms contribute to the both. In contrast, in the preruminant calf, the major absorption of digestion products occurs in the intestinal mucosa. The importance of GI digestive secretions is unquestionable; besides providing enzymatic degradation, they also provide the optimal environment for digestion processes (dilution, pH, temperature, etc.), protection of the GI epithelium, and regulation of GI functions. This chapter synthesizes recent studies on GI secretions in preruminant and ruminant calves, and also summarizes the mechanisms of their regulation.

Digestive secretions - some general aspects

Several experimental approaches have been developed to study the production of digestive secretions. The secretion of GI glands (parotid, pancreatic) and bile can be assessed by cannulating the ducts draining out the digestive juices into the lumen of the GI tract. In a chronic study, this approach gives a chance to precisely record the kinetics of flow and the daily output of digestive juices for many weeks. In calves and lambs it is possible to record secretion during the transition from preruminant to ruminant, as was done with the secretion of pancreatic and abomasal juices (Pierzynowski *et al.*, 1992; Zabielski *et al.*, 1997a; Guilloteau and Le Calvé, 1977). In the case of pancreas duct cannulation or gastric (abomasal) pouch, an important fact is that the juice is pure and non-activated (Zabielski *et al.*, 1997a; Guilloteau, 1986). The cannulation technique is also used to investigate the mucosal secretion of the sac-like (e.g., abomasum) and canal-like (e.g., small intestine) structures. In this case a small pouch can be formed from a part of the stomach or a loop formed from an intestine segment to collect pure secretion. To overcome problems related to maintenance of an intestinal loop, indirect methods employing various marker substances can be used instead (Andrews *et al.*, 1992; Wilsen *et al.*, 1993). In recent decades, digesta and tissue samples obtained by biopsy or after sacrificing the experimental animal are the most commonly used. This experimental approach is more acceptable to society than chronic cannulation, although paradoxically it requires more animals to be sacrificed to get information equivalent to that obtained in cannulation studies, especially when a kinetic process is investigated. The most reliable biological material for analyses, and the best results, can be obtained when all these techniques are used in a sensible combination.

PRODUCTION AND COMPOSITION

Salivary glands produce saliva, which is of little importance for digestion in the preruminant calf (apart from digestion of lipids), but plays an important role in the ruminant calf. The reticulo-rumen, the major site of microbial fermentation in the ruminant calf, and the omasum do not produce their own digestive secretions, thus the processes of degradation of nutrients in the forestomach rely on the microbial ecosystem. The abomasum in calves corresponds to the chemical stomach in monogastric animals and produces gastric (abomasal) juice. The pancreas plays a crucial role in digestion of all components present in the small intestinal lumen by supplying it with pancreatic juice. The small intestine produces intestinal juice and, more importantly, the brush border enzymes that complete the digestion of nutrients in the gut lumen. Moreover, there are a number of lysosomal enzymes that are important in the intracellular digestion that takes a place in the large lysosomal vacuoles (Figure 9.1) of the enterocytes (Baintner 1986, Fujita et al., 1990, Biernat et al. 1999). This kind of digestion occurs in lower vertebrates (e.g. salmon) throughout their lifetime, whereas in mammals it is observed only during the first few weeks of life. The function of intracellular digestion in mammalian neonates is to support digestion in the gut lumen until the lumen is adequately developed. The large intestine is of limited importance for digestion in the preruminant calf, but it can have a role in the ruminant calf. Thus, in calves the GI secretions are produced mostly by the salivary glands, the abomasum, the pancreas and the small intestine.

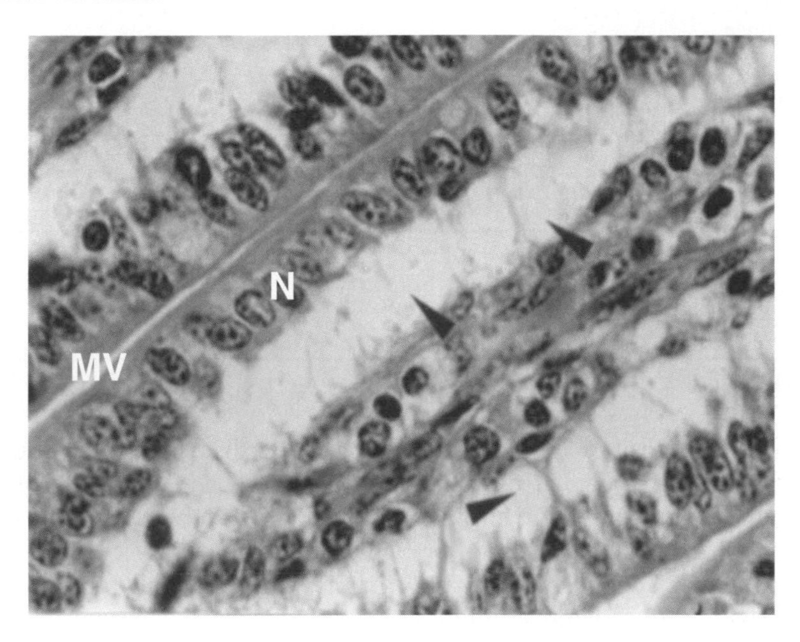

Figure 9.1 Histological sections of villi in the mid-jejunum of a neonatal calf. Nearly all enterocytes contain large lysosomal vacuoles (arrowheads) which move the nuclei (N) toward the apical pole of the cell (MV – microvilli).(Biernat et al, unpublished data)

It is interesting to compare the secretions produced by the preruminant calf (considered as a monogastric animal) with those of the ruminant calf, the young pig (monogastric animal) and the horse (a species which can be considered semi-ruminant). In the comparison, total daily secretions by the GI tract are important, but there are differences between the animal species. Salivary secretions are important in the ruminant calf and the horse; gastric juice secretions are important in the pig and the ruminant calf; and pancreatic secretions are important in the pig. Interestingly, in the preruminant calf the total volume of daily GI secretion is 19 to 25 litres (Table 9.1), whereas water intake is only 8 to 9 litres per day (Guilloteau and Le Huërou-Luron, 1996). This involves water metabolism and considerable water exchange at the level of the GI tract.

Table 9.1 TOTAL OUTPUT OF SECRETIONS AND CHEMICAL COMPOSITION OF PANCREATIC AND GASTRIC JUICES IN PRERUMINANT (PRE) AND RUMINANT (RU) CALVES (RUCKEBUSH, 1977; GUILLOTEAU, 1986; GUILLOTEAU *et al*, UNPUBLISHED DATA).

Total output (l/d)	*Preruminant*	*Ruminant*
Total secretions	19 -25	36 - 64
Saliva	2 - 4	20 -40
Gastric juice	2 -5	7 - 11
Pancreatic juice	1	1 - 2
Bile	7	1 - 2
Intestinal juice	7 -8	7 - 8

Chemical composition of pancreatic and gastric juices (g/kg juice)

	Dry Matter	*Lipids*	*Ash*	*Nitrogen*	*Amino Acids*
Pancreatic juice	10.2	0.1	6.0	0.3	ND
Gastric juice (Pre)	10.5	0.2	7.2	0.4	85
Gastric juice (Ru)	9.6	0.2	6.6	0.6	21

ND : Not Determined

The chemical composition of dry matter (DM) in gastric and pancreatic juices obtained from preruminant calves is shown in Table 9.1. Dry matter is composed almost entirely of minerals and nitrogenous matter. Digestive secretions contain similar elements, but their ratios differ. Thus, minerals supply electrolytes and nitrogenous matter supplies urea (mainly in the ruminant) and proteins, which can be divided into several compounds: enzymes, mucins, regulatory and antimicrobial proteins, protein structures of extruded cells and others (Table 9.2).

ROLE OF DIGESTIVE SECRETIONS

Role of different components

In the preruminant calf, digestive enzymes are either secreted into the GI lumen with saliva, gastric and pancreatic juices, attached to the brush border of the apical pole of the enterocytes (brush border enzymes), or placed inside the enterocytes (lysosomal enzymes). Thus, the term "digestive secretions" is not

Table 9.2 MAJOR COMPONENTS OF DIGESTIVE SECRETIONS IN CALVES (MONTAGNE, 2000).

		Proteins						
	Water	*Enzyme*	*Mucins*	*AP*	*Other*	*Urea*	*Electrolytes*	*Ext.cells*
Saliva	**	*	*	*	*	**	**	*
Gastric juice	**	**	*	*	*	tr	**	*
Pancreatic juice	**	**	?tr	*	*	tr	**	*
Bile	**	-	*	?	*	tr	**	*
Intestine juice	*	(1)	*	?	*	*	*	**

1 enzymes are retained in brush border and inside cells
AP: antimicrobial proteins Ext.cells : extruded cells tr : trace

strictly correct; it would be more accurate to speak of "digestive productions". Moreover, the secreted juices contain little enzymes extruded from cells. There are complementarities between enzymes from the proximal and distal gut (Guilloteau *et al.*, 1995). For example, different enzymes involved in protein digestion in preruminant calf are shown in Table 9.3. Firstly, abomasal enzymes, pancreatic enzymes (e.g. trypsin, chymotrypsin and elastases) and some intestinal enzymes (e.g. prolinase) are endopeptidases, i.e. they cut the bonds between amino acids inside a protein chain. Secondly, pancreatic enzymes (e.g. carboxypeptidases) and some intestinal enzymes (e.g. aminopeptidases) are exopeptidases, i.e. they cut off single amino acids at the extremities of a protein chain. Thirdly, each enzyme is more or less specialized in cutting a protein chain according to the nature of the amino acid bond.

Table 9.3 DIGESTION OF PROTEINS IN PRERUMINANT CALVES.

Compartment	*Enzymes*		*Substrates*
	1 *Endopeptidases*	*2* *Exopeptidases*	
Abomasum (gastric juice)	Chymosin pepsins		Casein K proteins (milk or milk replacer)
Pancreas (pancreatic juice)	Trypsins Chymotrypsins (ABC) Elastases (I, II)	Carboxy- peptidases (A, B)	proteins polypeptides peptides
Small intestine			
membrane		Aminopeptidases (A, N) Dipeptidases IV	peptides, tri or dipeptides
intracellular	Iminopeptidase (or prolinase)		dipeptides
	↓2 ↓1 ↓2 CC — CC — CC — CC		Protein

Enzymes require an optimal environment for their function, especially optimal pH. Therefore, it is important to look at the electrolyte composition of each secretion. The total anions are similar to the total cations in each GI secretion, but the proton of hydrochloric acid is predominant in the abomasum, whereas bicarbonates are predominant in the small intestine (Figure 9.2). In the ruminant calf (with a developed forestomach), the high pH of saliva maintains rumen pH at about 6.5-7.5, thereby providing an optimal environment for rumen microorganisms. In the abomasum, the hydrochloric acid of abomasal juice reduces the pH of digesta. In the ruminant calf, abomasal pH is between 2.5 and 3.5, which allows only the proteolytic action of pepsin. In the preruminant calf, however, the pH value is between 2.5 and 5.5, allowing coagulation of milk proteins and their hydrolysis by chymosin and pepsin. In the proximal duodenum, the pH increases to nearly neutral, due to the high pH (between 7 and 8 units) of pancreatic juice, bile and intestinal juice, providing optimal conditions for the enzymes of pancreatic juice and of the brush border. Thus we can see here a second complementarity between enzymes and electrolytes, which allows enzymes to function in a given segment of the GI tract.

Mucins are glycoproteins produced by the goblet cells in the GI epithelium. The density of goblet cells increases along the small intestine, from a few per cent of cells in the duodenum to about 20 per cent in the ileum. Mucins are the main component of mucus, and they are responsible for structural and functional properties of mucus. Besides the filtering function (selective diffusion of nutrients), the main role of mucus is to protect the mucosal epithelium against physical and chemical (enzymatic, acidic, toxic, etc.) alterations, and to control the host-microbial relationships. Studies on these topics in farm animals are scarce. It has been recently estimated that the mucin output in the ileal digesta is similar in the preruminant calf (4.0 g of DM intake) and in piglet (3.9 g of DM intake) fed with milk proteins (Montagne *et al.*, 2000a; Piel *et al.*, personal communication).

A number of antimicrobial proteins are produced by the epithelial cells of the skin, respiratory and GI tract (Zasloff, 2002). Pancreatic juice also has antibacterial activity (Rubinstein *et al.*, 1985). Laubitz *et al.*, (2003) identified a heat resistant polypeptide of 14 kDa in porcine pancreatic juice that is active against *Escherichia coli* strain AB1157. The polypeptide was active within a pH range normally seen in pancreatic juice; *in vitro* it reduced the number of living bacterial cells in overnight culture by a factor of 10,000, from 3 x 10^8 cells/ml for controls to 4 x 10^4 cells/ml for samples with pancreatic juice. Mass spectroscopy analysis of the active fraction showed high similarity with the pancreatic spasmolytic polypeptide. Antibacterial polypeptides controlling bacterial homeostasis have also been found in the pancreatic juice of the preruminant calf, though of lower activity compared with that in piglets (Laubitz and Grzesiuk – personal communication). It remains unclear whether the antibacterial polypeptide functions only in the pancreas duct system or also after it is released into the gut lumen.

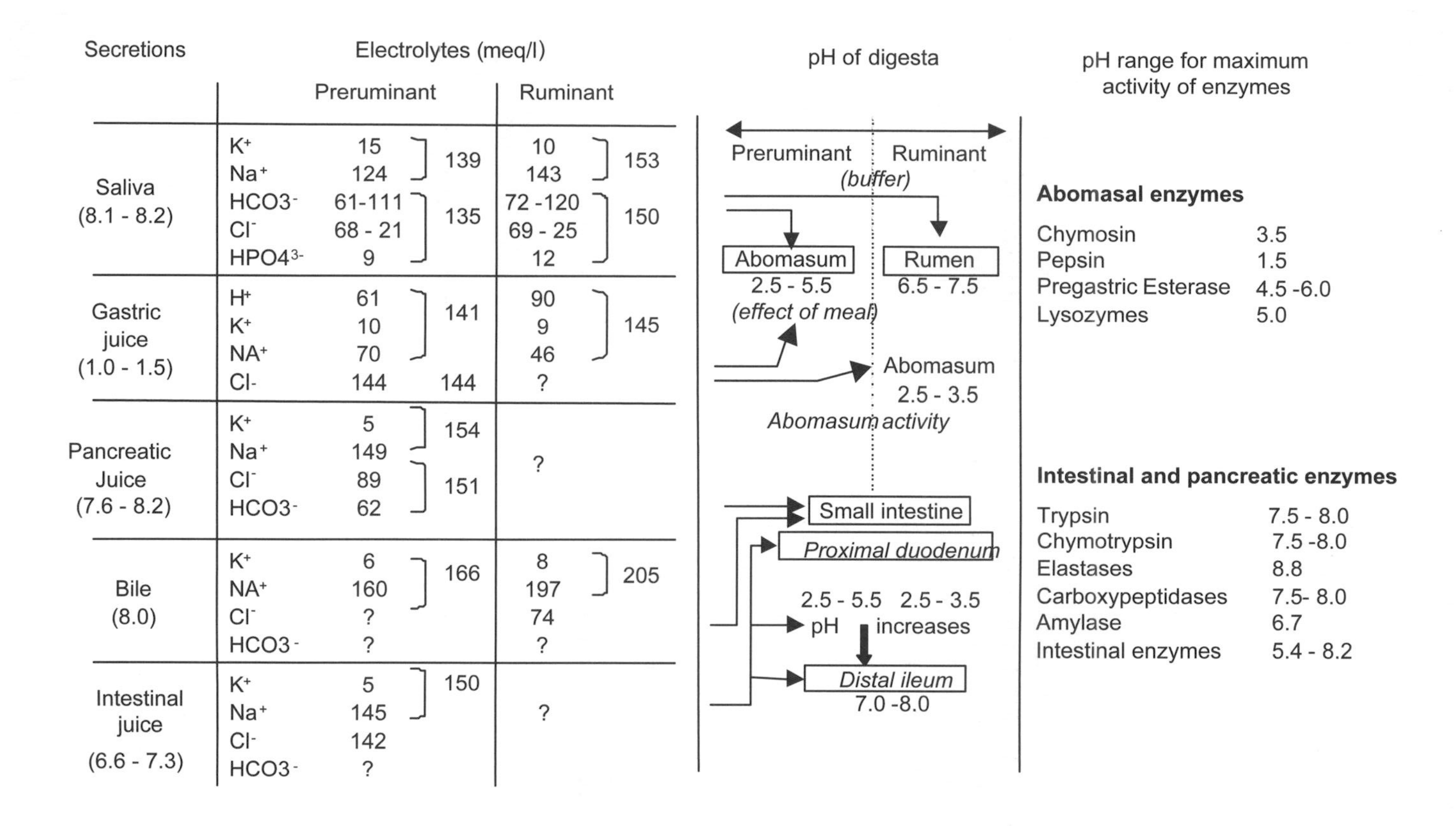

Figure 9.2 Electrolyte composition of digestive secretions, pH of digesta along the digestive tract and pH range for maximum activity of enzymes in preruminant and ruminant calves.

There are several other phenomena naturally present in the GI tract to protect the mucosa. Firstly, some enzymes are secreted into the GI lumen in the inactive form of proenzymes (e.g. gastric pepsinogen and pancreatic trypsinogen). They are activated in the GI lumen inside the digesta. Other enzymes are attached to the brush border or are located inside the cells. Secondly, the intestinal microflora help the host to antagonize colonisation by pathogenic bacteria and to protect against dangerous substances arriving in the lumen. Thirdly, the immune system is involved because, at intestinal sites, lymphocytes are directed to the lamina propria and mucosal epithelium to produce antibodies, to eliminate pathogens and to contribute to the normal activity of epithelial cells.

The endogenous contributions provided by GI secretions are not negligible. For instance, we have measured production of minerals by the abomasum of the preruminant calf and they represent significant quantities when compared with dietary intake. Endogenous protein productions are more important for pancreatic than abomasal enzymes (Table 9.4). The sum of the values obtained in different parts of the GIT tract before the ileo-caecal valve, is higher than the value measured at the end of ileum. A significant proportion of digestive secretions is digested and absorbed along with nutrients. It was estimated that in the preruminant calf, more than 70 % of endogenous nitrogen production is absorbed before the end of ileum (Montagne *et al.*, 2000b). However, mucin protein is resistant to enzymatic digestion and is poorly digested prior the large intestine (Hoskins, 1984); this probably represents one of the most significant contributions to flow of endogenous protein out of the small intestine (Fuller and Cadenhead, 1991).

Table 9.4 CONTRIBUTION OF DIGESTIVE SECRETIONS TO ENDOGENOUS PRODUCTION IN PRERUMINANT CALVES(GUILLOTEAU, 1986; CORNOU-LE DRÉAN, 1997; MONTAGNE, 2000).

Part of GI tract		*Nutrient*	*Proportion of nutrient ingested (% minerals or nitrogen)*
Abomasum	Minerals	Na+ Cl-	100 - 150
		K+	10 - 30
	N	total (N)	3.0 -7.0
		enzymes	0.2 - 0.8
Pancreas	N	total (N)	2.0 -5.0
		enzymes (80% of N)	1.6 - 4.0
At the end of the Small intestine	N	total (CP)	3.1 -6.8
		mucins	3.4 - 6.8

N : Nitrogen CP : Crude protein

To summarize, there are important differences in chemical composition of GI secretions along the GI tract. On the one hand this permits distinct phases of digestion but, on the other hand, these actions are complementary.

Effect of digestive secretions on digestibility in the preruminant calf

Abomasal secretion

When the young calf is given whole milk or a milk replacer based on skim milk protein powder, a tough coagulum forms in the abomasum due to chymosin and/or hydrochloric acid. After ingestion of a meal, this coagulum soon shrinks, holding back caseins and lipids, which are released slowly and steadily from the abomasum. When casein is replaced by protein substitutes that cannot coagulate, all components of the digesta leave the abomasum at the same speed. Thus, substituting skim milk in milk replacers by other sources of proteins increases the rate at which lipids and proteins are emptied from the abomasum (Toullec and Guilloteau, 1989). To estimate the effect of abomasal secretion, preruminant calves fitted with a catheter in the proximal duodenum were given two diets having very different kinetics with respect to gastric emptying. The proteins of these diets were supplied either almost solely by skim milk powder (milk diet) or mainly by a fish concentrate (fish diet). The milk replacers were either ingested by the animals or infused slowly or quickly into the duodenum to simulate previously-observed gastric emptying rates for the proteins (Guilloteau *et al.*, 1975). Bypassing the mouth and the abomasum decreased the digestibility of each diet when infusion simulated the gastric emptying rate of its protein (slow infusion of the milk diet, quick infusion of the fish diet). Infusion at a higher rate (quick infusion of the milk diet) increased this effect, while infusion at a lower rate (slow infusion of the fish diet) suppressed it (Guilloteau *et al.*, 1981). The effect could be a result of removal of abomasal enzymes, because infused diets by-passed the abomasum, or it could be related to flow rate, i.e. the time of enzyme and substrate co-incubation. Moreover, the effect was related to the nature of dietary proteins, since the decrease was more profound for fish than milk proteins.

Pancreatic secretion

Neonates secrete low but measurable amounts of pancreatic juice from the first day of life (Figure 9.3). Secretion responds to feeding, showing a clear cephalic phase associated with plasma pancreatic polypeptide (PP) elevation. However, no gastric or intestinal phase is observed. The first cycles of the periodic pancreatic secretion (PPS) associated with duodenal migrating motor complex (MMC) and plasma fluctuations of PP appear on the second or third day of life. Pancreatic protein and trypsin outputs follow day-by-day increases in juice volume in the pre- and postprandial periods (Zabielski *et al.*, 2002). This suggests that the exocrine pancreas in the first few days of life is capable of modifying its enzyme secretion, which supports earlier findings in human neonates fed milk formulae of different composition (Zoppi *et al.*, 1972).

In calf neonates the response to food occurs within approximately 20 minutes, and is followed by a marked decrease in pancreatic secretion, often

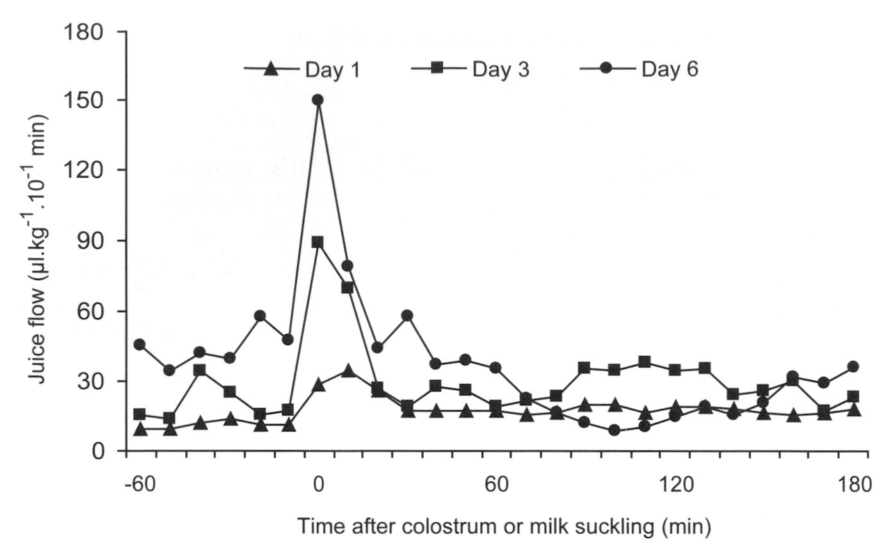

Figure 9.3 Kinetics of pancreatic juice secretion in the neonatal calf during the first 6 days of life. Note presence of a weak but significant cephalic phase in day 1, day-by-day increase in the cephalic phase, and lack of gastric and intestinal phases. (Zabielski *et al*, 2002).

below the preprandial level (Zabielski *et al*., 2002), as observed in older preruminant calves (McCormick and Stewart, 1967). It seems that although the exocrine pancreas is capable of responding to colostrum/milk suckling in calves, its overall secretory capacity is limited. These observations are in agreement with the increase in digestive utilisation of milk and whey proteins from birth to the fifth week of age (Grongnet *et al.,* 1981). Taking into account the characteristics of gastro-duodenal digesta flow pattern in preruminant calves, the cephalic phase of pancreatic secretion has to provide the duodenum with pancreatic enzymes to digest the soluble substances leaking from the abomasum during suckling (Guilloteau *et al*., 1979). Casein is readily coagulated by abomasal chymosin, and a large proportion of ingested milk is consequently transported in small portions into the small intestine by the following MMCs. Since this process is relatively slow, there is no need for the gastric and intestinal phases of pancreatic secretion to proceed early during the postprandial period, as is normally observed in adult monogastric animals.

To evaluate the global action of pancreatic enzymes, the digestibility of nitrogen and fats was measured in preruminant calves fed milk replacer based on skim milk powder. Pancreatic juice was collected by a chronic pancreatic duct catheter. When the juice was collected without being reintroduced into the duodenum, digestibility of nitrogen significantly decreased by 44 % and lipids by 46 % (Guilloteau *et al*., unpublished data). These decreases were far more dramatic than that obtained with digesta by-passing the abomasum. This study clearly confirmed that pancreas enzymes play a crucial role in food digestion. Interestingly, the effect in calves was related to the nature of dietary proteins, since the decrease was more important for milk replacer based on soyabean proteins than skim milk powder proteins (Guilloteau *et al.*, unpublished data). In the same calves, pooled pancreatic juice was infused for several days at different doses to evaluate the digestibility of nutrients. This was achieved using an automatic system for recording and reintroduction of pancreatic juice (and pancreatic proteins) into the duodenum (Le Dréan *et al.*, 1997). With increasing intra-duodenal load of pancreatic proteins, digestibility of dietary proteins and fats increased progressively until a limit was reached (Guilloteau *et al.*, 1999). The limits (95 per cent for proteins) corresponded to the digestibility values found in non-operated calves, further confirming that pancreatic enzymes play a key role in enzymatic digestion. Maximal digestibility was obtained when at least 25 mg of pancreatic protein per kg body weight was reintroduced into the proximal duodenum per day. However, more pancreatic juice reintroduction was necessary for milk replacer based on soyabean protein than on bovine milk protein (Guilloteau *et al.*, unpublished data). These observations could partly explain why the digestion of soyabean proteins is more difficult that those of bovine milk proteins, and also why some calves could have enzyme deficiencies with soyabean proteins but not with milk proteins.

Arrival of host secretions and digesta in the GI lumen

For feed digestion to be efficient, digestive juices should enter the lumen of digestive tract at the same time as the substrates in the digesta. We have shown that the daily pancreatic protein productions were similar in the preruminant calf fed milk replacer based on skim milk powder (milk diet) or soyabean concentrate (soyabean diet), in particular for total trypsin output. However, there was a great difference in the kinetics of the secretions during the 7-hour postprandial period (Le Dréan *et al.*, 1997). With the soyabean diet, pancreatic enzyme secretion was more abundant just after feeding (Figure 9.4A). Moreover, no coagulation of digesta in the abomasum was observed, so the digesta might leave the abomasum much faster just after the soyabean meal, as shown in Figure 9.4B (Guilloteau *et al.*, 1979). Thus, pancreatic secretion seems to be controlled by the characteristics of gastric emptying and the arrival of substrates into the proximal duodenum.

The ratio of trypsin output to duodenal dietary proteins, calculated just after feeding, showed higher values for the soyabean diet than for the milk diet (Figure 9.4). We think this would improve digestibility of soyabean proteins according to results with the reintroduction of different doses of pancreatic juice (see above). However, it seems that adaptation was not sufficient, because the ratio of duodenal trypsin to duodenal dietary proteins was significantly lower for the soyabean diet than for the milk diet. Finally, with the soyabean diet, enzyme secretion was not sufficient to achieve a ratio of enzyme to substrate in the GI lumen similar to that observed with the milk diet. In modern soyabean milk replacers, enzyme inhibitors (e.g., trypsin inhibitors) and glycoproteins (e.g., lectins) are no longer a problem, but the remaining soyabean proteins are more resistant to enzyme degradation than those of milk (Lallès, 1993), which might explain their lower digestibility than milk proteins in the preruminant calf.

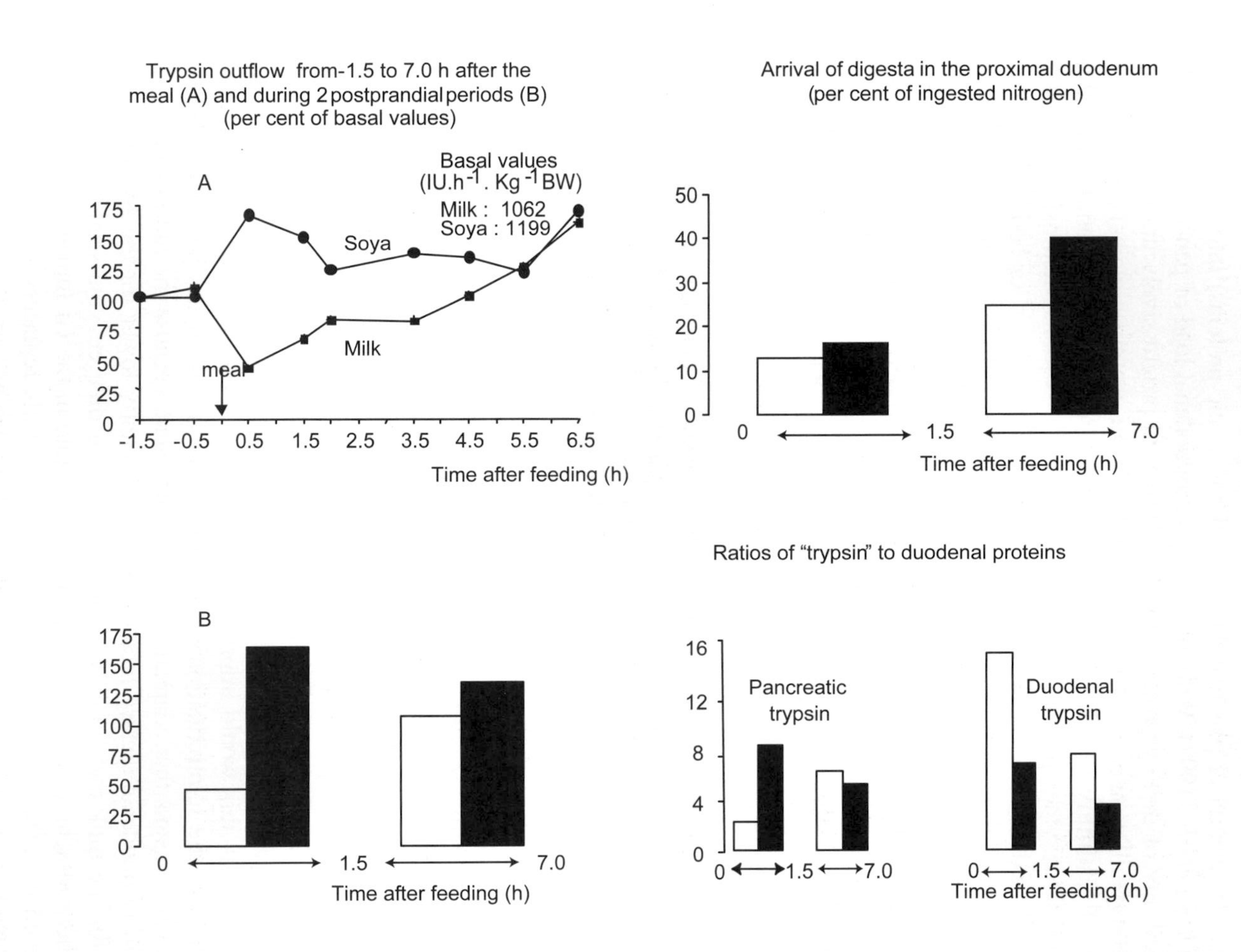

Figure 9.4 Comparison of trypsin production and flow of digesta proteins after meals based on milk (☐) or soya (■) proteins in preruminant calves (Le Dréan et al, 1998; Guilloteau *et al*, 1979; Guilloteau and Le Huërou, 1996).

Consequences of endogenous secretions

To maximise nitrogen digestibility, it is necessary to optimise enzyme activity in the GI tract as well as to minimise endogenous protein losses in the small intestine. When preruminant calves were fed milk replacer based on skim milk powder, total output of protein in the distal ileum was 14.3 g per kg DM intake (Montagne and Lallès, 2000). Thus, the true digestibility of milk protein was almost total, due to a very low level of dietary protein in the digesta. Endogenous protein represented about 45 per cent of the total protein output. Among this endogenous production, a large proportion came from mucins (Montagne *et al.*, 2000a). Since little digestion of mucin protein is supposed to occur before the large intestine, the abomasal, intestinal and, to lesser extent, salivary mucins support digestive processes in the large intestine. When dietary milk protein was replaced by colostrum protein, ileal flow of mucins was increased more than twofold (Montagne *et al.*, 2000a). The actual reason is not precisely known. For example, immunoglobulins from the colostrum diet, which largely escaped digestion (Grongnet *et al.*, 1986, and unpublished data), might have stimulated endogenous nitrogen loss by increasing especially the secretion of mucins. On the other hand, feeding calves with milk replacers containing plant proteins (potato or soyabean proteins without anti-nutritional factors) also increased ileal flow of endogenous proteins as well as that of mucins (Montagne *et al.*, 2001). Plant storage proteins are resistant to digestion and can stimulate endogenous nitrogen losses. Thus, the origin of dietary proteins, and also their technological manipulation as shown for soyabean, can influence endogenous protein production. Indeed, these data could be used to evaluate the nutritive value of a diet in terms of true digestibility. Finally, as mucus constitutes the interaction between the epithelium and the lumen, mucin could be used as a marker of interaction between the epithelium and dietary components (Montagne and Lallès, 2000). So, it provides a promising approach to study the impact of origin of dietary proteins and treatment of dietary ingredients, including dairy products, on these interactions. Thus, the "health value" could be added to the nutritive value of the diet.

Factors affecting digestive secretions

FACTORS AFFECTING DIGESTIVE SECRETIONS IN THE PRERUMINANT CALF

Newborn calf

In newborns, it is difficult to measure digestive secretions in the GI lumen and such studies are scarce. Thus, in the majority of studies the tissue content, in particular the enzyme tissue content, is analysed. On a live-weight basis, the enzyme content of the GI tissues was lower in the calf for chymotrypsin, colipase, amylase and pepsin, in comparison with lamb GI tissue, but the opposite is found for lipase. These data also depended on breed and sex of experimental animals (Guilloteau *et al.*, 1985). Newborn calves that are born after 277 d of gestation seem to have more lactase that those born at term. However, aminopeptidase A activity is not modified. The activity of aminopeptidase A seems more dependent on postnatal feeding (Bittrich *et al.*, 2004). After birth, the quantity of colostrum ingested and glucocorticoid treatments modify the tissue content of several pancreatic and intestinal enzymes (Blatter *et al.*, 2001; Sauter *et al.*, 2003). It is also interesting that the profiles of abomasal chymosin, and the profiles of pepsin and pancreatic chymotrypsin and elastase II, are opposite during the first 3 weeks of life (Le Huërou *et al.*, 1992a, Gestin *et al.*, 1997b). After birth, chymosin activity is high enough to coagulate colostrum and milk, but this enzyme has a low hydrolysis activity. In contrast, the activity of the other proteolytic enzymes is low and then increases progressively. These conditions favour absorption of intact immunoglobulins during the first days of postnatal life.

Postnatal development

In calves the secretion of pancreatic juice significantly develops with age (McCormick and Stewart, 1967; Ternouth and Buttle, 1973; Pierzynowski *et al.*, 1992; Zabielski *et al.*, 2002). First of all, it seems important to mention that three enzymes are typical in the young calf fed milk or milk replacer based on skim milk powder. The enzymes concerned are chymosin, produced by the abomasum, elastase II, produced by the pancreas, and lactase, produced by the small intestine (Figure 9.5). All three of these enzymes are present at a high level in the GI tissues from birth, further increase during the first 2 days of life, and then decrease with age (Guilloteau *et al.*, 1983; Le Huërou *et al.*, 1992a, Gestin *et al.*, 1997b).

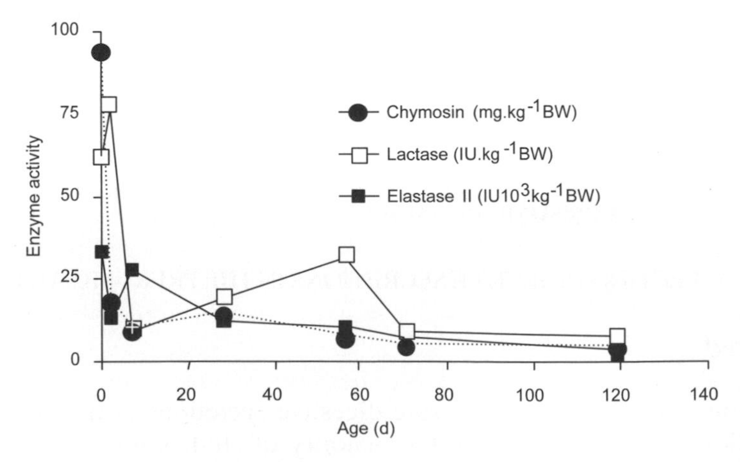

Figure 9.5 Change in enzyme activities with age in preruminant calves (Guilloteau *et al*, 1984; Le Huërou *et al*, 1992b, Gestin *et al*, 1997b)

Chymosin is specialised in the coagulation of milk in the abomasum by cutting the bonds between phenylalanine and methionine of κ-casein. Pepsin plays a similar role, but with much lower efficacy. Both enzymes are responsible for protein hydrolysis in the abomasum. We have shown that abomasal enzymes in the preruminant calf contribute to the hydrolysis of about 24 % of ingested proteins when the animals are fed with milk or milk replacer based on skim milk powder. However, abomasal hydrolysis is less important when the proteins come from sources other than milk (Guilloteau *et al.*, 1975 and 1979). We studied the secretion of gastric juice in preruminant calves using a gastric pouch technique. In general, the quantities of abomasal secretions produced per day increased with age. Total production of chymosin relative to body weight decreased up to 4 months of age and remained low after that. The same trend was observed for daily quantities of juice and electrolytes secreted by the abomasum. However, pepsin activity seems to remain constant (Guilloteau, 1986 and unpublished data). This form of data expression (per day and per kg of body weight) has two advantages: it reduces individual variations within a group of calves, and it is better related to the quantity of food ingested daily by the calf. For chymosin activity, we have found that the quantity of enzyme secreted is always sufficient to coagulate the amount of ingested milk protein, even in older preruminant animals that apparently secrete less chymosin (Guilloteau *et al.*, 1983) suggesting either that chymosin is secreted in great excess or that the other enzymes, e.g., pepsin, are also involved in protein coagulation.

Elastase II produced by the pancreas is the second typical (marker) enzyme for the development of the GI tract in preruminant calves. We have studied *in vitro* the effects of elastase I and II on ß-lactoglobuline, a globular protein. The action of pepsin plus pancreatic trypsin and chymotrypsin, studied simultaneously with that of elastase I or elastase II, did not seem to change the extent of hydrolysis. However, the nature and the number of peptides obtained was modified. The most interesting finding was a difference in the reduction of residual antigenicity: high by elastase II and lower by elastase I (Gestin *et al.*, 1997a). These observations suggested a technological application, i.e. to include elastase II during processing to reduce the allergenicity of milk formulas prepared for human infants. More generally, elastase II cleaves globular proteins like ß-lactoglobulins in milk, but also cleaves storage proteins like soyabean that are resistant to other proteolytic enzymes. This is particularly interesting when such proteins are included in the diet.

Lactase is produced in the intestinal epithelial cells, and is the third typical (marker) enzyme in the young ruminant. Its activity is particularly high during the first week of postnatal life. It is low after that, but still sufficient to digest lactose in milk. In contrast, even if the level of pancreatic amylase increases, it is not possible to incorporate large quantities of starch into the diet of preruminant calves (Toullec and Lallès, 1995). Finally, insufficient production of colipase compared with lipase could be the cause of poor utilisation of lipids in older preruminant calves.

Effect of rearing factors

The kinetics of pancreatic protein secretion, in relation to the timing of a meal in preruminant calves fed with milk replacer based on skim milk powder is presented in Figure 9.6. A similar profile for pancreatic juice and trypsin flow was obtained (Le Dréan et al., 1999). The flow increased just before the meal and over the next 15 minutes. This increase reflects a cephalic phase of digestion due to excitation by environmental stimuli in the farm conditions, since an unexpected meal given earlier in the day, without environmental disturbances, did not induce such a preprandial increase in pancreatic flow. This adaptation seems to be important to prepare the GI tract for digestion of diet and will be discussed further.

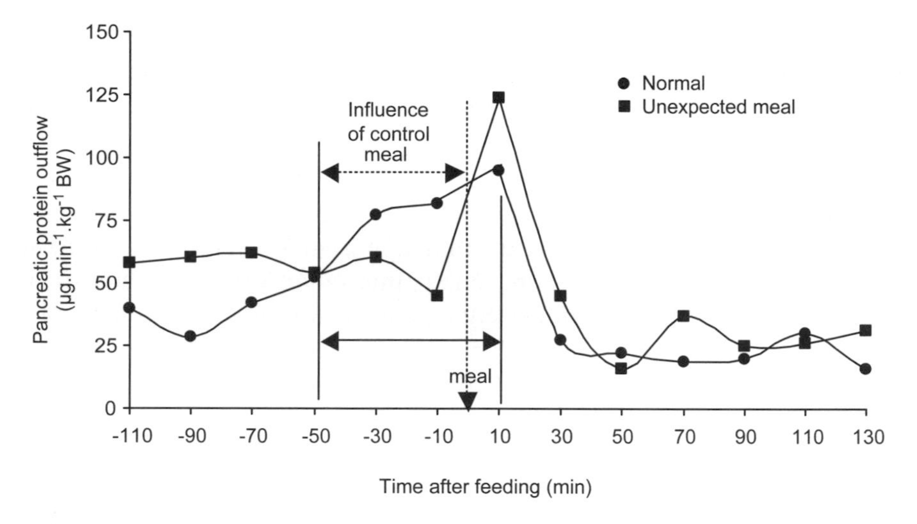

Figure 9.6 Pancreatic protein outflow, after a meal based on milk proteins in preruminant calves under normal [control] conditions or with an unexpected meal (Le Dréan *et al.*, 2000).

Another factor is the season of rearing. We showed that lipase, and especially the chymotrypsin content in pancreatic tissue, was lower in veal calves slaughtered in March compared with those slaughtered in September (Guilloteau and Le Huërou-Luron, 1996). Results from calves reared over the winter-spring period correspond with observations on the rearing conditions - greater faeces production, possibly due to a reduction in digestion of the diet and more digestive problems. Pathophysiological factors need to be taken into account, so during our nutrition studies we carried out some observations in calves suffering from digestive disorders. In regard to the secretion of abomasal juice in calves with a gastric pouch, a marked sustained increase in gastric juice pH due to reduced acid secretion was recorded 1 to 2 days before the clinical signs of gastric disorders (Guilloteau *et al.*, unpublished data). In calves chronically cannulated for pancreatic juice collection, it was observed that juice secretion sharply decreased during the 3 to 4 days following a rise in body temperature or during diarrhoea. Chymotrypsin content was reduced in the pancreas tissue, and the colipase/lipase ratio was disturbed in calves suffering from digestive problems in comparison with the healthy animals (Guilloteau *et al.*, unpublished data). Accordingly, the administration of LT colitoxin into the duodenum resulted in an immediate reduction in pancreatic juice secretion accompanied by gut motility disturbance and a rise in rectal temperature by 0.5° C (Zabielski *et al.*, 1999). Interestingly in that study, no clinical signs of diarrhoea were observed and pancreatic secretion and gut motility were normalized within several hours following administration of LT.

FACTORS AFFECTING DIGESTIVE SECRETIONS IN THE RUMINANT CALF (EFFECT OF WEANING)

Weaning is the period of transition from a liquid to a solid diet. It can also be defined by the separation of the young from its mother. Generally, it is a progressive period that results in a number of anatomical and physiological changes. The rumen and other forestomachs develop and become functional during this time, and some digestive processes (mainly of lipids and carbohydrates) depend on microorganisms that colonise the reticulo-rumen. In contrast to a preruminant calf, microbial products in the ruminant calf make an important contribution to the digesta arriving at the abomasum, constituting approximately 56 % (from 7 to 88 %) of total protein. Nevertheless, a significant proportion of dietary components can arrive at the abomasum undigested, and constitute approximately 38 % (from 10 to 60 %) of total protein in the abomasal digesta (Toullec and Lallès, 1995). These contributions, along with endogenous protein (6%), are digested in the abomasum as in the preruminant.

Weaning affects the pancreatic secretions. Early weaning (15 - 42 days of age) of calves leads to decreased preprandial pancreatic protein output and trypsin activity without affecting the pancreatic response to food (Pierzynowski *et al.*, 1991). After weaning, daily pancreatic secretion doubles and pancreatic protein secretion increases about three-fold (Zabielski *et al.*, 1997b). Pancreatic amylase activity apparently depends on feed type; it is low while the calf receives milk, and increases several-fold when the calf starts to consume solid food. The highest trypsin activity in the exocrine secretion was observed during the first two weeks of life, when the animals received cows' milk. After their diet was changed to a milk replacer, trypsin activity decreased and only started to increase 3 to 4 weeks later, in parallel with the increase in the amount of solid food ingested (Pierzynowski *et al.*, 1991).

In ruminants, digesta entering the abomasum has a higher dry matter content, due to the ingestion of a solid diet and the re-absorption of water in the omasum. Consequently, gastric outflow of digesta is lower and more regular in ruminants than in preruminants, and overall digestive secretions increase (Table 9.1). Saliva contains not only water, but also urea and rumen buffers to ensure optimal environmental conditions for maximal action of the micro-organisms. Nitrogen can be reused by the way of urea recycling in saliva. The secretion of abomasal juice, in terms of its nitrogen and mineral components, is more important in ruminants than in preruminants. The total production is greater and the chemical composition is different (Table 9.1). It contains less dry matter, mainly due to less ash, and more nitrogen, but only 20 % is in form of amino acids suggesting that the ruminant calf produces less gastric enzymes than the preruminant calf. Abomasal pepsin secretion increases during weaning, and abomasal chymosin secretion decreases. Chymosin activity remains at a low level and never disappears in the adult ruminant, which contrasts with observations in most other mammalian species. Under experimental conditions, it was possible to reverse food transition (from solid to liquid diet), thereby transforming ruminant calves into preruminants. In these calves, abomasal pepsin was maintained at the same level, whilst chymosin tended to increase (Guilloteau, 1986), suggesting a great adaptative capacity in relation to diet in calves.

The lysozyme family of enzymes is also produced by the abomasum. Lysozymes are produced from birth, and their levels decrease with age in preruminant calves. After weaning, lysozyme levels increase again (Le Huërou *et al.*, 1992a). In fact, lysozymes are typical for herbivorous species, in particular ruminants like bovine and ovine species (Dobson *et al.*, 1984). In the preruminant calf, these enzymes provide antibacterial control of the GI tract, and participate in hydrolysing peptidoglycans, which are components of bacterial cell walls that are present in milk replacers when part of the dietary protein comes from unicellular microorganisms (Guilloteau and Toullec, 1980; Guilloteau *et al.*, 1980). The role of lysozymes is, however, markedly increased in ruminants since microbial proteins entering the abomasum represent a major proportion of total protein.

In ruminants, pancreatic ribonuclease also seems to play a major role, since micro-organisms are rich in nucleic acids. Thus pancreatic ribonuclease content increases after weaning (Le Huërou *et al.*, 1992a). Pancreatic lipase appears to be high in the ruminant calf, but its action appears to be limited by low colipase level, and the same is observed after weaning. This may limit digestion when too many lipids enter the proximal intestine. Finally, the ruminant has a good enzymatic system to digest carbohydrates, since pancreatic amylase, as well as intestinal maltase and isomaltase significantly increase after weaning (Le Huërou *et al.*, 1992b). However, this is not true for structural carbohydrates that escape microbial digestion in the forestomach.

Besides differences in the composition of digestive secretions, their pattern of secretion is also changed in ruminant animals. Generally, in ruminants secretion is more regular, thereby reducing the differences between preprandial and postprandial levels (Le Dréan *et al.*, 1997). This phenomenon is related to the more regular and continuous outflow of digesta, firstly from the rumen and secondly from the abomasum, but also to the increased number of meals. There are very few data about the other digestive secretions, such as mucins and antibacterial proteins. For example, due to the physical effects of fibre (Montagne *et al.*, 2003), it is expected that excretion of mucins is will be higher in ruminants than in preruminants.

Some aspects of regulation of digestive secretion

The digestive secretions are regulated by complex neurohormonal mechanisms (involving the gut regulatory peptides, e.g., cholecystokinin (CCK), and the vagal nerves) which, in general, seem to be similar in the preruminant and ruminant calf although the relative importance of particular mechanisms may vary according to the stage of development of the forestomach. Food intake and the ingested food components provide the major input into the neurohormonal mechanisms that control the digestive secretions. These mechanisms are a part of the brain-gut axis that regulates various GI functions involving food intake, motility, secretion, absorption, cell protection and many others (Dockray 2003, Konturek *et al.*, 2003 and 2004). Within the axis, a dorsal vagal complex plays an important role in coordination and integration with the signals coming from behind the GI tract.

In ruminants, a more regular digesta flow is responsible, as mentioned above, for more uniform GI secretions during the pre- and post-prandial periods. Also, the effects of different feeds result in less dramatic modifications of digestive secretions. However, studies with exogenous nutrients administered directly into the abomasum or small intestine showed that in ruminants the regulatory mechanism characteristics of preruminants (and monogastric animals) still exist, but are masked by the luminal regulations caused by high digesta flow in the GI tract (Zabielski and Pierzynowski, 2001). Besides the regulation of the exocrine pancreas, little is known about the regulation of the other digestive secretions in preruminant and ruminant calves.

CEPHALIC PHASE

Milk is a weak stimulant of pancreatic secretion in preruminant calves (Pierzynowski *et al.*, 1991, Zabielski *et al.* 1997b) if we compare the effects of feeding in calves and adult monogastric animals such as dogs. Only a brief increase in pancreatic secretion is observed in calves during and after suckling milk, whereas in dogs there is an elevation in pancreatic secretion lasting several hours (Konturek *et al.*, 1986). Apparently the majority of this effect is due to cephalic stimulation via the vagal nerves in preruminant calves, since it can be abolished by blocking vagal conductivity by cooling the vagi, and infusions of atropine, a muscarinic non-specific receptor antagonist (Pierzynowski *et al.*, 1992) or 4-DAMP, a muscarinic M_3-specific receptor antagonist (Lesniewska *et al.*, 1996). Moreover, a similar cephalic stimulation of pancreatic secretion was observed after giving an empty feeding-bucket to a calf for couple of minutes (Lesniewska and Zabielski, unpublished data). The administration of tarazepide, a CCK_1 receptor antagonist, into the duodenum reduced pancreatic protein output following milk feeding (Zabielski *et al.*, 1998a), suggesting that CCK released by the duodenal mucosa contributed to the response. The cephalic phase of pancreatic secretion is present in suckling calves fed liquid food (milk or milk replacer) and, surprisingly, it is more evident in the morning than in the evening (Zabielski *et al.*, 1997b). The mechanism behind this effect remains unclear.

The cephalic phase of secretion increases during the first few days of life (Figure 9.3). The greatest response is observed when the calves are 4 weeks of age and decreases thereafter. The cephalic phase is not significant in ruminant calves receiving solid food. However, at the same time, the inter-digestive or daily secretions increase with age (Pierzynowski *et al.*, 1991, Zabielski *et al.*, 1997b). This phenomenon may be explained by a continuous flow of acidic digesta through the abomasum and small intestine during the preprandial period in ruminating calves that might permanently stimulate the exocrine pancreas to a certain extent. Therefore, additional cephalic stimulation during solid food ingestion can not substantially increase the secretion of juice over the already stimulated preprandial level. This was also observed in adult sheep with blocked flow of digesta into the duodenum (Kato *et al.*, 1984). Under these conditions pancreatic secretion was stimulated much more by food ingestion, than it was in sheep with undisturbed flow of digesta. Moreover, it was found that giving milk to weaned early ruminant calves can still produce a significant increase in pancreatic protein secretion, suggesting the involvement of different stimulatory mechanisms during suckling liquid food and ingestion of solid food (Lesniewska and Zabielski – unpublished observations).

The other two phases of pancreatic secretion classically described by Pavlov, the gastric and intestinal phases, are not manifested in preruminant calves. As shown originally by McCormick and Stewart (1967), the peak secretion associated with milk suckling is preceeded by a 2 to 3 hour profound reduction in pancreatic secretion below the preprandial level. This phenomenon may be related to: (1) a rapid increase in duodenal pH due to the arrival of large amounts of milk, thereby reducing some of the secretion that depends on secretin, or (2) a decreased vagal input after food ingestion as evidenced by a postprandial drop in plasma PP, an established marker of vagal tone (Toullec *et al.*, 1992, Zabielski *et al.*, 1998b). The third possible reason for this unusual event could be a low secretory capacity of the digestive organs in neonates (Pierzynowski and Zabielski 1999). To support this, a continuous intravenous infusion of a pharmacological dose of CCK-8 was shown to stimulate pancreatic secretion only for about 30 minutes and then the secretion fell despite continued stimulation (Zabielski *et al.*, 1995b).

PERIODIC PANCREATIC SECRETION AND ITS ASSOCIATION WITH DUODENAL MOTILITY

Pancreatic secretion in preruminant calves is closely related to duodenal motility and manifests clear-cut oscillations in concert with the MMC that appears in the duodenum (Zabielski *et al.*, 1993). Pancreatic secretions, namely the volume flow, bicarbonate output and pancreatic enzyme activity, are low (secretory trough) when a phase 1 of MMC (characterised by lack of action potentials and muscle contractions) is present in the duodenum, and high (secretory peak) when a phase 2 of MMC (manifested by irregular action potentials and intense muscle contractions) is recorded. Secretion returns to a nadir during phase 3 of duodenal MMC. Boldyreff (1911) was the first to describe this basic phenomenon of the GI tract involving gastric and intestinal motility and secretion, and biliary and pancreatic secretions. It was demonstrated later in humans (Vantrappen *et al.*, 1979) and other animal species (DiMagno *et al.*, 1979, Abello *et al.*, 1988, Onaga *et al.*, 1993, Kiela *et al.*, 1996). Our contribution was, however, to indicate it and follow its development in cattle from the neonate to ruminant (Zabielski *et al.*, 1993, Zabielski *et al.*, 1997a, Zabielski *et al.*, 2002).

The periodic pattern in the preruminant calf is transiently disturbed by feeding (1-2 h) as discussed in relation to the cephalic phase. In the ruminant calf the effect of feeding is even less expressed (Zabielski *et al.*, 1997b). Thus, the regulation of pancreatic secretion (and possibly the other GI secretions) could be considered as the regulation of the periodic activity of the GI tract in regard to cycle length and amplitude, according to nutritional and neurohormonal stimulations (Zabielski *et al.*, 1995a). Using electromagnetic flow probes implanted in the duodenum of preruminant calves, Girard and Sissons (1992) demonstrated that the duodenal MMC is well synchronized with the rate of digesta flow. During the phase 1 of MMC, the flow of digesta was halted; it increased during the phase 2, and was maximal just before the phase 3 of MMC.

NEUROHORMONAL CONTROL OF GI SECRETIONS

In preruminant calves, as in monogastric animals, the control of GI function is under the brain-gut axis constituting of a concerted neurohormonal action of vagal nerves and gut regulatory peptides (Konturek *et al.*, 2003). In the calf, our understanding of the neurohormonal control of the exocrine pancreas is the most advanced. The vagal nerves are responsible for generation of the periodic pancreatic activity since atropine and 4-DAMP can completely block this activity, and cold vagal blockade can almost completely block the periodic pancreatic fluctuations in the preruminant calf (Zabielski *et al.*, 1993 and 1995a, Zabielski and Naruse, 1999). The cephalic phase of pancreatic secretion is also entirely abolished by means of these blockades, as are the pancreatic responses to intra-duodenal stimulation with soyabean extract and hydrochloric acid (Pierzynowski *et al.*, 1992, Zabielski *et al.*, 1992). It has also been shown that the action of secretin and CCK on the secretion of pancreatic juice is dependent on the vagal nerves. The effect of secretin is blocked, and that of CCK markedly decreased, by cold vagal blockade (Zabielski *et al.*, 1992). Normal responses to secretin and CCK reappear soon after the vagi are rewarmed.

Local duodeno-pancreatic reflexes regulate the amount of secreted pancreatic juice, but appear to be less relevant regulators of the periodic activity of the exocrine pancreas in calves. Duodenal perfusion with lidocaine and intra-duodenal installation of capsaicin did not abolish the PPS cycles, but decreased the secretion per cycle; the response to intra-duodenal HCl stimulation and diversion of the gastro-duodenal contents, produced similar effects (Zabielski *et al.*, 1995a). These effects were, however, considerably weaker than those of the cold vagal blockade. Experiments in adult sheep and cows (Magee, 1961; Pierzynowski *et al.*, 1988) showed that, in contrast to preruminant calves, pH of the duodenal contents and duodenal digesta flow are essential regulators of pancreatic secretion. Therefore, it seems that the local reflexes are present but immature in preruminant calves and develop after weaning (Zabielski and Pierzynowski 2001).

In calves, concentrations of gut regulatory peptides in blood plasma change during postnatal development (Guilloteau *et al.*, 1992, Toullec *et al.*, 1992) and the composition of food (protein source) is important (Guilloteau *et al.*, 1986, Le Huërou-Luron *et al.*, 1998, Zabielski *et al.*, 1998b). Earlier studies on the presence of CCK_1 and CCK_2 receptors in the calf pancreas (Le Meuth *et al.*, 1993) were recently revisited with the immunohistochemistry technique (Morisset *et al.*, 2003). Gastrin is a newly proposed stimulator that might act via a CCK_2 and CCK_1 receptor in preruminant calves (Le Dréan *et al.*, 1999, Zabielski *et al.*, 2004). However, the CCK_1 and CCK_2 receptors located in the mucosa of the duodenum and proximal jejunum (Figure 9.7), rather than those on the pancreatic acini, are the targets for gastrin under physiological conditions (Zabielski *et al.*, 2002 and 2004). In contrast to gastrin, CCK and secretin are well-established stimulators of pancreatic juice secretion, and PP and somatostatin inhibitors. The role of other gut-regulatory peptides (e.g., VIP, pituitary adenylate cyclase activating polypeptide (PACAP), motilin) seems to be less relevant (Guilloteau *et al.*, 2002). The gut-regulatory peptides seem to work through a complex mechanism, in which an indirect effect, i.e. not related to specific receptors located on the pancreatic acinar cells, plays an important role. At least in the case of CCK, gastrin, secretin and VIP, the intermediation of vagal nerves has been demonstrated (Zabielski *et al.*, 1992, 1994 and 2004). In the case of CCK (Figure 9.7), in addition to a direct receptor effect on pancreatic acinar cells, an indirect mechanism that controls pancreatic juice secretion via a long vago-vagal reflex has been found in the duodenal mucosa of preruminant calves (Zabielski *et al.*, 1995b, and 1998a). These observations are compatible with findings in rats (Li and Owyang, 1992, Owyang, 1996) and pigs (Kiela *et al.*, 1996). Our studies in preruminant calves indicated that the CCK_1 receptor is present on the nerve terminals in the proximal small intestinal mucosa (Zabielski *et al.*, 2002). Consequently, selective block of this receptor by CCK_1 receptor antagonist reduces the secretion of pancreatic juice, and intra-duodenal administration of CCK stimulates the secretion of pancreatic juice in a dose-related manner via an atropine-sensitive pathway (Zabielski *et al.*, 1995a and 1998a). In conclusion, it seems that a brain-gut axis involving duodenal CCK and vagal nerves plays a crucial role in regulating pancreatic exocrine secretion (Konturek *et al.*, 2003; Zabielski, 2003). Recently, evidence has grown for the role of the brain-gut axis in controlling other GI functions involving gastric secretion, and gastric and pancreatic cytoprotection. This stimulates further research, as does new evidence concerning regulation of GI functions in ruminant species.

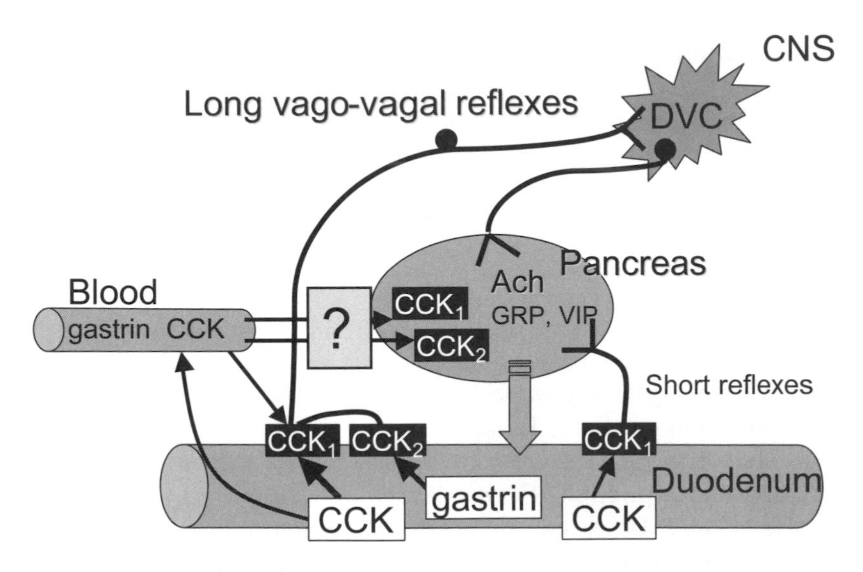

CNS : Central nervous system; DVC : Dorsal-vagal complex; CCK : Cholecystokinin; CCK_1 and CCK_2 : Receptors to CCK and/or gastrin of type 1 (or A) and type 2 (or B), respectively; Ach : Acetylcholine; GRP: Gastrin releasing peptide; VIP : Vasoactive intestinal polypeptide

Figure 9.7 Neurohormonal control of pancreatic juice secretion involving CCK, luminal gastrin and vagal nerves as an example of brain-gut axis regulation. Role of circulating CCK and gastrin in physiological doses in the regulation of pancreatic secretion is small (from our experiments and bibliographical data reported in the text).

General conclusion

Digestive secretions are key factors for digestion in the preruminant as well as the ruminant calf. Three major points are emphasised: (1) there are complementary actions between digestive secretions, which must be present in sufficient quantities and at the correct time in given points of the GI tract; (2) these digestive secretions contribute to the nutritional value of the diet, and mucin assay could be a good marker to evaluate the health value of the diet; (3) there is great adaptation of the digestive secretions in relation to environmental factors, which is linked to complex regulation mechanisms. Understanding digestive secretions might help to suggest alternative solutions to the utilisation of antibiotics as growth promoters, which will be banned in European Union from January 1, 2006. Currently intensive research is being performed to seek alternatives, e.g., acids, exogenous enzymes, prebiotics, probiotics and many others. Another attractive idea is to model the relationships between digestive secretions and requirements during critical periods of calf rearing. This review we hope will give some impact to this topic.

Acknowledgements

The authors would like to thank J. Chevalier who kindly agreed to produce the illustrations for this review.

References

Abello J., Laplace J.P. and Corring T. (1988). Biliary and pancreatic secretory component of the migrating myoelectric complex in the pig. Effect on intraduodenal pH. *Reproduction Nutrition Development,* **28**, 953-967.

Andrews F.M., Jenkins C.C., Frazier D. and Blackford J.T. (1992). Gastric secretion in foals: measurement by nasogastric intubation with constant infusion and aspiration. *Equine Veterinary Journal*, **13 (Sup.)**, 75-79.

Baintner K. (1989). Intestinal absorption of macromolecules and immune transmission from mother to young. CRC Press International, Florida.

Biernat M., Zabielski R., Sysa P., Sosak-Swiderska B., Le Huërou-Luron I. and Guilloteau P. (1999). Small intestinal and pancreas microstructures are modified by an intraduodenal CCK-A receptor antagonist administration in neonatal calves. *Regulatory Peptides,* **85**, 77-85.

Bittrich S., Philipona C., Hammon H.R., Romé V., Guilloteau P., Blum J.W. (2004). Preterm as compared with full-term neonatal calves are characterized by morphological and functional immaturity of the small intestine. *Journal of Dairy Science,* **87**, 1786-1795.

Blattler U., Hammon H.M., Morel C., Philipona C., Rauprich A., Romé V., Le Huërou-Luron I., Guilloteau P. and Blum J.W., 2001. Feeding colostrum, its composition and feeding duration variably modify proliferation and morphology of the intestine and digestive enzyme activities of neonatal calves. *Journal of Nutrition,* **131**, 1256-1263.

Boldyreff, W. (1911). Einige neue Seiten der Tatigeit des Pancreas. *Ergebnisse der Physiologie,* **11**, 121-217.

Cornou-Le Dréan G. (1997). Basal and postprandial pancreatic secretion in calf. Implication of cholecystokinin and gastrin, and of their receptors (in french). *PhD thesis*, Université Rennes I, N° 1813, 172pp.

DiMagno E.P., Hendricks J.C., Go V.L.W. and Dozois R.R. (1979). Relationships among canine fasting pancreatic and biliary secretions, pancreatic duct pressure, and duodenal phase III motor activity - Boldyreff revisited. *Digestive Diseases Science*, **24**, 689-693.

Dobson D.E., Prager E.M. and Wilson A.C. (1984). Stomach lysosyme of ruminants. I. Distribution and catalytic properties. *Journal of Biology and chemistry*, **259**, 11607-11616.

Dockray G.J. (2003). Luminal sensing in the gut: an overview. *Journal of Physiology and Pharmacology*, **54 (Sup. 4)**, 9-17.

Fujita M., Reinhart F. and Neutra M. (1990). Convergence of apical and basolateral endocytic pathways at the apical late endosomes in absorptive cells of suckling rat ileum in vivo. *Journal of Cell Science,* **97**, 394.

Fuller M.F. and Cadenhead A. (1991). Effect of the amount and composition of the diet on galactosamine flow from the small intestine. In *Proceeding of 5th International Symposium on Digestive Physiology in Pigs-1991*, pp 330-333: Edited by M.W. Verstegen, J. Huisman and den L.A. Hartog.

Gestin M., Desbois C., Le Huërou-Luron I., Romé V., Le Dréan G., Lengagne T., Roger L., Mendy F. and Guilloteau P. (1997a). In vitro hydrolysis by pancreatic elastases I and II reduces ß–lactoglobulin antigenicity (in French). *Le Lait,* **77**, 399-409.

Gestin M., Le Huërou-Luron I., Peiniau J., Thioulouse E., Desbois C., Le Dréan G., Feldman D., Aumaitre A. and Guilloteau P., (1997b). Method of measurement of pancreatic elastase II activity and postnatal development of proteases in human duodenal juice and bovine and porcine pancreatic tissue. *Digestive Disease and Science,* **42**, 1302-1311.

Girard C. L. and Sissons J. W. (1992). The role of migrating myoelectric complexes in the regulation of digesta transport in the preruminant calf. *Canadian Journal of Physiology and Pharmacology,* **70**, 1142-1147.

Grongnet J.F., Grongnet-Pinchon E., Levieux D., Piot M. and Lareynie J. (1986). Newborn calf intestinal absorption of immunoglobulins extracted from colostrum. *Reproduction Nutrition and Développement,* **26**, 731-743.

Grongnet J.F., Patureau-Mirand P., Toullec R. and Prugnaud J. (1981).Utilisation of milk and whey proteins by the young preruminant calf. Influence of age and whey protein denaturation (in French). *Annales de Zootechnie*, **30**, 443-464.

Guilloteau P. (1986). Protein digestion in preruminant calf (in French). *PhD thesis*, Université P. et M. Curie, Paris VI, 224pp.

Guilloteau P., Biernat M., Wolinski J. and Zabielski R. (2002). Gut regulatory peptides and hormones of the small intestine. In *Biology of the Intestine in Growing Animals*, pp 271-324. Edited by R. Zabielski; B. Gregory, B.

Weström,. Elsevier Science, Amsterdam.

Guilloteau P., Chayvialle J.A., Toullec R., Grongnet J.F., Bernard C., 1992. Early-life patterns of plasma gut regulatory peptide levels in calves: Effects of the first meals. *Biology of the Neonate,* **61**, 103-109.

Guilloteau P., Corring T., Chayvialle J.A., Bernard C., Sissons J.W., Toullec R., 1986. Effect of soya protein on digestive enzymes, gut hormone and anti-soya antibody plasma levels in the preruminant calf. *Reproduction Nutrition and Développement,* **26,** 717-728.

Guilloteau P., Corring T., Garnot P., Martin P., Toullec R., Durand G., (1983). Effect of age and weaning on enzyme activities of abomasum and pancreas of the lamb. *Journal of Dairy Science,* **66**, 2373-2385.

Guilloteau P., Corring T., Toullec R., Robelin J. (1984). Enzyme potentialities of the abomasum and pancreas of the calf. I- Effect of age in the preruminant. *Reproduction Nutrition and Développement,* **23**, 161-173.

Guilloteau P., Corring T., Toullec R., Villette Y. and Robelin J., (1985). Abomasum and pancreas enzymes in the newborn ruminant : effects of species, breed, sex and weight. *Nutrition Report International,* **31**, 1231-1236.

Guilloteau P., and Le Calvé J.L. (1977). Technique de réalisation d'une poche abomasale chez le veau en vue due l'obtention de sue gastrique pur. *Annales de Biologie Animale Biochimie Biophysique*, **17**, 1047-1060.

Guilloteau P., Le Huërou-Luron I., Romé V. and Plodari M. (1999). Nutrient absorption is related to quantity of pancreatic enzyme secretion : preliminary results. *South African Journal of Animal Science,* **29**, 241-242.

Guilloteau P. and Le Huërou-Luron I. (1996). Pancreatic secretions and their regulation in the preruminant calf. In *Veal perspectives to the year 2000*, pp 169-189. Fédération de la Vitellerie Française, Paris.

Guilloteau P., Le Huërou-Luron I., Malbert C.H. and Toullec R. (1995). Digestive secretions and their regulation (in French). In *Nutrition des Ruminants Domestiques*, pp489-526. Edited by R. Jarrige, Y. Ruckebusch, C. Demarquilly, M.H. Farce and, M. Journet. INRA Editions, Paris.

Guilloteau P., Paruelle J.L., Toullec R. and Mathieu C.M. (1975). Utilisation of proteins by the preruminant calf. III - Influence of replacement of milk powder proteins by those of fish, on the gastric emptying (in French). *Annales de Zootechnie,* **24**, 243-253.

Guilloteau P., Patureau-Mirand P., Toullec R. and Prugnaud J., (1980). Digestion of milk protein and methanol grown protein in the preruminant calf. II - Amino acid composition of ileal digesta and faeces and blood levels of free amino acids. *Reproduction Nutrition and Développement,* **20**, 615-629.

Guilloteau P., Toullec R., (1980). Digestion of milk protein and methanol grown bacteria protein in the preruminant calf. I - Kinetics and balance in the terminal small intestine and faecal balance. *Reproduction Nutrition and Développement,* **20**, 601-613.

Guilloteau P., Toullec R., Sauvant D. and Paruelle J.L. (1979). Utilisation of proteins by the preruminant calf. III - Influence of replacement of milk powder proteins by those of soya or faba bean, on gastric emptying (in French). *Annales de Zootechnie,* **28**,1-17.

Guilloteau P., Toullec R., Patureau-Mirand P. and Prugnaud J. (1981). Importance of the abomasum in digestion in the preruminant calf. *Reproduction Nutrition and Dévelopment,* **21**, 885-899.

Hoskins L.C. (1984). Mucin degradation by enteric bacteria: ecological aspects and implications for bacterial attachment to gut mucosa. Edited by E.C. Boedeker. CRC Press, Boca Raton.

Kato S., Usami M., Ushijima J. (1984). The effect of feeding on pancreatic exocrine secretion in sheep. Japanese Journal of Zootechnical Science, **55**, 973-977.

Kiela P., Zabielski R., Podgurniak P., Midura M., Barej W., Gregory P.C. and Pierzynowski S.G. (1996). Cholecystokinin-8 and Vasoactive Intestinal Polypeptide stimulate exocrine pancreatic secretion via duodenally mediated mechanisms in the conscious pig. *Experimental Physiology* **81,** 375-384.

Konturek S.J., Thor P.J., Bilski J., Bielañski W., Laskiewicz J. (1986). Relationships between duodenal motility and pancreatic secretion in fasted and fed dogs. *American Journal of Physiology*, **250**, G570-G574.

Konturek S.J., Zabielski R., Konturek J.W., Czarnecki J. (2003). Neuroendocrinology of the pancreas; role of brain-gut axis in pancreatic secretion. *European Journal of Pharmacology,* **481**, 1-14.

Konturek S.J., Konturek J.W., Pawlik T., Brzozowski T. (2004). Brain-gut axis and its role in the control of food intake. *Journal of Physiology and Pharmacology*, **55**, 137-154.

Lallès J.P. (1993). Nutritional and antinutritional aspects of soyabean and field pea proteins used in veal calf production. *Livestock Production Science*, **34**, 181-202.

Laubitz D., Zabielski R., Woliński J., Nieminuszczy J. and Grzesiuk E. (2003). Physiological and chemical characteristics of antibacterial activity polypeptide of pancreatic juice. *Journal of Physiology and Pharmacology,* **54**, 283-290.

Le Dréan G., Le Huërou-Luron I., Chayvialle J.A., Philouze-Romé V., Gestin M., Bernard C., Toullec R. and Guilloteau P. (1997). Kinetics of pancreatic exocrine secretion and plasma gut regulatory peptide release in response to feeding in preruminant and ruminant calves. *Comparative Biochemistry and Physiology,* **117A**, 245-255.

Le Dréan G., Le Huërou-Luron I., Gestin M., Desbois C., Romé V., Bernard C., Dufresne M., Moroder L., Gully D., Chayvialle J.A., Fourmy D. and Guilloteau P., (1999). Exogenous CCK and gastrin stimulate pancreatic exocrine secretion via CCK-A but also CCK-B/gastrin receptors in the calf. *Pflügers Archiv – European Journal of Physiology,* **438**, 86-93.

Le Dréan G., Le Huërou-Luron I., Gestin M., Romé V., Bernard C., Chayvialle J.A., Fourmy D. and Guilloteau P. (2000). Pancreatic secretory response to feeding in the calf : CCK-A receptors, but not CCK-B/gastrin receptors are involved. *Canadian Journal of Physiology and Pharmacology,* **78**, 813-819.

Le Dréan G., Le Huërou-Luron I., Gestin M., Romé V., Plodari M., Guilloteau P., Bernard C. and Chayvialle J.A. (1998). Comparison of the kinetics of pancreatic secretion and gut regulatory peptides in the plasma of preruminant calf fed milk or soyabean protein. *Journal of Dairy Science,* **81**, 1313-1321.

Le Huërou I., Guilloteau P., Wicker C., Mouats A., Chayvialle J.A., Bernard C., Burton J., Toullec R., Puigserver A. (1992b). Activity distribution of seven digestive enzymes along small intestine in calves during development and weaning. *Digestive Diseases and Science,* **37**, 40-46.

Le Huërou-Luron I., Gestin M., Le Dréan G., Romé V., Bernard C., Chayvialle J.A., Guilloteau P. (1998). Source of dietary protein influences kinetics of plasma gut regulatory peptide concentration in response to feeding in preruminant calves. *Comparative Biochemistry and Physiology,* **119A**, 817-824.

Le Huërou-Luron I., Guilloteau P., Wicker-Planquart C., Chayvialle J.A., Burton J., Mouats A., Toullec R. and Puigserver A. (1992a). Gastric and pancreatic enzyme activities and their relationship with some gut regulatory peptides during postnatal development and weaning in the calf. *Journal of Nutrition,* **122**, 1434-1445.

Le Meuth V., Philouze-Romé V., Le Huërou-Luron I., Formal M., Vaysse N., Gespach C., Guilloteau P., Fourmy D. (1993). Differential expression of A- and B-subtypes of cholecystokinin/gastrin receptors in the developing calf pancreas. *Endocrinology,* **133**, 1182-1191.

Lesniewska V, Ceregrzyn M, Pierzynowski S.G. and Zabielski R. (1996). Effect of M_3 muscarinic antagonist (4-DAMP) administration on non-stimulated and cholecystokinin-8 stimulated secretion of pancreatic juice and duodenal motility in conscious calves. *Regulatory Peptides,* **64**, 108.

Li Y., Owyang C., (1992). Cholecystokinin at physiologic levels stimulates pancreatic enzyme secretion via an afferent vagal pathway originating from the duodenum. *Regulatory Peptides,* **40**, 196.

Magee D.F. (1961). An investigation into the external secretion of the pancreas in sheep. *Journal of Physiology*, **158**, 132-143.

Magee D.F. and Naruse S. (1983). Neural control of periodic secretion of the pancreas and the stomach in fasting dogs. *Journal of Physiology,* **344**, 153-160.

McCormick R.J. and Stewart W.E. (1967) Pancreatic secretion in the bovine calf. *Journal of Dairy Science*, **50**, 568-571.

Montagne, L. (2000). Effect of dietary protein on endogenous protein mucin, and mucosa of the small intestine, in preruminant calf. *PhD thesis*, Ecole Nationale Supérieure Agronomique de Rennes, N° 13-B-119, 200pp.

Montagne, L., Pluske, J. R., & Hampson, D. J. (2003) A review of interactions between dietary fibre and the intestinal mucosa, and their consequences on digestive health in young non-ruminant animals. *Animal Feed Science and Technology*, **108**, 95-117.

Montagne L., Toullec R. and Lallès J.-P. (2000a) Calf intestinal mucin: isolation, partial characterization, and measurement in ileal digesta with an enzyme-linked immunosorbent assay. *Journal of Dairy Science*, **83**, 507-517.

Montagne L. and Lallès J.P. (2000). Digestion of plant protein sources in the preruminant calf. Quantification of endogenous protein and importance of mucin (in French). *Productions Animales*, **13**, 315-324.

Montagne L., Toullec R. and Lallès J.P. (2001). Intestinal digestion of dietary and endogenous proteins along the small intestine of calves fed soyabean or potato. *Journal of Animal Science,***79**, 19-30.

Montagne L., Toullec R., Formal M. and Lallès J.P. (2000b) Influence of dietary protein level and origin on the flow of mucin along the small intestine of the preruminant calf. *Journal of Dairy Science*, **83**, 2820-2828.

Morisset J., Bourassa J., Tessier P., Lainé J., Lessard M., Romé V., Guilloteau P. (2003). Revisiting cholecystokinin control of pancreatic secretion in calf. *Regulatory Peptides*, **111**,103-109.

Onaga T., Zabielski R., Mineo H. and Kato S. (1993). The temporal coordination of interdigestive pancreatic exocrine secretion and intestinal migrating myoelectric complex in rats. *Proceedings of XXXII Congress of the International Union of Physiological Sciences*, Glasgow, Great Britain, 95.1/P, p 83.

Owyang C. (1996). Physiological mechanisms of cholecystokinin action on pancreatic secretion. *American Journal of Physiology* **271**, G1– G7.

Pierzynowski S.G., Barej W., Mikolajczyk M., Zabielski R. (1988). The influence of light fermented carbohydrates on the exocrine pancreatic secretion in cows. *Journal of Animal Physiology and Animal Nutrition*, **60**, 234-238.

Pierzynowski S.G., Zabielski R. (1999). Gastrointestinal enzymology in young animals – feeding strategies. In *Proceedings of the 5th International Feed Production Conference- 1998*, pp 215-236. Edited by G. Piva, Facolta di Agraria U.C.S.C., Piacenza.

Pierzynowski S.G., Zabielski R., Podgurniak P., Kiela P., Sharma P., Weström B., Kato S., Barej W. (1992). Effect of reversible cold vagal blockade and atropinization on exocrine pancreatic function during liquid food consumption in calves. *Journal of Animal Physiology and Animal Nutrition*, **67**, 268-273.

Pierzynowski S.G., Zabielski R., Weström B., Mikolajczyk M. and Barej W. (1991). Development of the exocrine pancreatic function in chronically cannulated calves from the preweaning period up to early rumination. *Journal of Animal Physiology and Animal Nutrition,* **65**, 165-172.

Rubinstein E., Mark Z., Haspel J., Ben-Ari G., Dreznik Z., Mirelman D., Tadmor A. (1985). Antibacterial activity of the pancreatic fluid. *Gastroenterology,* **88**, 927-932.

Ruckebusch Y. (1977). In *Physiologie, Pharmacologie, Thérapeutique Animales* Edited by Y Ruckebusch. 424p. Maloine, Paris.

Sauter S.N., Roffler B., Philipoma C., Morel C., Romé V., Guilloteau P., Blum J.W. and Hammon H.M. (2003). Intestinal development in neonatal calves: Effects of glucocorticoids and dependance on colostrum feeding. *Biology of the Neonate*, **85**, 94-104.

Ternouth J.H. and Buttle H.L. (1973). Concurrent studies of the flow of digesta in the duodenum and of exocrine pancreatic secretion of calves. The collection of the exocrine pancreatic secretion from a duodenal cannula. *British Journal of Nutrition*, **29**, 387-97.

Toullec R., Chayvialle J.A., Guilloteau P., Bernard C., 1992. Early-life patterns of plasma gut regulatory peptide levels in calves. Effects of age, weaning and feeding. *Comparative Biochemistry and Physiology,* **102**, 203-209.

Toullec R. and Guilloteau P. (1989). Research into the digestive physiology of milk-fed calf. In: *Nutrition and Digestive Physiology of Monogastric Farm Animals*, pp 35-37. Edited by E.J. Van Weerden and J. Huisman. PUDOC, Wageningen.

Toullec R. and Lallès J.P. (1995).Digestion in abomasum and in small intestine (in French). In *Nutrition des Ruminants Domestiques*, pp 528-581. Edited by R. Jarrige, Y. Ruckebusch, C. Demarquilly, M.H. Farce and, M. Journet. INRA Editions, Paris.

Wilsen O., Hellstrom P.M. and Johansson C. (1993). Meal energy density as a determinant of postprandial gastrointestinal adaptation in man. *Scandinavian Journal of Gastroenterology*, **28**, 737-743.

Vantrappen G.R., Peeters T.L. and Janssen J. (1979). The secretory component of the interdigestive migrating motor complex in man. *Scandinavian Journal of Gastroenterology,* **14**, 663-667.

Zabielski R., Barej W., Leœniewska V., Pierzynowski S.G. (1999). Pancreas and upper gut dysfunctions around weaning in pigs and calves. In: *Production Diseases in Farm Animals*, pp 134-144. Edited by Th. Wensing. Wageningen Pers, Wageningen.

Zabielski R. and Naruse S. (1999). Neurohormonal regulation of the exocrine pancreas during postnatal development. In *Biology of the Pancreas in Growing Animals*, pp151-192. Edited by S.G. Pierzynowski and R. Zabielski. Elsevier, Amsterdam.

Zabielski R. (2003). Luminal CCK and its neuronal action on exocrine pancreatic secretion. *Journal of Physiology and Pharmacology,* **54**, 81-94.

Zabielski R., Dardillat C., Le Huërou-Luron I., Bernard C., Chayvialle J.A., Guilloteau P. (1998b). Periodic fluctuations of gut regulatory peptides in phase with the duodenal migrating myoelectric complex in preruminant calf: effect of different sources of dietary protein. *British Journal of Nutrition,* **79**, 287-296.

Zabielski R., Kiela P., Lesniewska V., Krzemiński R., Mikolajczyk M. and Barej W. (1997b) Kinetics of pancreatic juice secretion in relation to duodenal migrating myoelectric complex in preruminant and ruminant calves fed twice daily. *British Journal of Nutrition* **78**, 427-442.

Zabielski R., Kiela P., Onaga T., Mineo H., Gregory P.C., Kato S. (1995a). Effect of neural blockades, gastrointestinal regulatory peptides and diversion of gastroduodenal contents on periodic pancreatic secretion in the preruminant calf. *Canadian Journal of Physiology and Pharmacology* **73**, 1616-1624.

Zabielski R., Lesniewska V., Borlak J., Gregory P.C., Kiela P., Pierzynowski S.G. and Barej W. (1998a). Effects of intraduodenal administration of tarazepide on pancreatic secretion and duodenal EMG in neonatal calves. *Regulatory Peptides,* **78**, 113-123.

Zabielski R., Lesniewska V., Guilloteau P. (1997a). Collection of pancreatic juice in experimental animals. *Reproduction, Nutrition and Development,* **37**, 385-399.

Zabielski R., Morisset J., Podgurniak P., Romé V., Biernat M., Bernard C., Chayvialle J.A., Guilloteau P. (2002). Bovine pancreatic secretion in the first week of life; potential involvement of intenstinal CCK receptors. *Regulatory Peptides,* **103**, 93-104.

Zabielski R., Normand V., Romé V., Woliński J., Chayvialle J.A., Guilloteau P. (2004). The role of luminal gastrin in the regulation of pancreatic juice secretion in preruminant calves. *Regulatory Peptides*, **119**, 169-176.

Zabielski R., Pierzynowski S.G. (2001). Development and regulation of pancreatic juice secretion in cattle. State-of-the-art. *Journal of Animal and Feed Sciences* **10**, 25-45.

Zabielski R., Kato S., Pierzynowski S.G., Mineo H., Podgurniak P., Barej W. (1992). Effect of intraduodenal HCl and soyabean extract on pancreatic juice secretion during atropinization and cold vagal blockade in calves. *Experimental Physiology,* **77**, 807-817.

Zabielski, R., Onaga T., Mineo H., Kato S., Pierzynowski, S.G. (1995b). Intraduodenal cholecystokinin octapeptide (CCK-8) can stimulate pancreatic secretion in the calf. *International Journal of Pancreatology,* **17, 3**, 271-278.

Zabielski R., Onaga T., Mineo H., Kato S. (1993). Periodic fluctuations in pancreatic secretion and duodenal motility investigated in neonatal calves. *Experimental Physiology,* **78**, 675-684.

Zasloff M. (2002). Antimicrobial peptides of multicellular organisms. *Nature*, **415**, 389-395.

Zoppi G., Andreotti G., Pajno-Ferrara F., Njai D.M., Gaburro D. (1972). Exocrine pancreas function in premature and full term neonates. *Pediatric Research*, **6**, 880-886.

10

MANAGEMENT OF DISEASES IN CALVES AND HEIFERS

ROGER W. BLOWEY
Wood Veterinary Group, Gloucester

Introduction

The majority of calf disorders occur in young calves. Nationally mortality is still much higher than should be acceptable, ranging between 2 and 5% of all live-born calves. One of the major problems remains an inadequate intake of colostrum, and this has been covered in Chapter 8. Enteric problems are frequently seen in milk-fed calves and a few of the more common issues will be described here. Pneumonia is a risk from weaning until four months of age, and this will be discussed in some detail. Disease risk is reduced in older animals, although management factors associated with worm control are necessary in grazing animals. The management of the periparturient heifer becomes of major importance in relation to the survival and productivity of that heifer in her first lactation. There is an enormous wastage of first lactation heifers due to poor management in the periparturient transition period, and this will be discussed at some length.

Enteric disorders

Enteric disorders can be subdivided into infectious and managemental. Of the infectious disorders, bacterial conditions such as E. coli, salmonella and increasingly commonly, clostridial infections, cause problems in the first three days of life. At a slightly later age, rotavirus, coronavirus and cryptosporidia become important. In all of these conditions management has a significant impact. Calves can withstand low levels of challenge and should develop an immune response. In calves where the immune response is poor, both rotavirus and coronavirus lead to a sloughing of epithelial cells on the tips of the villi and this produces a poor reabsorption of fluid from the intestine. The overall effect of this is excessively fluid faeces, because fluid uptake is disrupted and the animal becomes dehydrated. There is also a loss of ability to digest milk, which is why, with rotavirus especially, in the early stages the faeces passed tend to be voluminous and pale in colour, associated with large quantities of undigested milk. The initial source of infection will be either from the dam or, in pen systems that have been used repeatedly, infection will result from previous or adjacent calves using the same environment. It is in these latter circumstances that the level of challenge builds up and calves can change from being simply *infected* at 7 to 10 days of age, to becoming *affected*. Control is therefore based on hygiene, preferably with an all in/all out system, and continued feeding of colostrum. Many calf rearers still do not use the full potential of Day-One colostrum; simply using it as a feed rather than as a "medication". Ideally calves need to be fed a small quantity of colostrum (around 5% of total milk intake) for the first three weeks of life. To facilitate this, the farm needs to store colostrum taken from cows on Day One or Two after calving. This should be stored separately from "other milk", which will be milk from Day Three onwards, mastitic milk or milk from cows under antibiotic treatment.

Cryptosporidia is a protozoan parasite that is also transmitted from dam to offspring or horizontally from calf to calf in the environment. Cryptosporidia can cause excessive straining, passage of mucoid faeces and quite intense abdominal pain, resulting in marked weight loss and, in severe cases, death. For this reason, it can sometimes be confused with coccidiosis. It is similarly a disease of hygiene and management, although recently the drug *halofuginome* (*Halocur – Intervet)* has been used quite successfully as a prophylactic.

TREATMENT OF ENTERIC CONDITIONS

Treatment will not be discussed in detail, but it is worth reiterating the main principles:

- withholding milk and replacing it with electrolytes. Electrolyte solutions positively promote the uptake of fluid through the damaged intestinal wall (Scott et al., 2004)

- however, only withhold milk for a maximum of two days. If milk is withheld for any longer than this, then the epithelial cells responsible for secreting lactase and other digestive enzymes become dysfunctional. When milk is reintroduced, the milk in itself can cause scouring because it cannot be digested.

- milk should be reintroduced, even if the calf continues to scour, although in this instance it is preferable to feed milk twice daily and electrolytes twice daily.

- nursing, i.e. keep the calf warm, dry and in a facility where it cannot infect others.

Calf pneumonia

Pneumonia remains a key problem in calf rearing and is something that continues to be difficult to control. There is a range of infections involved, including RSV (respiratory syncitial virus), IBR (infectious bovine rhinotracheitis), PI_3 (parainfluenza 3), BVD (bovine viral diarrhoea), mycoplasmas and bacterial infections, such as Pasteurella and Haemophilus, both of which can be secondary to viral damage or can cause diseases in their own right. Recently Pasteurella haemolytica, biotype A1, has become increasingly important as a primary pathogen. This produces an intense pleurisy and pericarditis in young calves, and leads to increased mortality before vaccination is applicable (Andrews, 2000).

Environment - Ventilation will obviously have an effect, but calves can still develop pneumonia in what appears to be an extremely well ventilated building, with low stocking density. The poorly ventilated, low roofed building, which leads to condensation dripping onto the calves' backs is particularly dangerous. Wherever possible, calves should be allowed to run into outside yards, as this both reduces stocking density and usually provides good ventilation. The amount of sheltered air space should not be restricted or overcrowding will occur during inclement weather (calves do not like driving rain, although they do not object to cold dry weather).

It is vital that bedding is changed regularly. If left to build up and ferment, straw bedding can reach a temperature of 40 °C, generating high humidity, which can predispose to the survival of micro-organisms in the atmosphere. Fermenting bedding also gives off noxious gases that can adversely effect the tracheal mucociliary escalator mechanism, which is important in the natural repulsion of inhaled infections. The frequency of cleaning out will depend on stocking density, but ideally sheds should be cleaned out every four weeks.

Age and source-matching - Animals of different ages will have different levels of colostral immunity and hence show a variable response to exposure to infection. Because of this, it is preferable to keep calves in relatively fixed age groups. Similarly, calves from different sources may be carrying different pathogens and again are best reared as separate groups within their own air space. This is difficult to achieve, but in the pig industry, where pneumonia is probably an even greater problem, the benefits of this have been considerable. Age grouping alone will not totally eliminate pneumonia, as it can still be a problem on some farms that have totally closed units.

Vaccination – Vaccines have improved in recent years, but there is still no single vaccine that is effective against all pathogens. As antibiotics are generally effective in treatment, then presumably vaccination against bacterial components, especially Pasteurella, must be logical, and many Pasteurella vaccines are available.

Importance of the oesophagel groove

The function of the oesophageal groove is to transport milk from the oesophagus into the abomasum without it being spilled into the rumen. Milk that enters the rumen ferments, leading to colic, bloat, scour and poor growth. Affected calves have a compromised immune system and are more susceptible, therefore, to a range of other enteric and pneumonic conditions. Achieving good groove closure in calves fed milk is vital and it is important that those feeding calves understand the significance of groove closure. The calves need to know that they are to be fed, so the sound, sight and smell of food preparation are important. Calves that are fed at irregular intervals, those that are fed warm milk some days, cold others, and those that are fed abruptly, fail to achieve good groove closure; this results in digestive upsets and other infectious disorders.

Clostridial disorders

A range of clostridial diseases affect calves and growing heifers, and they seem to have increased over the past few years. Disorders such as tetanus, blackleg and enterotoxaemia are becoming much more common. Vaccines have improved, however, and as they are relatively inexpensive, they represent good value for money.

The periparturient heifer

The main disease periods for the heifer from birth to calving are firstly in the younger animal, where scouring and pneumonia are seen, and secondly in the transition heifer. Management of diseases during the transition period is critical. An enormous number of heifers are wasted because they are inadequately integrated into the herd. This integration should be social, nutritional and environmental (Blowey 2002). Calving itself is a major risk for many diseases, especially mastitis, lameness and fertility. Prior to calving the natural keratin teat seal at the teat end begins to dissolve and the teat becomes susceptible to the entry of new infections. As parturition approaches, the suppression of the heifer's immune system allows many of these new infections to proliferate, which is why peracute toxic mastitis increases in animals close to calving, even though the colostrum present contains large amounts of antibody. Much of the mastitis in the first 100 days after calving is known to originate from infections in the dry period. Lameness is also an issue. At calving horn growth stops but hoof wear dramatically increases due to longer times spent standing to be milked, standing to feed and, if there has been no cubicle training, waiting to learn where to lie. At the same time the heifer may be mixed with dry cows and the resulting increased social interaction leads to further hoof wear. Relaxation of ligaments within the foot leads to increased movement of the pedal bone, which leads to bruising of the corium and subsequent lameness (Tarlton, 2004). The changes are most pronounced where heifers are not adequately integrated, i.e. where they are not acclimatised to the housing or feeding, or the social hierarchy of the herd that they are about to join. This results in depression of food intake and an increase in a range of post-partum disorders such as ketosis and left displacement of the abomasum.

References

Andrews, A.H. (2000) *The Health of Dairy Cattle*, p. 11. Blackwell Scientific, Oxford.

Blowey, R.W. (2002) *A Veterinary Book for Dairy Farmers*. Old Pond Publishing, Ipswich.

Scott, R.R., Hall, G.A., Jones, P.W. and Morgan, J.H. (2004) *Calf Diarrhoea in Bovine Medicine*, 2nd Ed, p185. Blackwell Scientific, Oxford.

Tarlton, J. (2004) *Proceedings 13th International Symposium on Disorders of the Ruminant Digit*, Maribor, Slovenia, p. 88. http://www.ruminantlameness.com

11

WELFARE IMPLICATIONS OF DAIRY CALF AND HEIFER REARING

ALISTAIR B LAWRENCE, CATHY M DWYER, SUSAN JARVIS AND DAVID ROBERTS
Sustainable Livestock Systems, Scottish Agricultural College, Edinburgh, EH9 3JG

Introduction

Dairy production is dependent on production of calves to replace cull cows at the end of their productive lives. The main management focus of these calves has been over their immediate health although increasingly there is interest in the longer-term implications of rearing on health and fertility (e.g. Offer *et al.*, 2003). However, concern has also been raised about the welfare implications of dairy calf and heifer management practices (e.g. FAWC, 1997). Animal welfare continues to grow in importance at a national and international level (e.g. OIE, 2004), making it increasingly relevant for livestock industries to consider carefully the standards they apply with respect to farm animal welfare at all stages of the production chain.

Animal welfare is a broad concept potentially encompassing scientific, ethical and economic perspectives. The UK's Farm Animal Welfare Council (FAWC) sees animal welfare as including 'physical and mental state and ... good animal welfare implies both fitness and a sense of well-being' (FAWC, 2004). Considering this broad conception of animal welfare it is possible to see that many of the chapters in this book have relevance to welfare, at least in the sense that they deal with aspects of physical health or fitness. There is a need to be cautious, however, in assuming that practices that have evolved to suit particular production systems or even to protect health, are necessarily the optimal solution in terms of overall welfare. Calf and heifer rearing practices, as part of high output systems of dairy production, are extremely divergent from the biology of the species and in that sense alone require careful consideration for their potential impact on welfare.

This paper will examine the rearing of calves and heifers from the perspective of behavioural development. Behavioural science can provide an insight into the animals' biological requirements independent of constraints imposed by certain management systems. Behavioural development is currently of considerable interest as it has become established that early experiences during mammalian development can have long-term implications for both mental and physical health.

Pre-natal effects

The possibility of pre-natal events affecting the long-term health and welfare of dairy calves has not received a great deal of attention, despite obvious possibilities for the calf foetus to be affected in a variety of ways during its uterine development. There is growing evidence that the pre-natal environment can have important influences on long-term physical and behavioural development. For example, pre-natal stress experienced by the pregnant mother can affect the capacity of her off-spring to adapt effectively to environmental challenges in later life (e.g. Braastad, 1998). Exposure of cows to repeated transportation stress during gestation altered their calves' physiological response to stress at both 10 and 150 days post-partum, indicating that pre-natal stress can affect the capacity of calves to adapt to stress in later life (Lay *et al.*, 1997).

Prenatal nutrient supply can also have important impacts on brain development and postnatal behaviour, depending on which component of the diet is lacking (e.g. energy, protein or trace element deficiency), the timing of the malnutrition in relation to stage of development, and the duration and severity of the nutrient restriction (Morgane *et al.*, 1993). Nutritional effects on brain development occur during periods of rapid brain growth and during early organisational events, such as neurogenesis, cell migration and differentiation; thereafter the brain is protected in older animals even during starvation. Although some of the neuronal deficits can be recovered later, either by improved postnatal nutrition or provision of a stimulating environment, deficits in brain development appear to be permanent. These can affect spatial learning and memory in the adult animal (Dauncey and Bicknell, 1999), and are thus likely to be important for the welfare of the adult. However, another consequence of early prenatal malnutrition is long-lasting changes in brain neural receptor function, which affect emotional responses leading to alterations in motivation and anxiety (Strupp and Levitsky, 1995). In sheep, for example, the offspring of ewes undernourished in pregnancy showed greater reactivity to management conditions such as restraint or novelty than offspring of well-fed ewes (Erhard *et al.*, 2004). These changes may be of greater consequence for animal welfare than learning or memory deficits.

Another important area where prenatal nutrition can have direct effects on welfare is through its effects on neonatal survival. For example, animals with low birth weight have severely limited body reserves and difficulty in maintaining body temperature after birth. They also show retarded behavioural development, taking longer to feed or suck from their mothers (Dwyer, 2003), and compromised physiological function (reviewed by Ashworth and Antipatis, 2001) which also influences survival. Deficits in play behaviour and social interactions are also seen in pre-natally malnourished animals as juveniles and adults (Almeida *et al.*, 1996). These results suggest that animals undernourished during early development may be less able to cope with environmental and social stresses, perhaps due to the alterations in emotional responses described above. Additional effects of prenatal nutrition include impacts on long term health and fertility, through for example impacts on the developing gonads, which may affect culling decisions and cow longevity in the herd.

It seems that the role of pre-natal nutrition as a factor affecting not only immediate neonatal behaviours but also longer-term adaptive capacities in dairy cattle has not been considered. The potential for alterations of foetal nutrition as a result of genotype x nutrition interactions in dairy cows suggests this as an area worthy of future research work.

Post-natal life

The single most significant event in the post-natal development of the dairy calf is its early removal from the cow, often at less than 24 hours after birth. Weaning is usually seen as the abrupt point at which the mother finally stops providing the calf with milk. However, it is probably more accurate to see weaning as a gradual process that begins from the point that the mother first begins to restrict access to the udder and hence her supply of milk to her offspring (Martin, 1984). In sheep, using Martin's concept, it is likely that weaning begins within 1-2 weeks after birth when the ewe first begins to control both the initiation and termination of suckling bouts, whilst the final cessation of suckling (under unconstrained conditions) does not occur until the lamb is 6-7 months of age (e.g. O'Connor, 1990; Pickup and Dwyer, 2002). It is likely that natural weaning in cattle occurs at least as late as 6 months into the calf's life (Wood-Gush *et al.*, 1986), indicating the extent to which early-weaning as practiced in the dairy industry diverges from the biological norm for the species.

Attention on the early removal of the calf from the cow has tended to focus on the acute responses of calf. Most experiments use treatment groups which are separated by only small degrees given the length of biological weaning in the cow (e.g. separation <24 hours vs. after 4 days; Lidfors, 1996). The trend in the data suggests that the longer the time the calf spends with the cow the greater its acute behavioural response to weaning (e.g. Weary and Chua, 2000; Flower and Weary, 2001). Recent data suggest that calves are able to discriminate their mothers calls even when separated after 24 h when tested within 48 h of birth (Marchant-Forde *et al.*, 2002). In sheep there is evidence that this early preference of the neonate for its mother is mediated by direct effects of colostrum on the brain neuroendocrine systems involved in social attachment (e.g. Nowak *et al.*, 2001), and that lambs learn the sounds of their mothers' voices in utero (Vince *et al.*, 1985). A study of abrupt weaning in beef calves indicated acute physiological and immune responses consistent with weaning acting as a significant stressor (Hickey *et al.*, 2003). Available data therefore suggest that weaning results in a generalised 'stress' response and that calves reared with the cow for longer show a greater response relative to those weaned within 24 hours. These data on acute responses support the views expressed in the FAWC (1997) report on dairy cow welfare that: 'Concern has been expressed that removal of the calf at such an early age is not in the welfare interests of the cow or calf. Others argue that if it is not possible to keep a calf with its mother for six months, as in a beef suckler herd, then the least cruel act is to separate them as soon as possible (Webster, 1994). Leaving a calf with its mother until weaning and separation occur naturally may be an ideal but is not practicable within the present modern dairy industry. As the industry is currently set up, early separation is the least stressful option'.

This view of early calf separation holds, provided that there are only short-term (acute) issues to be addressed with no longer-term 'hidden' penalties to early weaning. There is currently considerable scientific interest in the longer-term effects of maternal deprivation (of which early weaning is one form), stemming from a series of observations that link exposure to early maternal deprivation to later mental and physical health issues. In humans, early maternal loss has been identified as a risk factor for later abnormalities in mood (Mortensen *et al.*, 2003), with females perhaps being especially susceptible (Mireault *et al.*, 2001). Experimental studies in rodents and primates have explored the long-term effects of early adverse experiences in the form of maternal separation or loss, abuse or neglect, and social deprivation. The evidence suggests that early disruption to the maternal-offspring relationship is associated with alterations in a range of functions, including emotional and behavioural regulation, neuroendocrine responsiveness to stress, and cognition (e.g. Sanchez *et al.*, 2001; but see Pryce and Feldon (2003) for a re-interpretation of these data). The prevailing view from bio-medical research would therefore suggest that very early weaning, as experienced by the calf, could be a risk factor for emotional and stress hyper-reactivity in the longer-term. There is also evidence to suggest that heightened stress responses can have detrimental effects on other functions, such as immunity (e.g. Kusnecov and Rossi-George, 2002) and reproduction (e.g. Turner *et al.*, 2002). Thus a longer-term emotional/ stress hyper-reactivity in calves, induced by early weaning, could be related to production problems such as increased disease and reduced fertility.

This is clearly an important hypothesis, in that exposure to very early weaning could be instrumental in reducing the capacity of the calf to adapt to later environmental challenges such as post-weaning housing. However, our understanding of the longer-term impact of early weaning in calves is limited by a number of factors.

EXTRAPOLATION FROM EXISTING DATA

The evidence for significant effects of early maternal separation on later development stem mainly from research on rodents and primates. In extrapolating from this research to very early weaning in calves it is relevant that maternal separation has been used as a collective term for a variety of extremely different experimental manipulations (Lehmann and Feldon, 2000). Treatments can vary from 15 minutes of maternal separation to much longer periods, but in general involve returning the offspring to the mother once the prescribed period of separation is complete. In addition to the variability of what is termed maternal separation, there is the difficult question of what constitutes an appropriate control for maternal separation in different species (Pryce and Feldon, 2003). In research on rats the chosen control group is generally non-handled pups which under laboratory conditions remain in constant close contact with the mother. However, if given a choice, rat mothers will absent themselves from the nest for periods of up to 3 hours (e.g. Huot *et al.*, 2000). Rat pups should therefore be adapted to regular and relatively long periods of maternal separation. Maternal separation research in rats which uses periods of separation of less than 3 hours, may in fact be replicating the adapted response of rat pups to 'normal' maternal behaviour (Pryce and Feldon, 2003). Cattle are a 'hider' species, where the mother leaves the calf concealed whilst grazing (e.g. Bailey *et al.*, 1989). In cattle also, therefore, the calf will be adapted to the periods of maternal separation when the cow is grazing. Nonetheless, as the evidence of acute responses to weaning indicates (e.g. Flower and Weary, 2003), the calf is responsive to sufficiently prolonged periods of separation.

In terms of relevant data from cattle, there has been little controlled experimentation investigating the longer-term impact of early weaning on calf adaptation and health. For example, there are no specific data on the longer-term effects of very early weaning on later emotional and stress responses of calves. In pigs we have found that early weaning at 21 days results in a more 'anxious' behavioural profile than later weaning at 42 days (Wemelsfelder et al., 2003). In relation to health, Weary and Chua (2000) observed a non-significant tendency for calves weaned at 4 days to have fewer bouts of diarrhoea over the first 3 weeks of life relative to calves weaned at 6 h or 1 day; all calves received bottle-fed colostrum within 24 h. Calves allowed to remain with the cow for longer also tend to gain more weight, based on data from a number of studies (see Table 1 of Flower and Weary (2003)). The mechanisms involved in these responses to variation in weaning age have not been firmly identified, and may be due to direct effects of consuming either more colostrum or milk as opposed to any mediating effect of emotional or stress reactivity.

OTHER CONFOUNDING FACTORS

A major problem in determining the specific impact of early weaning *per se* is that it is confounded with a number of other factors that could also exert important effects on calf development.

Changes in nutrition and feeding behaviour

Early separation of the calf from the cow involves both loss of maternal contact (e.g. grooming) and changes both to the form of the milk diet and also how the calf receives the milk. Calves are often fed milk on a regime where they are offered a set amount of milk in a restricted number of meals. Such prescribed feeding regimes may impose an effective restriction on milk intake, resulting in calves experiencing heightened feeding motivation and compounding their response to early weaning. Thomas *et al.* (2001) found that new-born weaned calves offered 5 l of milk per day in 2 meals, vocalised more and at a higher frequency than calves offered 8 l of milk per day in 4 meals. This indicates the potential importance of providing weaned calves with the opportunity of regulating their own intake patterns. Appleby *et al.* (2001) compared meal feeding (2 bucket-fed milk meals/ day, at 5% of body weight per meal) with *ad libitum* feeding (unrestricted access to milk from a teat). Over the first 4 weeks of life, *ad libitum* milk feeding resulted in a greater weight gain (as a result of an increased milk intake), and a reduced number of days on which calves showed diarrhoea (Appleby *et al.*, 2001). *Ad libitum* milk intake has been shown to have no detrimental long-term effect on solid food intake (Jasper and Weary, 2002). These advantages of *ad libitum* access to milk through a teat appear to result from this method of feeding being a much better match to the calves biological (behavioural and physiological) requirements, including allowing calves to express individual differences in feeding motivation and behaviour (Appleby *et al.*, 2001).

Post-natal housing

When replacement dairy calves are removed from the cow, they are placed in a range of housing systems from individual pens and outdoor hutches to group housing systems (Howard, 2003). The choice of individual penning will often relate in part to the perceived need to control spread of disease through tactile contact, such as sharing of teats, grooming or behaviours such as cross-sucking. In biomedical research social isolation is used as an adverse early experience that is suggested to result in similar detrimental effects on brain-endocrine-immune regulation as found with maternal separation (e.g. Kanitz *et al.*, 2004). In calves individual housing without social contact in 'barren' pens resulted in greater behavioural reactivity and a higher incidence of 'abnormal' behaviours such as tongue-rolling (Veissier *et al.*, 1997). Individually housed calves have also been shown to have a greater cortisol response to an exogenous challenge with adrenocorticotrophic hormone (ACTH) than pair housed control calves, indicating a greater exposure to chronic stress (Raussi *et al.*, 2003). Thus, for the calf the combination of early separation from the cow and individual housing involving degrees of social isolation, could potentially reinforce each other with respect to longer-term detrimental effects on calf development. Indeed, the need for individual housing to protect against disease might be required partly as a result of early weaning increasing the calves susceptibility to disease challenges.

The relatively few studies that have looked at the potential long-term effects of housing are somewhat equivocal. Earlier work (e.g. Creel and Albright, 1998) tended to focus on the effects of isolation (with no physical contact) versus individual (physical contact allowed) housing on later behaviour and production. This work suggested either that calves reared in isolation had increased milk production as cows (e.g. Donaldson, 1970) or that there was no effect of housing on later production (Arave *et al.*, 1992). Creel and Albright (1988) suggested that effects of early rearing conditions on later production are mediated via changes in sensitivity of the hypothalamic-pituitary-adrenal axis, as they found heifers reared in isolation to be more behaviourally and physiologically reactive to acute stressors. Morgensen *et al.* (1999) compared group, isolation and individual housing and found that group housed calves were less ready to approach humans, but that overall there was no obvious effect of housing on later production. Recent epidemiological evidence suggests that there may be increased health risks associated with larger group sizes combined with the use of automatic feeding systems (Svensson *et al.*, 2003).

Human-animal interactions

The importance of human-animal interactions both in animal welfare and production is now well established. Many studies have shown that the quality of the human-animal interface can affect the behaviour, physiology and productivity of farm animals (e.g. Hemsworth and Coleman, 1998). In dairy cattle, as with other species, cow behaviours indicative of fear of humans are negatively associated with production levels (e.g. Breuer *et al.*, 2000). Fear of humans could affect milk production by activating stress-sensitive endocrine systems that interfere with oxytocin release (e.g. Lawrence *et al.*, 1992) and milk let down (e.g. Negro and Marnct, 2003).

In terms of human-animal interactions during early life, there is evidence that early exposure to humans can mediate against early adverse experiences. The combination of stroking and feeding (but not stroking alone) of rat pups during a 24-hour separation from the mother removed adverse effects of the maternal deprivation on brain development (van Oers *et al.*, 1999). We might have evidence of a similar effect in piglets where pigs early weaned at day 12 showed similar levels of anxiety to pigs weaned at day 42 and significantly less anxiety than pigs weaned at day 21 (Wemelsfelder et al., 2003; see also above). The day-12 pigs were removed from their mothers and kept in a nursing unit where they might have formed an association between the stockperson and delivery of milk from the nursing unit. In the early work demonstrating apparently improved production in calves reared in visual and tactile isolation (e.g. Arave *et al.*, 1985), it was suggested that the isolated calves might have 'imprinted' on the stockperson and thus adapted more easily to milking at a later date.

In conclusion, the early weaning and individual housing of dairy calves would appear from data on other species including humans, to be substantial risk factors for poor health and welfare in later life. However, both insufficient data and the presence of several confounding factors make it impossible at this stage to weigh properly the longer-term impact of these early experiences. In the absence of a proper understanding of the risks associated with early calf management, it is still worth considering interventions that could reduce any long-term effects of early weaning and individual housing.

POST-NATAL INTERVENTIONS

Feeding practices

We indicated earlier that there is evidence of potential benefits to keeping calves with cows for longer than the conventional 24 hours. However, set against these benefits is the greater response to weaning of later-weaned calves. One line of future research could be to investigate how the advantages of later weaning in calves can be achieved, whilst limiting the acute effects of later separation on the calf.

As mentioned earlier, weaning involves both loss of maternal contact and an abrupt change to the calf's nutritional environment. In beef production a promising line of study has been to separate these two components by imposing a restriction on sucking before separation of cow and calf, so called Two Step Weaning (e.g. Haley *et al.*, 2001). This is achieved by fitting calves with an anti-sucking device four days before actual separation from the cow (Phase 1). The device prevents sucking, but allows grazing and drinking, and results in little observable response by the cow-calf pair. Following Phase 1, the cow-calf pair are physically separated (Phase 2) and there is a clear reduction in behavioural evidence of weaning stress in calves exposed to Two Step Weaning; they vocalise and walk significantly less, but eat more than conventionally weaned calves. The principle of Two Stage Weaning is not directly applicable to young calves because they have no source of nutrition other than sucking. However, the impact of weaning on any age of calf is likely to be lessened if the post-weaning feeding environment is as similar as possible to the pre-weaning phase. From this perspective, post-weaning feeding of calves should ideally allow expression of individual differences in sucking pattern with little restriction on intake of milk, thus limiting confounding of disruption of sucking patterns and hunger (e.g. Thomas *et al.*, 2001; Appleby *et al.*, 2001) with effects of separation from the cow. The advent of automatic milking devices would appear to be able to meet the behavioural and physiological needs of calves because milk supply can be programmed to individual calf requirements both in terms of total amount of milk delivered and also frequency of milk meals. In addition, automatic feeders, by providing milk via a teat, should help limit the occurrence of non-nutritive sucking (see next point).

Post-natal housing

Group housing would appear more appropriate to the longer-term development of a social animal such as the calf (e.g. FAWC, 1997) although, as mentioned earlier, other confounding issues such as early exposure to humans, need to be taken into account. One behavioural problem that emerges in group housing is that of sucking other calves, a behaviour that rarely occurs in calves reared by cows (e.g. Krohn, 2001), and dairy producers may be hesitant to group house calves for fear of stimulating cross-sucking of pen-mates, which may develop into later inter-sucking and milk stealing (e.g. Keil *et al.*, 2000). Both cross- and inter-sucking might also be a route for disease transmission, thus mediating the higher incidence of disease and mortality reported in group versus individual housing.

However, as explained earlier, a major cause of non-nutritive sucking such as cross-sucking is bucket-feeding (e.g. de Passille, 2001), which provides a positive stimulus with little negative feedback to sucking behaviour. Thus bucket feeding might in itself necessitate the use of individual housing to prevent non-nutritive sucking by calves. In a study of group-housed veal calves drinking milk from a bucket (de Passille and Caza, 1997), cross-sucking of calves occurred immediately after the bucket-fed meal and was significantly reduced by provision of (non-nutritive) rubber teats. In addition to provision of teats, slowing the rate of milk supply to allow nutritive sucking to last longer (de Passille *et al.*, 1992) and providing hay at the end of milk meals (Haley *et al.*, 1998) also help reduce non-nutritive sucking. Suppression of cross-sucking by controlling motivation for non-nutritive sucking might also lead to reduction of later inter-sucking if, as has been suggested, cross- and inter-sucking share a similar motivational basis (e.g. Spinka, 1992).

Stockperson-calf interactions

We discussed earlier that positive early interactions between stockpersons and calves might also help in mitigating the longer-term effects of practices such as early weaning. Confounding good handling practices with feeding might be a particularly effective route to establishing a long-term reduction in fear of humans. Producers who use group housing for calves with automatic feeding systems might need to pay particular attention to providing human contact in conjunction with feeding, even if supply of milk is largely under automatic control.

Environmental enrichment

There has been considerable attention given to the enrichment of environments for farm animals. Much of this attention has focused on pigs and poultry as a result of them being housed in environments that are highly confining or lacking stimuli. There are broad interpretations of what can be defined as environmental enrichment, but it is probably best seen as providing stimuli that facilitate species-typical behaviour (e.g. Duncan, 2001). Appropriate environmental enrichment is often found to reduce behavioural reactivity and fear (e.g. Jones 1996; Veissier *et al.*, 1997), occurrence of problematic behaviours such as stereotypies (e.g. Spoolder *et al.*, 1995), and injurious behaviour directed at pen-mates such as tail-biting in pigs (e.g. Beattie *et al.*, 2001).

In the sense that environmental enrichment facilitates expression of species-typical behaviour, then in calves allowing for expression of sucking, social contact and access to objects for oral investigation as we have previously discussed can all be regarded as providing appropriate environmental enrichment. Space in which to move and to play might also be regarded as an important aspect of environmental enrichment in calves. Increasing space allowance from 1.5 m^2 to 3 and 4 m^2 per calf has been found to increase the occurrence of locomotory play both in the home pen and in an open arena (Jensen and Kyhn, 2000). Play is an important if somewhat overlooked behavioural component of welfare, as it can be used as an indicator of high quality environments and good health (e.g. Lawrence, 1987). Providing sufficient space for play might not only have direct benefits on welfare but also allow expression of this useful indicator of health and welfare.

Conclusions

In this chapter we have focused primarily on behavioural issues that emerge from current practices of calf rearing. We have used the perspective of behavioural development and existing knowledge from other species to emphasise that early events in the calf's life (early weaning and individual housing) might have long-term implications for its later health and welfare. Extrapolation from other studies to calves is limited by a lack of data and the presence of important confounding factors (e.g. the post-weaning feeding regime), which can also have potentially large effects on adaptation to the post-weaning production system. Indeed some elements of post-weaning management (group housing; positive human-animal interactions) might act to reverse adverse effects arising from early weaning.

However, we have presented sufficient evidence to suggest that practices such as early weaning and individual housing need to be more carefully examined in terms of their long-term impact on calf health and welfare. A practice such as early weaning can attract adverse public reaction, is not at first sight compatible with the development of sustainable (e.g. organic) dairy systems and, as we have highlighted, might be the cause of production problems (e.g. disease) through increasing the sensitivity of the calf to environmental challenges. At the same time, even in the absence of a complete understanding of the importance of the factors affecting the long-term health and welfare of calves, there are potential interventions that are worth considering. These are based on meeting the calf's biological needs and include extending the length of the suckling period, allowing individual expression of sucking behaviour and feeding behaviour, providing for both appropriate social and human contact in early life, and facilitating the expression of positive indicators of welfare such as play behaviour. In addition to expected benefits for calf welfare, it is also relevant to consider whether these interventions might also reduce hidden costs of current practices by increasing the adaptation of calves to the production environment and thereby have a positive overall effect on the net margin of the business.

Acknowledgements

The Scottish Agricultural College receives financial assistance from the Scottish Executive Environment and Rural Affairs Department.

References

Almeida, S.S., Tonkiss, J. and Galler, J.R. (1996) Prenatal protein malnutrition affects the social interactions of juvenile rats. *Physiology and Behavior*, **60**, 197-201.

Appleby, M.C., Weary, D.M. and Chua, B. (2001) Performance and feeding behaviour of calves on ad libitum milk from artificial teats. *Applied Animal Behaviour Science*, **74 (3)**, 191-201.

Arave, C.W., Mickelsen, C.H. and Walters, J.L. (1985) Effect of early rearing experience on subsequent behavior and production of Holstein heifers. *Journal of Dairy Science*, **68 (4)**, 923-929.

Arave, C., Albright, J., Armstrong, D., Foster, W. and Larson, L. (1992) Effects of isolation of calves on growth, behaviour and first lactation yield of Holstein cows. *Journal of Dairy Science*, **75**, 3408-3415.

Ashworth, C.J. and Antipatis, C. (2001) Micronutrient programming of development throughout gestation. *Reproduction*, **122 (4)**, 527-535.

Bailey, D.W., Rittenhouse, L.R., Hart, R.H. and Richards, R.W. (1989) Characteristic of spatial memory in cattle. *Applied Animal Behaviour Science*, **23**, 331-340.

Beattie, V.E., Sneddon, I.A., Walker, N. and Weatherup, R.N. (2001) Environmental enrichment of intensive pig housing using spent mushroom compost. *Animal Science*, **72**, 35-42.

Braastad, B.O. (1998) Effects of prenatal stress on behaviour of offspring of laboratory and farmed mammals. *Applied Animal Behaviour Science*,

61 (2), 159-180.
Breuer, K., Hemsworth, P.H., Barnett, J.L., Matthews, L.R. and Coleman, G.J. (2000) Behavioural response to humans and the productivity of commercial dairy cows. *Applied Animal Behaviour Science*, **66 (4)**, 273-288.
Creel, S.R. and Albright, J.L. (1988) The effects of neonatal social-isolation on the behavior and endocrine function of Holstein calves. *Applied Animal Behaviour Science*, **21 (4)**, 293-306.
Dauncey, M.J. and Bicknell, R.J. (1999) Nutrition and neurodevelopment: mechanisms of developmental dysfunction and disease in later life. *Nutrition Research Reviews*, **12**, 231-253.
De Passille, A.M.B. (2001) Sucking motivation and related problems in calves. *Applied Animal Behavior Science*, **72 (3)**, 175-187.
De Passille, A.M.B. and Caza, N. (1997) Motivational and physiological analysis of the causes and consequences of non-nutritive sucking by calves. *Journal of Dairy Science*, **80 (suppl 1)**, 229.
Donaldson, S., Albright, J. and Black, W. (1972) Primary social relationships and cattle behaviour. *Proceedings of the Indiana Academy of Science*, **81**, 345-351.
Duncan, I.J.H. and Olsson, I.A.S. (2001) Environmental enrichment: from flawed concept to pseudo-science. *Proceedings of the 35th International Congress of the International Society for Applied Ethology*, pp 73. Edited by J.P. Garner, J.A. Mench and S.P. Heekin. ISAE, Davis, USA.
Dwyer, C.M. (2003) Behavioural development in the neonatal lamb: effect of maternal and birth-related factors. *Theriogenology*, **59 (3-4)**, 1027-1050.
Erhard, H.W., Boissy, A., Rae, M.T. and Rhind, S.M. (2004) Effects of prenatal undernutrition on emotional reactivity and cognitive flexibility in adult sheep. *Behavioural Brain Research*, **151**, 25-35.
Farm Animal Welfare Council (1997) *Report on the Welfare of Dairy Cattle*. FAWC, London, UK.
Farm Animal Welfare Council (2004) *Farm Animal Welfare Council web-pages*. http://www.fawc.org.uk/freedoms.htm.
Flower, F.C. and Weary, D.M. (2001) Effects of early separation on the dairy cow and calf: 2. Separation at 1 day and 2 weeks after birth. *Applied Animal Behaviour Science*, **70 (4)**, 275-284.
Flower, F.C. and Weary, D.M. (2003) The effects of early separation on the dairy cow and calf. *Animal Welfare*, **12 (3)**, 339-348.
Haley, D.B., Rushen, J., Duncan, I.J.H., Widowski, T.M. and De Passille, A.M. (1998) Effects of resistance to milk flow and the provision of hay on nonnutritive sucking by dairy calves. *Journal of Dairy Science*, **81 (8)**, 2165-2172.
Haley, D.B., Stookey, J.M., Clavelle, J.L. and Watts, J.M. (2001) The simultaneous loss of milk and maternal contact compounds distress at weaning in beef calves. *Proceedings of the 35th International Congress*

of the International Society for Applied Ethology, pp 41. Edited by J.P. Garner, J.A. Mench and S.P. Heekin. ISAE, Davis, USA.

Hemsworth, P.H. and Coleman, G.J. (1998) *Human-Livestock Interactions: The Stockperson and the Productivity and Welfare of Intensively-farmed Animals*. CAB International, Oxon UK.

Hickey, M.C., Drennan, M. and Earley, B. (2003) The effect of abrupt weaning of suckler calves on the plasma concentrations of cortisol, catecholamines, leukocytes, acute-phase proteins and in vitro interferon-gamma production. *Journal of Animal Science*, **81 (11)**, 2847-2855.

Howard, P. (2003) A review of calf health, welfare and rearing practices on UK dairy farms. *Cattle Practice*, **11**, 173-180.

Huot, R.L., Ladd, C.O. and Plotsky, P.M. (2000) Maternal deprivation. In *Encyclopedia of Stress, Vol. 1*, pp 699-707. Edited by G. Fink. San Diego, Academic Press.

Jasper, J. and Weary, D.M. (2002) Effects of ad libitum milk intake on dairy calves. *Journal of Dairy Science*, **85 (11)**, 3054-3058.

Jensen, M.B. and Kyhn, R. (2000) Play behaviour in group-housed dairy calves, the effect of space allowance. *Applied Animal Behaviour Science*, **67 (1-2)**, 35-46.

Jones, R.B. (1996) Fear and adaptability in poultry: Insights, implications and imperatives. *World's Poultry Science Journal*, **52 (2)**, 131-174.

Kanitz, E., Tuchscherer, M., Puppe, B., Tuchscherer, A. and Stabenow, B. (2004) Consequences of repeated early isolation in domestic piglets (Sus scrofa) on their behavioural, neuroendocrine, and immunological responses. *Brain Behavior and Immunity*, **18 (1)**, 35-45.

Keil, N.M., Audige, L. and Langhans, W. (2001) Is intersucking in dairy cows the continuation of a habit developed in early life? *Journal of Dairy Science*, **84 (1)**, 140-146.

Krohn, C.C. (2001) Effects of different suckling systems on milk production, udder health, reproduction, calf growth and some behavioural aspects in high producing dairy cows - a review. *Applied Animal Behaviour Science*, **72 (3)**, 271-280.

Kusnecov, A.W. and Rossi-George, A. (2002) Stressor-induced modulation of immune function: a review of acute, chronic effects in animals. *Acta Neuropsychiatrica*, **14 (6)**, 279-291.

Lay, D.C., Randel, R.D., Friend, T.H., Jenkins, O.C., Neuendorff, D.A., Bushong, D.M., Lanier, E.K. and Bjorge, M.K. (1997) Effects of prenatal stress on suckling calves. *Journal of Animal Science*, **75 (12)**, 3143-3151.

Lawrence, A. (1987) Consumer demand theory and the assessment of animal-welfare. *Animal Behaviour*, **35**, 293-295.

Lawrence, A.B., Petherick, J.C., McLean, K., Gilbert, C.L., Chapman, C. and Russell, J.A. (1992) Naloxone prevents interruption of parturition and increases plasma oxytocin following environmental disturbance in parturient sows. *Physiology and Behavior*, **52 (5)**, 917-923.

Lehmann, J. and Feldon, J. (2000) Long-term biobehavioral effects of maternal separation in the rat: Consistent or confusing? *Reviews of Neuroscience*, **11 (4)**, 383-408.

Lidfors, L.M. (1996) Behavioural effects of separating the dairy calf immediately or 4 days post-partum. *Applied Animal Behaviour Science*, **49 (3)**, 269-283.

Marchant-Forde, J.N., Marchant-Forde, R.M. and Weary, D.M. (2002) Responses of dairy cows and calves to each other's vocalisations after early separation. *Applied Animal Behaviour Science*, **78 (1)**, 19-28.

Martin, P. (1984) The meaning of weaning. *Animal Behaviour*, **32**, 1257-1259.

Mireault, G., Bearor, K. and Thomas, T. (2001) Adult romantic attachment among women who experienced childhood maternal loss. *Omega-Journal of Death and Dying*, **44 (1)**, 97-104.

Morgane, P.J., Austinlafrance, R., Bronzino, J., Tonkiss, J., Diazcintra, S., Cintra, L., Kemper, T. and Galler, J.R. (1993) Prenatal malnutrition and development of the brain. *Neuroscience and Biobehavioral Reviews*, **17 (1)**, 91-128.

Morgensen, L., Krohn, C.C. and Foldager, J. (1999) Long-term effect of housing method during the first three months of life on human-animal relationships in female dairy cattle. *Acta Agriculturae Scandinavia, Section A. Animal Science*, **49**, 163-171.

Mortensen, P.B., Pedersen, C.B., Melbye, M., Mors, O. and Ewald, H. (2003) Individual and familial risk factors for bipolar affective disorders in Denmark. *Archives of General Psychiatry*, **60 (12)**, 1209-1215.

Negro, J.A. and Marnet, P.G. (2003) Cortisol, adrenalin, noradrenalin and oxytocin release and milk yield during first milkings in primiparous ewes. *Small Ruminant Research*, **47 (1)**, 69-75.

Nowak, R., Breton, G. and Mellot, E. (2001) CCK and development of mother preference in sheep: a neonatal time course study. *Peptides*, **22 (8)**, 1309-1316.

O'Connor, C.E. (1990) Mother-offspring relationships in Scottish Blackface sheep. *PhD Thesis*, University of Edinburgh.

van Oers, H.J.J., de Kloet, E.R. and Levine, S. (1999) Persistent effects of maternal deprivation on HPA regulation can be reversed by feeding and stroking, but not by dexamethasone. *Journal of Neuroendocrinology*, **11 (8)**, 581-588.

Offer, J.E., Leach, K.A., Brocklehurst, S. and Logue, D.N. (2003) Effect of forage type on claw horn lesion development in dairy heifers. *Veterinary Journal*, **165 (3)**, 221-227.

Office International des Epizooties (OIE) (2004) *Global Conference on Animal Welfare: an OEI Initiatve*. OIE, Paris. ISBN 92-894-6614-6.

Pickup, H.E. and Dwyer, C.M. (2002) An investigation into postural communication between ewes and lambs and its role in maintaining ewe-lamb proximity. *Proceedings of the International Society for Applied Ethology*, pp 66. Edited by P. Koene. ISAE, The Netherlands.

Pryce, C.R. and Feldon, J. (2003) Long-term neurobehavioural impact of the postnatal environment in rats: manipulations, effects and mediating mechanisms. *Neuroscience and Biobehavioral Reviews*, **27 (1-2)**, 57-71.

Raussi, S., Lensink, B.J., Boissy, A., Pyykkonen, M. and Veissier, I. (2003) The effect of contact with conspecifics and humans on calves' behaviour and stress responses. *Animal Welfare*, **12 (2)**, 191-203.

Sanchez, M.M., Ladd, C.O. and Plotsky, P.M. (2001) Early adverse experience as a developmental risk factor for later psychopathology: Evidence from rodent and primate models. *Development and Psychopathology*, **13 (3)**, 419-449.

Spinka, M. (1992) Intersucking in dairy heifers during the first two years of life. *Behavioural Processes*, **28**, 41-50.

Spoolder, H.A.M., Burbidge, J.A., Edwards, S.A., Simmins, P.H. and Lawrence, A.B. (1995) Provision of straw as a foraging substrate reduces the development of excessive chain and bar manipulation in food restricted sows. *Applied Animal Behaviour Science*, **43 (4)**, 249-262.

Strupp, B.J. and Levitsky, D.A. (1995) Enduring cognitive effects of early malnutrition - A theoretical reappraisal. *Journal of Nutrition*, **125 (8)**, S2221-S2232.

Svensson, C., Lundborg, K., Emanuelson, U. and Olsson, S.O. (2003) Morbidity in Swedish dairy calves from birth to 90 days of age and individual calf-level risk factors for infectious diseases. *Preventative Veterinary Medicine*, **58 (3-4)**, 179-197.

Thomas, T.J., Weary, D.M. and Appleby, M.C. (2001) Newborn and 5-week-old calves vocalize in response to milk deprivation. *Applied Animal Behaviour Science*, **74 (3)**, 165-173.

Turner, A.I., Hemsworth, P.H. and Tilbrook, A.J. (2002) Susceptibility of reproduction in female pigs to impairment by stress and the role of the hypothalamo-pituitary-adrenal axis. *Reproduction Fertility and Development*, **14 (6)**, 377-391.

Veissier, I., Chazal, P., Pradel, P. and LeNeindre, P. (1997) Providing social contacts and objects for nibbling moderates reactivity and oral behaviors in veal calves. *Journal of Animal Science*, **75 (2)**, 356-365.

Vince, M.A., Billing, A.E., Baldwin, B.A., Toner, J.N. and Weller C. (1985) Maternal vocalizations and other sounds in the fetal lambs sound environment. *Early Human Development*, **11 (2),** 179-190.

Weary, D.M. and Chua, B. (2000) Effects of early separation on the dairy cow and calf 1. Separation at 6 h, 1 day and 4 days after birth. *Applied Animal Behaviour Science*, **69 (3)**, 177-188.

Webster, J. (1994) *A Cool Eye Towards Eden.* Blackwell Scientific, Oxford.

Wemelsfelder, F., Batchelor, C., Jarvis, S., Farish, M. and Calvert, S. (2003). The relationship between qualitative and quantitative assessments of pig behaviour. *Proceedings of the 37th International Congress of the ISAE*, Abano Terme, Italy.

Wood-Gush, D.G.M., Carson, K., Lawrence, A.B. and Moser, H.A. (1986) Parental behaviour in Ungulates. In *Parental Behaviour*, pp 85-115. Edited by W. Sluckin and M. Herbert. Blackwell, Oxford.

12

EARLY GROWTH EFFECTS ON SUBSEQUENT HEALTH AND PERFORMANCE OF DAIRY HEIFERS

JAMES K. DRACKLEY
Department of Animal Sciences, University of Illinois, Urbana, Illinois 61822 USA

Introduction

Interest has increased recently in the question of whether early-life events, particularly plane of nutrition and subsequent rates of growth during the milk-feeding period, might affect later-life health, productivity, and longevity. In many ways this is a unique question, one that probably would not have been asked a decade ago or, if it were asked, one that would be roundly dismissed. In the author's opinion, that the question is being asked (and, presumably, considered seriously) as a part of this international conference is attributable to two main factors: 1) the interest and research in the area of improving the plane of early nutrition to more "natural" levels, led by the pioneering work of M. E. Van Amburgh and associates at Cornell University (Diaz, Van Amburgh, Smith, Kelsey, and Hutten, 2001), and 2) the growing recognition of, and appreciation for, the considerable body of knowledge gained in other animal species and humans on the impacts of early-life nutrition on growth and health, both during the neonatal period and later in life (Lucas, 1998; Roberts and McDonald, 1998; Waterland and Garza, 1999; Burrin, Fiorotto, and Davis, 2001; Patel and Srinivasan, 2002).

This question is vitally important to establishing the value, both in economic terms and from an animal well-being standpoint, of increased liquid feeding programs. At the outset, however, it must be made clear that there are very few data available that directly link nutritional status in calves with later life health and productivity as cows. Available data will be discussed here, but most of the remaining discussion will be devoted to a speculative overview of potential physiological impacts of suboptimal early-life nutrition based on data from other species, which may serve as a foundation on which to develop testable hypotheses for dairy calves.

Definition of question

First, standards or definitions of what is meant by differences in early nutrition must be established. Current convention of calf rearing in North America and much of Europe is that calves are fed a limited amount of milk or milk replacer, typically 0.08 to 0.10 of body weight (BW) as liquid, in an effort to stimulate early consumption of solid feeds (starter). Because volatile fatty acids (particularly butyrate) from fermentation of concentrate-based ingredients are the stimulus for development of the ruminal epithelium, early consumption of starter dry matter (DM) is important for systems in which the goal is early weaning and the lowest-cost rearing program (Davis and Drackley, 1998). Consumption of this limited amount of solids from milk or milk replacer (typically 400 to 600 g/d) will support maintenance plus average daily gains (ADG) in the range of 200-300 g/d (for milk replacer) to 300-400 g/d (for whole milk) under thermoneutral conditions. However, under adverse environmental conditions, increased maintenance requirements for thermogenesis might result in reductions in BW gain or even cause BW loss. For example, a 45-kg calf consuming 500 g/d of a typical milk replacer powder will lose BW when the effective environmental temperature is below 5°C (National Research Council, 2001).

Restricted-feeding programs differ markedly from the natural feeding behaviour of calves allowed to suckle their dams or to consume liquid feed ad libitum. Calves allowed to suckle their mothers typically consume 6 to 10 meals per day, and may consume 0.16 to 0.24 of their BW daily as milk after 3 to 4 weeks of age (Hafez and Lineweaver, 1968). Recent studies have confirmed these patterns in Holstein calves. Flower and Weary (2001) showed that Holstein calves left with their dams weighed 59.9 kg at 14 days of age compared with 46.9 kg for calves fed milk from a bucket at a rate of 0.10 BW. Jasper and Weary (2002) reported mean milk intakes of 8.8 kg/d during the first 35 days of age when Holstein calves had free access to milk via an artificial teat, compared with 4.7 kg/d for calves fed milk at 0.10 BW. In that experiment, calves with ad libitum access to milk consumed over 9 kg/d by day 4 of life.

Studies with hand feeding of milk also show that ad libitum intakes of milk are in excess of 0.18 BW. For example, Khouri and Pickering (1968) fed a milk replacer twice daily to calves during the first 6 weeks of life at rates of 0.113, 0.139, 0.159, or 0.194 BW (ad libitum). The ADG during weeks 2 to 6 of life were 0.41, 0.50, 0.62, and 0.94 kg/d, respectively. Feed efficiencies (kg milk DM per kg BW gain) were 1.59, 1.47, 1.33, and 1.23, respectively. The latter values compare favourably with feed efficiencies for young pigs and lambs (Hodge, 1974; Greenwood, Hunt, Hermanson, and Bell, 1998; Kim, Heo, Odle, Han, and Harrell, 2001). Consequently, what has been referred to recently as "accelerated growth" by calves is, in fact, biologically normal growth. It is a management (i.e., economic) decision to feed smaller amounts of milk or milk replacer twice-daily to encourage dry feed intake.

As starter intake increases in restricted-milk programs, gains of BW increase and rates of ADG may approach those on more intensive milk feeding programs (Figure 12.1). Thus, differences in plane of nutrition are most pronounced during the first 2-3 weeks of life, when restricted-feeding falls far short of meeting nutrient requirements (National Research Council, 2001; see also chapter by Van Amburgh, this volume). Clearly, early growth is limited by restricted-milk feeding programs. Because of the close link between growth and development (see chapter by Brameld in this book), it is prudent to ask what other early-life developmental processes might be limited by such programs. Consequently, the frame of reference for the following discussion is the comparison between marginal nutrition during the first 2-3 weeks and more biologically appropriate early nutrition.

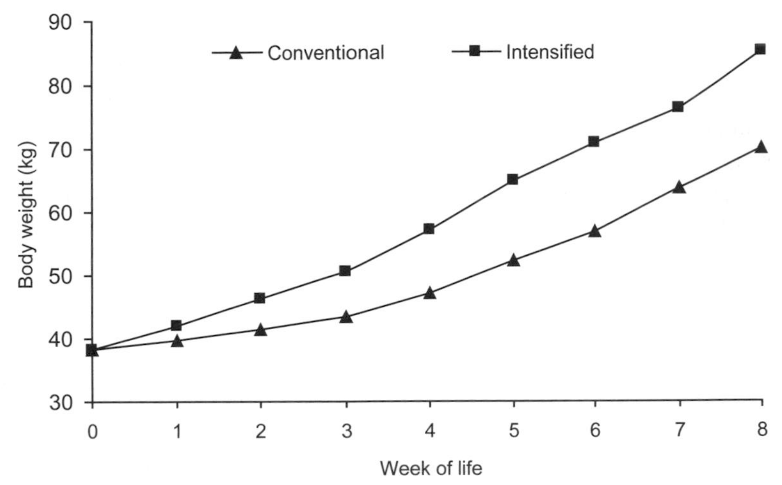

Figure 12.1 Example of differences in early growth between calves fed on a conventional restricted-feeding program (milk replacer powder fed at 0.0125 birth BW; calves weaned at 35 days) or on an intensified program (milk replacer fed at 0.02 birth BW for week 1, then 0.025 BW at week 2 during weeks 2-5; calves weaned at day 42). Calves had access to starter feed from week 1 of life. (B.C. Pollard and J.K. Drackley, unpublished data, 2002

Evidence for long-term effects of early life nutrition

IS THERE EVIDENCE THAT INCREASED EARLY NUTRITION MIGHT BE DETRIMENTAL?

Common concerns raised by those accustomed to restricted-feeding of calves include increased incidence of scouring and other early health problems, overfattening, impaired fertility, decreased milk production (perhaps mediated by impaired mammary development), bone or skeletal abnormalities at maturity, and, as a result of some combination of these factors, decreased herd life.

A common argument in favour of restricted liquid feeding and early weaning has been that scouring is decreased. Faecal consistency becomes less fluid as dry feed is consumed, primarily as a result of the bulking effect of dietary fibre. However, merely feeding more milk or more of a high-quality milk replacer does not cause scouring (Mylrea, 1966; Huber, Silva, Campos, and

Mathieu, 1984; Nocek and Braund, 1986; Appleby, Weary, and Chua, 2001; Diaz *et al.*, 2001; Weary and Jasper, 2002). The occurrence of calf scours, unless a poor-quality milk replacer containing damaged ingredients is fed, depends more on the load of pathogenic microorganisms in the calf's environment (Roy, 1980) and the degree of environmental stress on calves (Bagley, 2001). Calves fed larger amounts of milk replacer will have softer faeces, and that requires a shift of mindset by producers and advisors. Our own experiences with calves fed milk replacer at up to 0.18 BW indicated that average faecal scores were not significantly different, but days with elevated faecal score (softer faeces) were increased (Bartlett, 2001). Feeding milk replacer resulted in softer faeces than feeding similar amounts of whole milk, regardless of the macronutrient composition of the milk replacer (Bartlett, 2001).

Given the known adverse effects of allowing heifers to become overconditioned (Sejrsen, Purup, Vestergaard, and Foldager, 2000), concern exists that increased liquid feeding might allow calves to become too fat, in turn impairing subsequent reproduction or milk production. Our recent studies (Bartlett, 2001; Blome, Drackley, McKeith, Hutjens, and McCoy, 2003; Bartlett, McKeith, VandeHaar, Dahl, and Drackley, 2005) as well as those by Van Amburgh and associates (Diaz *et al.*, 2001; Tikofsky, Van Amburgh, and Ross, 2001) have shown clearly that body composition can be influenced by dietary composition in young dairy calves. Measurements of stature increase as the content of dietary crude protein (CP) is increased in isoenergetic diets (i.e., as the dietary protein to energy ratio is increased; Blome *et al.*, 2003; Bartlett *et al.*, 2005), indicating stimulation of skeletal growth. Whole-body deposition of protein (i.e., lean tissue) also increases linearly as dietary CP supply increases over a wide range of protein intakes (Figure 12.2). For calves fed isoenergetic diets at the same energy intake, whole-body protein deposition increases linearly as the protein to energy ratio increases, whilst fat deposition decreases linearly (Figure 12.3; Bartlett *et al.*, 2005). Fat deposition is increased more by dietary fat than by a similar amount of energy from lactose (Bartlett, McKeith, and Drackley, 2001a; Tikofsky *et al.*, 2001). Taken together, these studies indicate that overfattening should not be a concern with higher protein, moderate-fat milk replacers designed for enhanced growth rates; fat deposition might increase with larger amounts of whole milk fed for prolonged periods.

While controlled studies of early nutrition and subsequent reproductive characteristics have not been performed, field observations indicate no problems. Indeed, if growth from weaning to puberty and breeding are similar, there seems to be little biological basis to propose that differences in early life nutrition would affect fertility.

Over-feeding energy during the period of 3 months of age to puberty may negatively affect mammary development and milk production (Sejrsen *et al.*, 2000). Although concern has been raised that greater liquid feeding early in life also might affect mammary development, Danish researchers have found no evidence for effects of high growth rate during the first 2 months on mammary development (Sejrsen, Purup, Martinussen, and Vestergaard, 1998; Sejrsen *et al.*, 2000). Recent research from Michigan State University has shown that improved early nutrition actually stimulated mammary tissue development (Brown, VandeHaar, Daniels, Liesman, Chapin, and Weber Nielsen, 2002). Calves fed a milk replacer (CP 28.5 g/kg, fat 150 g/kg) for ADG of 666 g/d had 0.32 more parenchymal mass and 0.47 more parenchymal DNA than calves fed a conventional milk replacer (CP 200 g/kg, fat 200 g/kg) for ADG of 379 g/d.

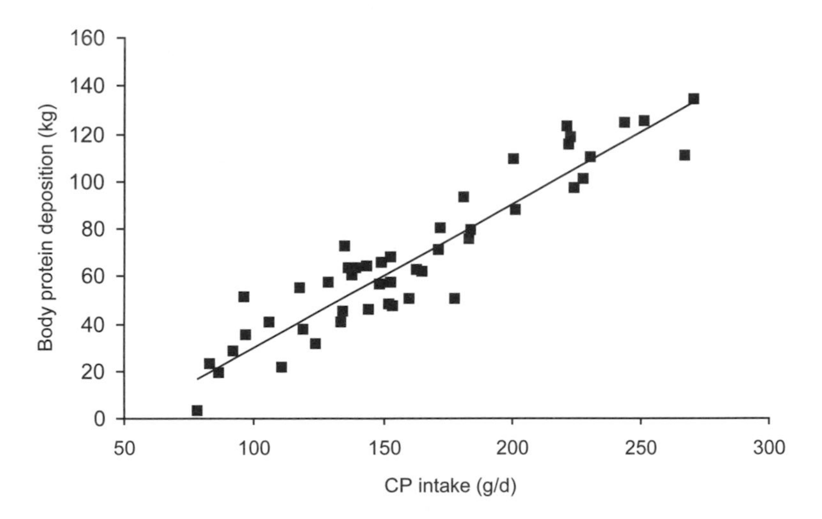

Figure 12.2 Relationship between dietary crude protein intake in milk replacer and whole-body protein deposition in Holstein calves fed milk replacer. (Bartlett *et al.*, 2005).

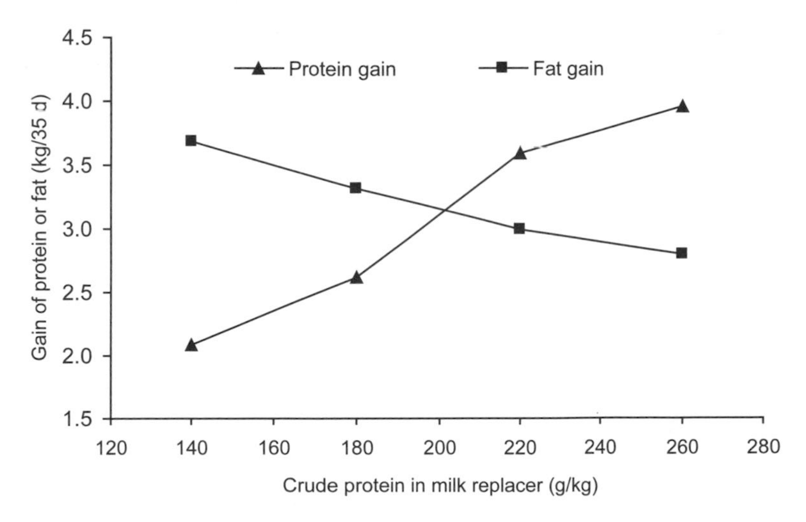

Figure 12.3 Relationships between whole-body protein deposition and whole-body fat deposition in Holstein calves fed milk replacers with increasing crude protein content. Reconstituted milk replacers were isoenergetic and were fed at a rate of 0.14 BW, adjusted weekly. The linear effect of crude protein content was significant ($P < 0.05$) for both. (Drawn from data in Bartlett *et al.*, 2005).

The author is aware of no evidence that might provide a basis for abnormal skeletal development as a result of greater milk intake in cattle, or any other mammalian species, during early life. One only has to consider the fact that beef heifers consume large amounts of milk early in life and live to greater average ages than do dairy cattle, with no evidence of bone problems. Although current biologically appropriate early nutrition programs have not been in existence long enough to make any evaluation of longevity, there is little basis on which to postulate negative effects if properly formulated and balanced diets are fed to meet nutrient requirements.

IS THERE EVIDENCE THAT INCREASED EARLY NUTRITION MIGHT BE BENEFICIAL?

As mentioned in the introduction, the possibility that differences in early-life nutrition might affect subsequent productivity or longevity has not been widely considered by researchers. Nevertheless, although data are limited there is more evidence that a higher plane of nutrition during early postnatal life may improve subsequent productivity than evidence for impairment.

Israeli researchers (Bar-Peled, Robinzon, Maltz, Tagari, Folman, Bruckental, Voet, Gacitua, and Lehrer, 1997) compared calves allowed to suckle nurse cows three times daily for 15 min per feeding during days 5-42 of life with calves fed a milk replacer (230 g protein/kg, 180 g fat/kg) in restricted amounts. All calves had free access to starter concentrate and hay. Suckled calves were changed to milk replacer at day 43, and the amount fed was decreased to be similar to controls by day 50. From day 51 until calving, management of the two groups was identical. Suckled calves consumed essentially no starter or hay during the treatment period, but consumed on average 0.14 more energy than calves fed limited amounts of milk replacer. Although ADG were greater during the treatment period (Table 12.1), suckled calves underwent a pronounced growth check at weaning. Consequently, suckled calves were actually nearly 10 kg lighter at 12 weeks of age. Nevertheless, suckled calves were 5 cm taller, calved 30 days earlier, and produced 453 kg more milk in first lactation than calves fed milk replacer in restricted amounts.

A Danish study compared calves fed 4.6 kg whole milk from birth to 56 days of life with calves allowed to suckle their dam for 30 min twice daily for 56 days (Foldager and Krohn, 1994). Suckled calves produced over 4.5 kg per day more milk in first lactation than the conventionally fed calves (1403 kg more for a 305-day lactation). A second Danish study compared calves fed 4.6 kg of whole milk for 42 days with calves fed ad libitum amounts of milk twice daily from a bucket for 42 days (Foldager, Krohn, and Mogensen, 1997). Calves fed the greater amounts of milk produced 489 kg more milk in first lactation than restricted-fed controls.

Table 12.1 GROWTH AND SUBSEQUENT PRODUCTION IN HOLSTEIN HEIFERS FED RESTRICTED AMOUNTS OF MILK REPLACER (CONVENTIONAL) OR ALLOWED TO SUCKLE COWS THREE TIMES DAILY DURING THE FIRST 42 DAYS OF LIFE (FROM BAR-PELED *et al.*, 1997).

Variable	*Conventional*	*SEM*	*Suckled*	*SEM*
n	20		20	
Live weight (kg)				
6 wk	61.9	3.2	73.4	4.7
12 wk	98.2	4.2	88.3	5.1
Live-weight gain (kg/d)				
0 to 6 wk	0.56	0.08	0.85*	0.11
7 to 12 wk	0.86	0.11	0.35*	0.35
0 to 12 wk	0.71	0.10	0.60	0.13
12 wk to conception	0.64	0.08	0.87*	0.09
Conception to calving	0.65	0.06	0.67	0.08
Age at conception (d)	426	13	394*	15
Calving age (d)	700	15	669*	12
Live weight at calving (kg)	507	24	544	30
Milk yield (kg/300 d)	9171	306	9624†	374
Wither height (cm)	134.4	1.9	139.7*	2.3

* $P < 0.05$
† $P < 0.10$

Our research group compared growth responses of Holstein heifer calves to dietary protein concentrations in milk replacer and starter designed for conventional restricted-feeding programs. Treatments were two protein contents in milk replacer (180 or 220 g/kg) and two protein contents in starter (180 or 220 g/kg) in a 2 x 2 factorial arrangement. Calves were assigned randomly to treatments (n = 15) at birth and received colostrum and transition milk for the first 2 days of life. Calves received milk replacer (solids reconstituted to 125 g/L) at a fixed allocation of 0.10 of birth BW divided into two feedings daily. During days 28 to 35, calves received one feeding of milk replacer; calves were weaned abruptly at day 35. Calves remained in individual hutches until day 49, and then were group-housed under common management and feeding until breeding and first calving.

Some data from this study are given in Table 12.2. Few interactions of milk replacer and starter CP content were detected, so only main effects are presented. Milk replacer containing the higher amount of CP resulted in greater ADG and heart girth increase. In contrast, higher starter CP content did not affect ADG, but resulted in greater concentrations of urea-N in plasma. A total of 47 of the original 60 heifers completed first lactations. BW at calving was 20 kg less for heifers fed the higher CP milk replacer as calves, although this difference was only marginally significant (P = 0.11). Despite being smaller on average, these heifers produced 508 kg more milk. It is interesting that this difference is of similar magnitude to the studies reported above, although the difference did not achieve statistical significance (P = 0.33). On the other hand, heifers that were fed the higher CP starter as calves produced an average of 540 kg less milk than those fed the lower protein starter; again, the difference was not statistically significant (P = 0.30). It is likely that, because of the immature rumen, nitrogen from CP degradation was in excess of available carbohydrate in the higher CP starter, as demonstrated by higher concentrations of urea-N in plasma. One might only speculate, therefore, whether the elevated ammonia or urea somehow resulted in developmental changes unfavourable for subsequent milk production.

Table 12.2 EFFECTS OF CRUDE PROTEIN (CP) CONTENT OF MILK REPLACER AND STARTER ON GROWTH AND SUBSEQUENT MILK PRODUCTION IN HOLSTEIN HEIFERS (DRACKLEY, J.K., BLOME, R.M., AND BARTLETT, K.S., UNPUBLISHED DATA).

	CP in milk replacer (g/kg)		*CP in starter (g/kg)*		
Variable	*180*	*220*	*180*	*220*	*SEM*
d 3-49					
n	30	30	30	30	—
Live-weight gain (kg/d)	0.52	0.58†	0.54	0.55	0.03
Wither height gain (cm)	18.5	20.3	18.6	19.7	0.9
Heart girth gain (cm)	13.5	15.4*	14.2	14.6	0.6
Plasma urea-N (mmol/L)	1.13	1.35*	1.09	1.39*	0.05
First lactation					
n	26	21	22	25	—
Calving age (mo)	25	26	26	26	—
Calving BW (kg)	581	561	570	572	8
Milk yield (kg)	11,354	11,862	11,878	11,338	366
Milk fat (kg)	410	425	423	412	14
Milk protein (kg)	350	362	367	346	12

* $P < 0.05$ within main effects.
† $P < 0.10$ within main effects.

Larger studies would be needed, of course, to confirm these tendencies. Given the magnitude of the differences between means and the degree of variability present in this experiment, over 200 cows per treatment would be necessary to have a 90% chance of demonstrating statistical significance at a probability of 0.05! While such large numbers of animals per treatment would be required in individual experiments to demonstrate effects in later life, data from smaller experiments could be combined for statistical analysis using techniques such as meta analysis. It is unfortunate, however, that most research studies on early life nutrition in calves have not measured subsequent milk production, reproductive efficiency, or longevity. Several studies are ongoing at the time of writing to determine longer-term effects of intensified milk replacer programs, but results are not yet available.

Potential mechanisms for long-term effects of early life nutrition

If improved early nutritional status does actually lead to enhanced productivity, health, and longevity, what might be the mechanisms? Given the close association between normal growth and developmental processes, the simple answer may be that more closely meeting nutrient requirements in early life may better support early developmental processes that in turn make for a better dairy cow. However, more defined mechanistic explanations or hypotheses would be more satisfying scientifically. Based on data obtained in other animal models, theories might be advanced that encompass 1) improved development or function of the immune system leading to better health status, 2) enhanced early mammary development, 3) altered endocrine development or function, 4) enhanced lean tissue development, and 5) metabolic programming or imprinting. Some of these theories have been referred to in preceding discussion, and the others are addressed in the following sections.

IMPROVED HEALTH STATUS

Poor health during early life is believed to have long-lasting effects on production and herd life. Epidemiological studies relating specific neonatal illnesses to later productivity generally have not found strong relationships between any specific illness or condition and subsequent survivability or productivity, although respiratory disease in calves increased the age at first calving (Correa, Curtis, Erb, and White, 1988). Perhaps most interesting is the report that early-life "dullness" in calves was a significant risk factor for shorter herd life. Calves that were characterized as having dullness before 90 days of age (defined as dull appearance, listlessness, droopy ears, and off feed) were 4.3 times more likely to die after 90 days of age (Curtis, White, and Erb, 1989) and 1.3 times more likely to leave the milking herd than herdmates (Warnick, Erb, and White, 1997). The authors speculated that this condition might reflect the combined effects of poor health and suboptimal nutrition.

Insufficiencies of protein or energy are well known to impair health and immune system function in other species (Woodward, 1998). What is the evidence that inadequate nutrition during early life decreases resistance to disease and compromises health and well being of calves? Williams, Day, Raven, and McLean (1981) compared calves fed two amounts of milk replacer solids (600 g/d and either 300 or 400 g/d) with either ad libitum or restricted access to calf starter. Calves fed the higher amount of milk replacer with ad libitum access to starter had the greatest ADG and least mortality. Other studies have shown that inadequate nutrition results in impaired immune responses in young calves. Griebel, Schoonderwoerd, and Babiuk (1987) fed neonatal calves either below maintenance or above maintenance intakes of milk replacer. Calves fed below maintenance lost BW and had higher (although nonsignificant) concentrations of cortisol in serum; lymphocytes isolated from these calves had decreased proliferative responses compared with adequately fed calves. Malnourished calves had lower primary antibody response to administration of K99 antigen.

Pollock, Rowan, Dixon, Carter, Spiller, and Warenius (1993) and Pollock, Rowan, Dixon, and Carter (1994) compared effects of weaning age (5, 9, or 13 weeks of age) and two levels of nutrition (400 g/d or 1000 g/d of milk replacer powder). Weaning at 5 weeks resulted in compromised lymphocyte responses (cellular immunity) at 10 weeks of age. The higher plane of nutrition, which was approximately twice maintenance, resulted in improved responses of cell-mediated immunity and decreased skin responses to antigen (Pollock *et al.*, 1993). In contrast, the high plane of nutrition resulted in decreased antibody titres to specific antigens, without changing total immunoglobulin concentration in serum (Pollock *et al.*, 1994). These results are consistent with recent evidence demonstrating that neonatal calves have vigorous antigen-specific cell-mediated immune responses but relatively weak antibody responses compared with adult cattle (Foote, Nonnecke, Waters, Fowler, Miller, and Beitz, 2003; Nonnecke, Foote, Smith, Pesch, and Van Amburgh, 2003).

Recent research (Nonnecke, Van Amburgh, Foote, Smith, and Elsasser, 2000; Nonnecke *et al.*, 2003) studied immune system characteristics in calves fed a milk replacer for greater rates of early growth (CP 300 g/kg, fat 200 g/kg; DM fed at 0.024 BW) or in calves fed a conventional milk replacer (CP 200 g/ kg, fat 200 g/kg) at slightly greater rates than industry standard (DM fed at 0.014 BW). Increased plane of nutrition did not affect total numbers of blood leucocytes, composition of the mononuclear leukocyte population, mitogen-stimulated DNA synthesis, or immunoglobulin M secretion. However, mononuclear leukocytes isolated from calves fed on the higher plane of nutrition produced more inducible nitric oxide and less interferon-γ than cells from conventionally fed calves. In another study by that research group, calves fed 568 g/d of conventional milk replacer powder (CP 220 g/kg, fat 200 g/kg) had lower antigen-induced proliferation of CD8 lymphocytes than calves fed 1136 g/d of a milk replacer (CP 280 g/kg, fat 200 g/kg) designed for intensified early nutrition (Foote *et al.*, 2003). Together, the available data support a role of nutritional status in at least some aspects of immune system function in young calves.

Improved neonatal nutritional status might be expected to affect the immune system via provision of deficient nutrients or energy, or by altering the endocrine environment that affects the developing immune system. Amino acid status, manifested particularly by glutamine availability, might be expected to affect both innate and specific immunity. Glutamine is a major fuel and biosynthetic amino acid for rapidly proliferating cells of the gut (Newsholme, Crabtree and Ardawi, 1985b), which might affect barrier function of the intestine in young calves. Moreover, glutamine is a major fuel for lymphocytes (Newsholme, Crabtree and Ardawi, 1985a) and other cells of the immune system (Calder and Yaqoob, 1999). Deficiency of glutamine results in immunosuppression in humans (Calder and Yaqoob, 1999) and mice (Kafkewitz and Bendich, 1983).

The anabolic hormones growth hormone (GH) and insulin-like growth factor-1 (IGF-1) play a direct role in integrating the growth, maintenance, repair, and function of the immune system in other species (Clark, 1997). Lymphocytes express receptors for both GH and IGF-1. In rodents, IGF-1 causes growth and maturation of B cells, increases size of the thymus and spleen, and increases antibody production by B cells (see review by Clark, 1997). Consequently, assuming that responses in calves are similar, increased concentrations of IGF-1 resulting from improved nutrition might be expected to enhance immunocompetence in calves.

Smith, Van Amburgh, Diaz, Lucy, and Bauman (2002) recently reported that calves fed for biologically appropriate growth during early life had greater concentrations of IGF-1 than did calves fed milk replacer at a conventional (restricted) rate. Plasma from calves in both groups showed increased IGF-1 concentrations in response to injection of bovine somatotropin at 5 weeks of age, but calves on the higher plane of nutrition also responded with increased growth rates. In our own experiments (Bartlett, 2001; Bartlett *et al.*, 2005), IGF-1 in plasma was increased by greater feeding rates and increased linearly as dietary CP increased (Figure 12.4). Patterns of change for plasma IGF-1 were nearly identical to those for ADG (Figure 4) and the two variables were highly correlated (r = 0.72). These results clearly show a functional IGF-1 system in young calves that is responsive to early nutritional status, although the relationship of the enhanced IGF-1 status to immune function and health remains to be determined.

The health status of young calves is probably affected by interactions between early nutrition and environment. Nutritional insufficiency might be especially problematic for immune function during cold or heat stress, when maintenance requirements for temperature regulation are increased. For example, our laboratory conducted an experiment to determine the value of supplementing milk replacer with energy sources for Jersey calves raised in hutches during winter (Drackley, Ruppert, Elliott, McCoy, and Jaster, 1996). To do so required establishment of an appropriate baseline feeding regimen. Preliminary investigations showed that Jersey heifer calves fed a conventional milk replacer at 0.08 BW did not maintain BW and had a high incidence of health problems. Calves fed the same milk replacer at 0.10 BW gained small amounts of BW but still were unhealthy. Only when calves were fed at a rate of 0.12 BW were they able to maintain health and BW gains.

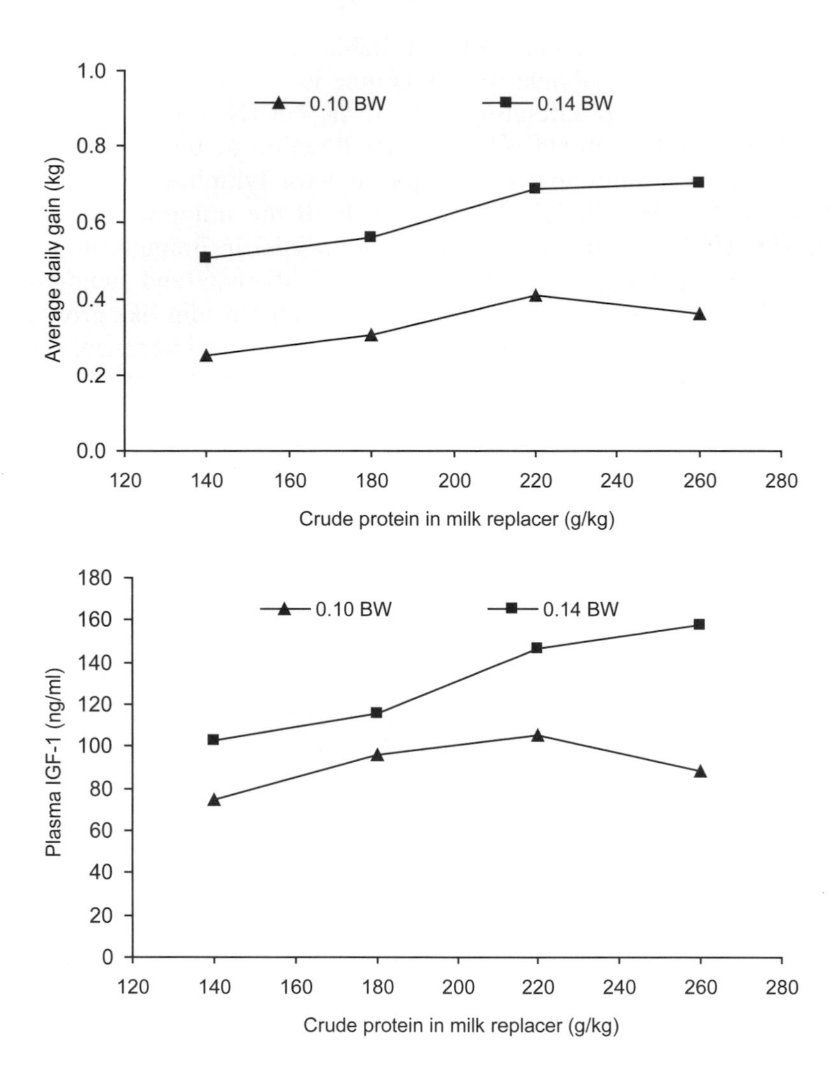

Figure 12.4 Average daily gains (top panel) and plasma IGF-1 concentrations (bottom panel) in Holstein calves fed milk replacers with increasing crude protein content at either 0.10 BW or 0.14 BW. The effects of feeding rate and the linear effect of increasing dietary protein content were significant ($P < 0.05$); for IGF-1, the interaction between feeding rate and the linear effect of increasing crude protein content was also significant. (Drawn from data in Bartlett *et al.*, 2005)

In other species, the degree of immune challenge present in the environment can have marked effects on growth and feed efficiency. For example, pigs subjected to challenge with bacterial lipopolysaccharide had decreased growth and feed efficiency compared with control pigs; growth of challenged pigs was lower than that of pair-fed littermates, indicating that decreased feed intake accounted for only about 2/3 of the growth depression (Dritz, Owen, Goodband, Nelssen, Tokach, Chengappa, and Blecha, 1996). Pigs raised in a clean environment with minimal exposure to pathogens had greater rates of gain, improved feed efficiencies, and a greater lean-to-fat ratio than pigs raised in the presence of a high degree of pathogen exposure (Williams, Stahly, and Zimmerman, 1997). Such responses are probably attributable both to increased nutrient demands by the immune system and to the anti-growth effects of cytokines produced by the activated immune system. Klasing and Calvert (1999) calculated that up to 60% of the impaired growth of chicks during an intense immune response could be explained by known processes, including synthesis of the acute phase proteins. Almost 7% of lysine intake was used by immune processes during a lipopolysaccharide challenge in chicks (Klasing and Calvert, 1999). Consequently, nutrient demands of the immune system may compete directly with growth processes for a limited supply of nutrients.

A fascinating discussion of the partitioning of nutrients among maintenance, immune function, and growth or other productive functions has been provided by Houdijk, Jessop, and Kyriazakis (2001). In particular, these authors raise the possibility that inadequate protein supply may limit function of the immune system, through the importance of glutamine (and potentially other amino acids) as discussed above. Given the importance of protein intake for growth, as demonstrated by our experiments (Bartlett, McKeith, and Drackley, 2001b; Blome *et al.*, 2003; Bartlett *et* al., 2005) in which increasing protein content of isoenergetic milk replacers markedly increased ADG and efficiency of gain at similar overall energy retention, relationships among nutrition, growth, and immune function in calves seem to be fertile areas for future research.

ALTERED ENDOCRINE DEVELOPMENT

In addition to the effects on the somatotrophic axis discussed above and reviewed elsewhere (Blum and Hammon, 1999), evidence exists for alterations in other aspects of the endocrine system as a consequence of differences in early nutrition. Insulin concentrations increase markedly in response to a higher plane of nutrition (Diaz *et al.*, 2001). Cortisol concentrations during the first week of life were higher, and prolactin concentrations were lower, in calves fed milk replacer compared with calves that remained on bovine colostrum (Hammon and Blum, 1998), although differences in nutrient concentration make interpretation difficult. Whether these early differences lead to permanent differences in hormonal secretion or responsiveness is not known, as is whether increased or decreased concentrations during the early life period might affect developmental processes to alter metabolism (discussed further below).

ENHANCED LEAN TISSUE DEVELOPMENT

As discussed earlier, appropriate nutrition during early life results in increased stature and BW, increased lean tissue deposition, and increased whole-body protein deposition (Figures 2 and 3). These changes are probably signalled by energy status via insulin and amino acid status via leucine or other branched-chain amino acids (Davis and Reeds, 1998; Meijer, 2003); signal transduction mechanisms for both insulin and amino acids converge to directly activate protein translation activity (Davis, Nguyen, Suyawan, Bush, Jefferson, and Kimball, 2000; Schmelzle and Hall, 2000; Meijer, 2003). Studies with rats (Winick and Noble, 1966), lambs (Greenwood *et al.*, 1998, 2000), and pigs (Campbell and Dunkin, 1983) that were protein malnourished during early life have demonstrated that growth is never fully restored in later life. Is it possible that changes in early growth observed with improved early protein nutrition are favourable for continued lean growth and subsequent productivity in dairy cattle?

Some evidence indicates that dairy cows are becoming leaner as selection for milk yield continues (Murphy, Chester-Jones, Crooker, Ziegler, Appleman, and Hansen, 1991), similar to effects of exogenous somatotropin administration (Brown, Taylor, DePeters, and Baldwin, 1989). Milk yield in established lactation is more sensitive to postruminal supplies of essential amino acids than to glucose (Ørskov, Grubb, and Kay, 1977). Potential advantages to maternal body protein balance for adaptations to lactation, lactation performance, and health have been postulated (Bell, Burhans, and Overton, 2000). Degradation of skeletal muscle protein appears to increase during the first 7-10 days postpartum (Overton, Drackley, Douglas, Emmert, and Clark, 1998), perhaps to provide amino acids for gluconeogenesis (Overton, Drackley, Ottemann-Abbamonte, Beaulieu, Emmert, and Clark, 1999) and immune function, as well as for milk protein synthesis. Glutamine is an important fuel for gut epithelial cells of ruminants as in other species (Okine, Glimm, Thompson, and Kennelly, 1995; Beaulieu, Overton, and Drackley, 2001). Muscle is also an important site of utilization of NEFA mobilized from adipose tissue, and the ketone bodies that are generated from NEFA in the liver. These data together support the hypothesis that muscle may be a critically important tissue to support whole-body metabolism and health during lactation and during metabolic adaptations.

Early postnatal development of muscle in cattle is characterized by hypertrophy, involving increases in both DNA and protein contents of muscle. Increased DNA results from proliferation of satellite cells. Although not well characterized for dairy calves, in pigs the rate of satellite cell proliferation and DNA incorporation into skeletal muscles is highest at birth and declines sharply with postnatal age (Mesires and Doumit, 2002). Greater DNA allows greater transcription of ribosomal RNA and greater protein synthetic capacity. Whether postnatal nutrition affects the rate or duration of the period of satellite cell incorporation is not known. However, IGF-1 is a known mitogen for satellite cells, and insulin increases proliferation of bovine satellite cells in vitro (Cassar-Malek, Langlois, Picard, and Geay, 1999). Consequently, improved nutritional status that results in greater insulin and IGF-1 concentrations (Diaz *et al.*, 2001;

Bartlett *et al.*, 2005) would be consistent with the possibility of greater postnatal satellite cell incorporation and greater long-term protein synthetic ability in muscle. In turn, this could confer advantages to production and longevity if the above-postulated effects of muscle mass and/or maternal protein balance are real.

METABOLIC PROGRAMMING OR IMPRINTING

Another way by which early nutrition might impart long-term effects on later-life events is the phenomenon of "metabolic programming" or "nutritional imprinting". This idea has been defined as the "early adaptations to a nutritional stress or stimulus that permanently change the physiology and metabolism of the organism and continue to be expressed even in the absence of the stimulus/ stress that initiated them..." (Patel and Srinivasan, 2002). An abundance of data, both experimental in animals and epidemiological in humans, provides evidence for the existence of such effects of early nutrition on responses in later life of the endocrine system, metabolism, and susceptibility to chronic disease (see reviews by Lucas, 1998; Waterland and Garza, 1999; Patel and Srinivasan, 2002).

The term "metabolic imprinting" encompasses "adaptive responses to specific nutritional conditions early in life that are characterized by 1) a susceptibility limited to a critical ontogenic window early in development, 2) a persistent effect lasting through adulthood, 3) a specific and measurable outcome (that may differ quantitatively among individuals), and 4) a dose-response or threshold relation between a specific exposure and outcome." (Waterland and Garza, 1999). In the context of this chapter, there obviously is much uncertainty about the existence of such effects in cattle, as "specific and measurable outcomes" are not available. Moreover, potential "critical developmental windows" in early postnatal life for subsequent productivity and longevity also remain to be determined. Nevertheless, clear evidence for imprinting effects does exist for cattle and other ruminants. For example, maternal androgen administration during a critical window of foetal development results in the genetically female offspring displaying masculine characteristics of structure, growth, and metabolism (Hansen, Drackley, Berger, and Grum, 1995; Hansen, Drackley, Berger, Grum, Cremin, Lin, and Odle, 1995; Reiling, Drackley, Grum, and Berger, 1997). Another example is the imprinting of "memory" of antigen exposure in cells of the immune system.

Much of the knowledge of early postnatal imprinting effects has been obtained with rodent models, which are born relatively much more immature than cattle. Consequently, foetal nutritional status might be more important to cattle and other ruminants (Greenwood and Bell, 2003). Maternal undernutrition in ewes, which led to lambs of smaller birth weight, resulted in more lasting differences in growth, body composition, and muscle development than did restrictions in postnatal nutrition (Greenwood *et al.*, 1998, 2000). Zanker, Hammon, and Blum (2001) searched for lasting effects of starvation on the first day of life in newborn calves. Although plasma concentrations of total protein and globulin protein were decreased in calves to 30 days of age, they found no differences in haematological, metabolic, or endocrine measurements or on growth performance that would suggest imprinting of later responses; however, they did not study calves beyond 30 days of age and the treatment was acute rather than chronic.

An excellent discussion of the potential biology responsible for metabolic imprinting has been provided by Waterland and Garza (1999), who proposed five candidate mechanisms: 1) induced variations in organ structure, 2) alterations in cell number, 3) clonal selection, 4) metabolic differentiation, and 5) hepatocyte polyploidization. The first of these, organ structure, might involve induced changes in vascularization or innervation as organs develop, or changes in the physical relationships between different cell types in an organ. Such changes may result in differences in the ability of cells to respond to external signals such as blood-born nutrients or hormones. The second mechanism, differences in cell numbers, might lead to permanent changes in the size or function of organ systems. Although in cattle the number of muscle cells is fixed at birth (Swatlund, 1994), effects on cellularity of other organs such as liver, adipose, and mammary are less clear. Clonal selection refers to subtle differences among proliferating cells that are induced by differences in early nutrition that in turn creates different microenvironments for the developing organ. Certain subpopulations of cells may then be favoured during subsequent cell turnover as cells compete for available nutrients.

Metabolic differentiation is the process by which all cells establish a stable profile of gene expression that in turn determines the "basal" metabolic activity in that cell. Metabolic differentiation thus depends on the profile of enzymes expressed in the cell, but also on the complex milieu of transcription factors, hormones, growth factors, endocrine receptors, and membrane transporters present during the developmental period. Phenomena whereby nutrition-induced changes become stably expressed and passed on to subsequent generations of cells in the organism are called "epigenesis". Potential molecular mechanisms underpinning epigenesis include the varying combinations of transcription factors and other DNA-binding proteins, alterations in chromatin structure, and DNA methylation.

The final potential mechanism of imprinting described by Waterland and Garza (1999) is hepatocyte polyploidy, which refers to cells that contain more than the normal copy number of chromosomes. Increases in hepatocyte polyploidy occur during postnatal development and persist into adulthood. Hepatocyte polyploidy may result in increased metabolic activity in the liver through increased basal and inducible expression of enzymes resulting from increased RNA and rates of RNA synthesis. Whether early postnatal nutritional status might affect the rate or extent of polyploidy in the neonatal calf is not known.

Concluding remarks

Few data are available to document conclusively what effects early nutrition might have on later life productivity, health, and longevity. Current conventional systems restrict nutrient intake during the milk-feeding period in an effort to encourage early intake of calf starter, allow earlier weaning, and decrease costs of heifer rearing. Programs of more intensive liquid feeding early in life result in nutrient intakes that more nearly meet the requirements of neonatal calves for normal growth and development. Consequently, from a biological perspective it would be difficult to argue that improving nutritional status of the young calf during the first few weeks of life should be anything but positive for subsequent productivity and longevity.

Malnutrition or sub-optimal nutrition has been well documented to affect growth and health status in the young of other species. To what extent can parallels be drawn between early development of dairy heifers and information on early-life events in other mammalian species? Knowledge from other species could be used as the basis for testable hypotheses in dairy cattle; however, experiments to test such hypotheses will probably require very large numbers of animals and will be long, difficult, and expensive to conduct.

References

Appleby, M.C., Weary, D.M., and Chua, B. (2001) Performance and feeding behaviour of calves on ad libitum milk from artificial teats. *Applied Animal Behaviour Science*, **74**, 191-201.

Bagley, C.V. (2001) Influence of nutrition and management on calf scours (bovine neonatal diarrhea). In *Proceedings 2001 Intermountain Nutrition Conference: Nutrition and Health for Farm Profitability*, pp 27-37. Publication No. 169, Utah Agricultural Experiment Station, Utah State University, Logan.

Bar-Peled, U., Robinzon, B., Maltz, E., Tagari, H., Folman, Y., Bruckental, I., Voet, H., Gacitua, H., and Lehrer, A.R. (1997) Increased weight gain and effects on production parameters of Holstein heifer calves that were allowed to suckle from birth to six weeks of age. *Journal of Dairy Science*, **80**, 2523-2528.

Bartlett, K.S. (2001) *Interactions of Protein and Energy Supply from Milk Replacers on Growth and Body Composition of Dairy Calves.* M.S. Thesis, University of Illinois, Urbana.

Bartlett, K.S., McKeith, F.K., and Drackley, J.K. (2001a) Effects of energy sources in milk replacers on growth and body composition of Holstein calves. *Journal of Dairy Science*, **84**, 1560. (Abstr.)

Bartlett, K.S., McKeith, F.K., and Drackley, J.K. (2001b) Effects of feeding rate and protein concentration in milk replacers on growth and body composition of Holstein calves. *Journal of Dairy Science*, **84**, 1560. (Abstr.)

Bartlett, K.S., McKeith, F.K., VandeHaar, M.J., Dahl, G.E., and Drackley, J.K. (2005) Growth and body composition of dairy calves fed milk replacers containing different amounts of protein at two feeding rates. *Journal of Animal Science* (in press).

Beaulieu, A.D., Overton, T.R., and Drackley, J.K. (2001) Substrate oxidation by isolated ovine enterocytes is increased by phlorizin-induced glucosuria. *Canadian Journal of Animal Science*, **81**, 585-588.

Bell, A.W., Burhans, W.S., and Overton, T.R. (2000) Protein nutrition in late pregnancy, maternal protein reserves and lactation performance in dairy cows. *Proceedings of the Nutrition Society*, **59**, 119-136.

Blome, R.M., Drackley, J.K., McKeith, F.K., Hutjens, M.F., and McCoy, G.C. (2003) Growth, nutrient utilization, and body composition of dairy calves fed milk replacers containing different amounts of protein. *Journal of Animal Science*, **81**, 1641-1655.

Blum, J.W., and Hammon, H. (1999) Endocrine and metabolic aspects in milk-fed calves. *Domestic Animal Endocrinology*, **17**, 219-230.

Brown, D.L., Taylor, S.J., DePeters, E.J., and Baldwin, R.L. (1989) Influence of sometribove, USAN (recombinant methionyl bovine somatotropin) on the body composition of lactating cattle. *Journal of Nutrition*, **119**, 633-638.

Brown, E.G., VandeHaar, M.J., Daniels, K.M., Liesman, J.S., Chapin, L.T., and Weber Nielsen, M.S. (2002) Increasing energy and protein intake of Holstein heifer calves increases mammary development. *Journal of Dairy Science*, **85**(Suppl. 1), 80. (Abstr.)

Burrin, D.G., Fiorotto, M.F., and Davis, T.A. (2001) Importance of neonatal nutrition: what have we learned from human and animal studies? In *Proceedings Cornell Nutrition Conference*, pp. 1-13. Cornell University, Ithaca, New York.

Calder, P.C., and Yaqoob, P. (1999) Glutamine and the immune system. *Amino Acids*, **17**, 227-241.

Campbell, R.G., and Dunkin, A.C. 1983. The influence of protein nutrition in early life on growth and development of the pig. 1. Effects on growth performance and body composition. *British Journal of Nutrition*, **50**, 605-617.

Cassar-Malek, I., Langlois, N., Picard, B., and Geay, Y. (1999) Regulation of bovine satellite cell proliferation and differentiation by insulin and triiodothyronine. *Domestic Animal Endocrinology*, **17**, 373-388.

Clark, R. (1997) The somatogenic hormones and insulin-like growth factor-1: Stimulators of lymphopoiesis and immune function. *Endocrine Reviews*, **18**, 157-179.

Correa, M.T., Curtis, C.R., Erb, H.N., and White, M.E. (1988) Effect of calfhood morbidity on age at first calving in New York Holstein herds. *Preventive*

Veterinary Medicine, **6**, 253-262.

Curtis, C.E., White, M.E., and Erb, H.N. (1989) Effects of calfhood morbidity on long-term survival in New York Holstein herds. *Preventive Veterinary Medicine*, **7**, 173-186.

Davis, C.L., and Drackley, J.K. (1998) *The Development, Nutrition, and Management of the Young Calf.* Iowa State University Press, Ames, Iowa.

Davis, T.A., and Reeds, P.J. (1998) The roles of nutrition, development and hormone sensitivity in the regulation of protein metabolism: an overview. *Journal of Nutrition*, **128**, 340S-341S.

Davis, T.A., Nguyen, H.V., Suyawan, A., Bush, J.A., Jefferson, L.S., and Kimball, S.R. (2000) Developmental changes in the feeding-induced stimulation of translation initiation in muscle of neonatal pigs. *American Journal of Physiology*, **279**, E1226-E1234.

Diaz, M.C., Van Amburgh, M.E., Smith, J.M., Kelsey, J.M., and Hutten, E.L. (2001) Composition of growth of Holstein calves fed milk replacer from birth to 105-kilogram body weight. *Journal of Dairy Science*, **84**, 830-842.

Drackley, J.K., Ruppert, L.D., Elliott, J.P., McCoy, G.C., and Jaster, E.H. (1996) Effects of increased solids in milk replacer on Jersey calves housed in hutches during winter. *Journal of Dairy Science*, **79**(Suppl. 1), 154. (Abstr.)

Dritz, S.S., Owen, K.Q., Goodband, R.D., Nelssen, J.L., Tokach, M.D., Chengappa, M.M., and Blecha, F. (1996) Influence of lipopolysaccharide-induced immune challenge and diet complexity on growth performance and acute-phase protein production in segregated early-weaned pigs. *Journal of Animal Science*, **74**, 1620-1628.

Flower, F.C., and Weary, D.M. (2001) Effects of early separation on the dairy cow and calf: 2. Separation at 1 day and 2 weeks after birth. *Applied Animal Behaviour Science*, **70**, 275-284.

Foldager, J., and Krohn, C.C. (1994) Heifer calves reared on very high or normal levels of whole milk from birth to 6-8 weeks of age and their subsequent milk production. *Proceedings of the Society for Nutrition and Physiology*, 3. (Abstr.)

Foldager, J., Krohn, C.C. and Mogensen, L. (1997) Level of milk for female calves affects their milk production in first lactation. In *Proceedings 48th Annual Meeting European Association for Animal Production.*

Foote, M.R., Nonnecke, B.J., Waters, W.R., Fowler, M.A., Miller, B.L., and Beitz, D.C. (2003) Effects of plane of nutrition on proliferation of lymphocyte subsets from neonatal calves vaccinated with *M. bovis* bacillus Calmette-Guerin (BCG). *FASEB Journal*, **17**, A1124. (Abstr.)

Greenwood, P.L., and Bell, A.W. (2003) Consequences of intra-uterine growth retardation for postnatal growth, metabolism and pathophysiology. *Reproduction Supplements*, **61**, 195-206.

Greenwood, P.L., Hunt, A.S., Hermanson, J.W., Bell, A.W. (1998) Effects of birth weight and postnatal nutrition on neonatal sheep: I. Body growth and composition, and some aspects of energetic efficiency. *Journal of Animal Science*, **76**, 2354-2367.

Greenwood, P.L., Hunt, A.S., Hermanson, J.W., and Bell, A.W. (2000) Effects of birth weight and postnatal nutrition on neonatal sheep: II. Skeletal muscle growth and development. *Journal of Animal Science*, **78**, 50-61.

Griebel, P.J., Schoonderwoerd, M., and Babiuk, L.A. (1987) Ontogeny of the immune response: effect of protein energy malnutrition in neonatal calves. *Canadian Journal of Veterinary Research*, **51**, 428-435.

Hafez, E.S.E., and Lineweaver, L.A. (1968) Suckling behaviour in natural and artificially fed neonate calves. *Zeitschrift fur Tierpsychologie*, **25**, 187-198.

Hammon, H.M., and Blum, J.W. (1998) Metabolic and endocrine traits of neonatal calves are influenced by feeding colostrum for different durations or only milk replacer. *Journal of Nutrition*, **128**, 624-632.

Hansen, L.R., Drackley, J.K., Berger, L.L., and Grum, D.E. (1995) Prenatal androgenization of lambs: I. Alteration of growth, carcass characteristics, and metabolites in blood. *Journal of Animal Science*, **73**, 1694-1700.

Hansen, L.R., Drackley, J.K., Berger, L.L., Grum, D.E., Cremin, J.D., Jr., Lin, X., and Odle, J. (1995) Prenatal androgenization of lambs: II. Metabolism in adipose tissue and liver. *Journal of Animal* Science, **73**, 1701-1712.

Houdijk, J.G.M., Jessop, N.S., and Kyriazakis, I. (2001) Nutrient partitioning between reproductive and immune functions in animals. *Proceedings of the Nutrition Society*, **60**, 515-525.

Huber, J.T., Silva, A.G., Campos, O.F., and Mathieu, C.M. (1984) Influence of feeding different amounts of milk on performance, health, and absorption capability of baby calves. *Journal of Dairy Science*, **67**, 2957-2963.

Jasper, J., and Weary, D.M. (2002) Effects of ad libitum milk intake on dairy calves. *Journal of Dairy Science*, **85**, 3054-3058.

Kafkewitz, D., and Bendich, A. (1983) Enzyme-induced asparagine and glutamine depletion and immune system function. *American Journal of Clinical Nutrition*, **37**, 1025-1030.

Khouri, R.H., and Pickering, F.S. (1968) Nutrition of the milk-fed calf I. Performance of calves fed on different levels of whole milk relative to body weight. *New Zealand Journal of Agricultural Research*, **11**, 227-236.

Kim, J.H., Heo, K.N., Odle, J., Han, I.K., and Harrell, R.J. (2001) Liquid diets accelerate the growth of early-weaned pigs and the effects are maintained to market weight. *Journal of Animal Science*, **79**, 427-434.

Klasing, K.C., and Calvert, C.C. (1999) The care and feeding of an immune system: an analysis of lysine needs. In *Protein Metabolism and Nutrition*, pp 253-264. Edited by G.E. Lobley, A. White, and J.C. MacRae. EAAP

Publication No. 96, Wageningen Press, Wageningen.
Lucas, A. (1998) Programming by early nutrition: an experimental approach. *Journal of Nutrition*, **128**, 401S-406S.
Meijer, A.J. (2003) Amino acids as regulators and components of nonproteinogenic pathways. *Journal of Nutrition*, **133**, 2057S-2062S.
Mesires, N.T., and Doumit, M.E. (2002) Satellite cell proliferation and differentiation during postnatal growth of porcine skeletal muscle. *American Journal of Physiology*, **282**, C899-C906.
Murphy, K.D., Chester-Jones, H., Crooker, B.A., Ziegler, D.M., Appleman, R.D., and Hansen, L.B. (1991) Dairy heifers from divergent genetic lines differ in dry matter intake, growth related parameters, and body composition. *Journal of Dairy Science*, **74**(Suppl. 1), 262. (Abstr.)
Mylrea, P.J. (1966) Digestion in young calves fed whole milk ad lib. and its relationship to calf scours. *Research in Veterinary Science*, **7**, 407-416.
National Research Council. (2001) *Nutrient Requirements of Dairy Cattle*. 7th Revised Edition. National Academy Press, Washington, DC.
Newsholme, E.A., Crabtree, B., and Ardawi, M.S. (1985a) Glutamine metabolism in lymphocytes: its biochemical, physiological and clinical importance. *Quarterly Journal of Experimental Physiology*, **70**, 473-489.
Newsholme, E.A., Crabtree, B., and Ardawi, M.S. (1985b) The role of high rates of glycolysis and glutamine utilization in rapidly dividing cells. *Bioscience Reports*, **5**, 393-400.
Nocek, J.E., and Braund, D.G. (1986) Performance, health, and postweaning growth on calves fed cold, acidified milk replacer ad libitum. *Journal of Dairy Science*, **69**, 1871-1883.
Nonnecke, B.J., Van Amburgh, M.E., Foote, M.R., Smith, J.M., and Elsasser, T.H. (2000). Effects of dietary energy and protein on the immunological performance of milk replacer-fed Holstein bull calves. *Journal of Dairy Science*, **83**(Suppl. 1), 135. (Abstr.)
Nonnecke, B.J., Foote, M.R., Smith, J.M., Pesch, B.A., and Van Amburgh, M.E. (2003) Composition and functional capacity of blood mononuclear leukocyte populations from neonatal calves on standard and intensified milk replacer diets. *Journal of Dairy Science*, **86**, 3592-3604.
Okine, E.K., Glimm, D.R., Thompson, J.R., and Kennelly, J.J. (1995) Influence of stage of lactation on glucose and glutamine metabolism in isolated enterocytes from dairy cattle. *Metabolism*, **44**, 325-331.
Ørskov, E.R., Grubb, D.A., and Kay, R.N. (1977) Effect of postruminal glucose or protein supplementation on milk yield and composition in Friesian cows in early lactation and negative energy balance. *British Journal of Nutrition*, **38**, 397-405.
Overton, T.R., Drackley, J.K., Douglas, G.N., Emmert, L.S., and Clark, J.H. (1998) Hepatic gluconeogenesis and whole-body protein metabolism of periparturient dairy cows as affected by source of energy and intake

of the prepartum diet. *Journal of Dairy Science*, **81**(Suppl. 1), 295. (Abstr.)

Overton, T.R., Drackley, J.K., Ottemann-Abbamonte, C.J., Beaulieu, A.D., Emmert, L.S., and Clark, J.H. (1999) Substrate utilization for hepatic gluconeogenesis is altered by increased glucose demand in ruminants. *Journal of Animal Science*, **77**, 1940-1951.

Patel, M.S., and Srinivasan, M. (2002) Metabolic programming: causes and consequences. *Journal of Biological Chemistry*, **277**, 1629-1632.

Pollock, J.M., Rowan, T.G., Dixon, J.B., Carter, S.D., Spiller, D., and Warenius, H. (1993) Alteration of cellular immune responses by nutrition and weaning in calves. *Research in Veterinary Science*, **55**, 298-306.

Pollock, J.M., Rowan, T.G., Dixon, J.B., and Carter, S.D. (1994) Level of nutrition and age at weaning: effects on humoral immunity in young calves. *British Journal of Nutrition*, **71**, 239-248.

Reiling, B.A., Drackley, J.K., Grum, L.R., and Berger, L.L. (1997) Effects of prenatal androgenization and lactation on adipose tissue metabolism in finishing single-calf heifers. *Journal of Animal Science*, **75**, 1504-1513.

Roberts, S.B., and McDonald, R. (1998) The evolution of a new research field: metabolic programming by early nutrition. *Journal of Nutrition*, **128**:400S.

Roy, J.H.B. (1980) Factors affecting susceptibility of calves to disease. *Journal of Dairy Science*, **63**, 650-664.

Schmelzle, T., and Hall, M.N. (2000) TOR, a central controller of cell growth. *Cell*, **103**, 253-262.

Sejrsen, K., Purup, S., Martinussen, H., and Vestergaard, M. (1998) Effect of feeding level on mammary growth in calves and prepubertal heifers. *Journal of Dairy Science*, **81**(Suppl. 1), 377. (Abstr.)

Sejrsen, K., Purup, S., Vestergaard, M., and Foldager, J. (2000) High body weight gain and reduced bovine mammary growth: physiological basis and implications for milk yield potential. *Domestic Animal Endocrinology*, **19**, 93-104.

Smith, J.M., Van Amburgh, M.E., Diaz, M.C., Lucy, M.C., and Bauman, D.E. (2002) Effect of nutrient intake on the development of the somatotropic axis and its responsiveness to GH in Holstein bull calves. *Journal of Animal Science*, **80**, 1528-1537.

Swatlund, H.J. (1994) *Structure and Development of Meat Animals and Poultry.* Technomic Publishing Company, Inc., Lancaster, Pennsylvania.

Tikofsky, J.N., Van Amburgh, M.E., and Ross, D.A. (2001) Effect of varying carbohydrate and fat content of milk replacer on body composition of Holstein bull calves. *Journal of Animal Science*, **79**, 2260-2267.

Warnick, L.D., Erb, H.N., and White, M.E. (1997) The relationship of calfhood morbidity with survival after calving in 25 New York Holstein herds. *Preventive Veterinary Medicine*, **31**, 263-273.

Waterland, R.A., and Garza, C. (1999) Potential mechanisms of metabolic imprinting that lead to chronic disease. *American Journal of Clinical Nutrition*, **69**, 179-197.

Williams, N.H., Stahly, T.S., and Zimmerman, D.R. (1997) Effect of level of chronic immune system activation on the growth and dietary lysine needs of pigs fed from 6 to 112 kg. *Journal of Animal Science*, **75**, 2481-2496.

Williams, P.E.V., Day, D., Raven, A.M., and McLean, J.A. (1981) The effect of climatic housing and level of nutrition on the performance of calves. *Animal Production*, **32**, 133-141.

Winick, M., and Noble, A. (1966) Cellular response in rats during malnutrition at various ages. *Journal of Nutrition*, **89**, 300-306.

Woodward, B. (1998) Protein, calories, and immune defenses. *Nutrition Reviews*, **56**, S84-S92.

Zanker, I.A., Hammon, H.M., and Blum, J.W. (2001) Delayed feeding of first colostrum: are there prolonged effects on haematological, metabolic and endocrine parameters and on growth performance in calves? *Journal of Animal Physiology and Animal Nutrition*, **85**, 53-66.

13

MAMMARY DEVELOPMENT AND MILK YIELD POTENTIAL

KRIS SEJRSEN
Department of Animal Nutrition and Physiology, Danish Institute of Agricultural Sciences, Foulum, Denmark

Introduction

The main objective of raising replacement heifers is to produce healthy cows that are able to fulfil their genetic potential for milk production. The first limiting factor for milk production is the synthetic capacity of the mammary glands. Since the majority of mammary gland development in cattle takes place before calving, it is essential that feeding and management during the rearing period secure optimal development of the mammary glands. It is obvious, therefore, that successful rearing of heifers is not a question of simply maximizing growth rate or body weight at calving; control of the growth pattern during the rearing period is also important for optimal mammary development and subsequent milk yield potential.

Normal mammary development

In mammals – including cattle – lactation is the last step in the reproductive cycle, supplying the newborn young with antibodies and essential nutrients. In line with the role of lactation in reproduction, mammary development occurs in distinct phases related to reproductive development during foetal life, puberty, pregnancy, and lactation.

The basic structures of the mammary glands are formed in foetal life. At birth the non-epithelial tissues and the outer shape of the glands are almost fully developed, but the epithelial tissue, in contrast, is still rudimentary. From birth to 2-3 months of age there is still no growth of the epithelial tissue.

At 2-3 months of age the glands start to grow at a faster rate than the rest of the body; the growth becomes allometric (Sinha and Tucker, 1969). In this phase there is rapid growth of the mammary fat pad and of the mammary ducts, but no alveoli are formed. The allometric growth phase of the mammary glands is closely linked to reproductive development. Most studies suggest that the allometric growth phase ends at onset of puberty or shortly thereafter. At this stage of development the mammary glands of heifers weigh about 2 to 3 kg, of which only 0.5 to 1 kg is parenchymal tissue containing the mammary ducts. The parenchymal tissue usually contains 10 to 20% epithelial cells, 40 to 50% connective tissue and 30 to 40% fat cells. In large dairy breeds average age at onset of puberty is 9-11 months, and average body weight is approximately 275 kg. Variation between animals in age and body weight at onset of puberty is wide, but puberty occurs at approximately the same live weight independent of feeding level (see Sejrsen and Purup, 1997).

From puberty to pregnancy, mammary development is isometric and relatively limited. During pregnancy, however, mammary growth is both qualitatively and quantitatively much more extensive than during puberty. In early pregnancy, growth of the mammary ducts continues and from mid pregnancy there is extensive lobulo-alveolar development. The lactating mammary glands can weigh as much as 25–30 kg, and lactating parenchyma consist of 40-50% epithelial cells (ducts and alveoli), 15-20% lumen, about 40% connective tissue and almost no fat cells (Harrison *et al.*, 1983).

In cattle, mammary development is essentially complete at calving. It is therefore likely that factors affecting mammary development in the rearing period can influence the milk yield capacity of heifers in their first lactation. Considering the pattern of mammary development, the rearing period can be divided into four separate phases: 1) the calf period from birth to 2-3 months of age; 2) the pre-pubertal period from 2-3 months of age to puberty; 3) the post-pubertal period from puberty to conception; and 4) pregnancy.

Effect of feeding on mammary development

As early as the 1940s, it was observed that mammary development was incomplete in heifers raised on a high feeding level (Herman and Ragsdale, 1946). This initial observation was later confirmed by Swanson (1960), Little and Kay (1979) and Harrison *et al.* (1983). Furthermore, it was noted that the glands of heifers raised on a high feeding level were clearly different in shape and size from glands of heifers raised on moderate feeding levels (Herman and Ragsdale, 1946; Swanson, 1960; Little and Kay, 1979; Foldager and Sejrsen, 1991).

In the early experiments, different feeding levels generally were maintained throughout the rearing period, without regard for the different phases of mammary development. Results from rearing experiments designed to examine mammary development suggested that the prepubertal period of development was especially critical for subsequent milk yield capacity of heifers (Sejrsen, 1978). In agreement, Sejrsen *et al.* (1982) found that the amount of mammary parenchyma was reduced by a high feeding level in the prepubertal period, but not in the post-pubertal period of virgin heifers. Subsequently, the effect of feeding level in the prepubertal period has been investigated in a large number of experiments (see Sejrsen *et al.*, 2000). The initial finding was confirmed in most of the subsequent experiments. Figure 13.1 shows cross sections of mammary glands from heifers fed moderate and high feeding levels in our most recent experiment (Sejrsen *et al.*, 1998). Similar to the lack of effect of feeding level in virgin postpubertal heifers, mammary development is unaffected by feeding level in pregnant heifers (Foldager and Sejrsen, 1991).

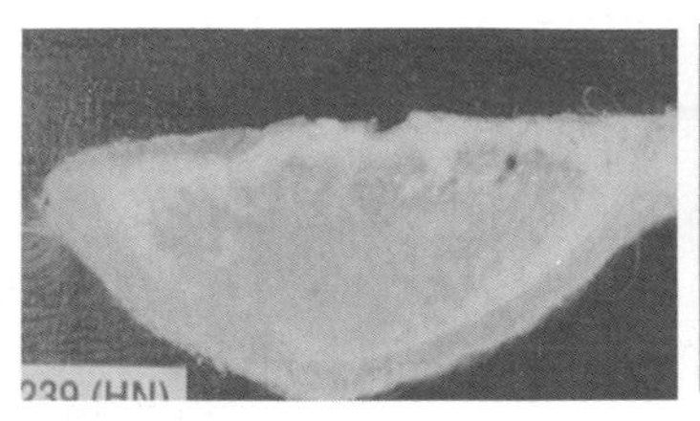

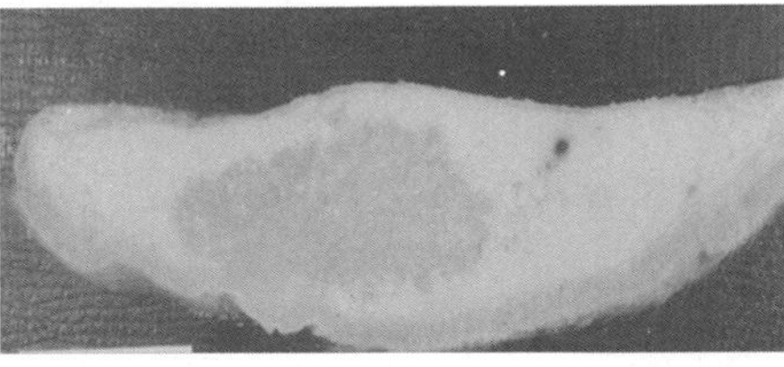

Moderate feeding level (ADG 700g/d)

High feeding level (ADG 1150 g/d)

Figure 13.1 Effect of feeding level in the prepubertal period on mammary development. The weight of the total gland was increased from 914 to 1971 g ($p < 0.05$), but the amount of parenchymal tissue was decreased from 440 to 245 g ($p < 0.05$) (Sejrsen *et al.*, 1998).

More recently, the effect of feeding level during the calf period has been investigated. Brown *et al.* (2002) observed a positive effect of high feeding level from 2 to 8 weeks of age on amount of mammary parenchyma per 100 kg body weight at 14 weeks of age. In our experiment, mammary growth measured at 220 kg live weight was not significantly affected by feeding level during the calf period from 0 to 6 weeks, but the effect tended to be positive (Sejrsen *et al.*, 1998). Petitclerc *et al.* (1999), however, observed a significant reduction in the amount of mammary parenchyma expressed in relation to body weight in calves fed on a high feeding level from 5 weeks to 4 months of age. It is difficult to draw a clear conclusion concerning the effect feeding level during the calf period, but the data of Petitclerc *et al.* (1999) suggest that the critical period before puberty starts before 4 months of age.

In many experiments, feeding level has been confounded with concentrate:roughage ratio, suggesting that the effects observed might have been due to differences in type of diet rather than feeding level. Capuco *et al.* (1995), for instance, observed that the negative effect of high feeding level on mammary growth was much more severe on a maize-based diet than on an alfalfa-based diet. We, however, observed no difference in mammary growth in heifers fed with two extreme barley straw:concentrate ratios (Sejrsen and Foldager, 1992).

The data of Capuco *et al.* (1995) suggest that the higher protein supply in the alfalfa diet might explain the different responses to feeding level. However, mammary growth was unaffected by protein level in two recent experiments (Whitlock *et al.*, 2002; Dobos *et al.* 2000). It has also been suggested that protein degradability might be important. Mäntysaari *et al.* (1995), however, found that the negative effect of feeding level was unaffected by the amount of by-pass protein. The role of fatty acid composition – including CLA - has also been investigated. McFadden *et al.* (1990) suggested that polyunsaturated fat may lead to increased pubertal mammary growth in sheep, but more recent data in cattle suggest that mammary growth in heifers is unaffected by fatty acid composition of the diet (Thibault *et al.*, 2003; Smith *et al.*, 2002). These data therefore suggest that diet composition has little or no effect on mammary development.

Hormonal regulation of mammary development

Regulation of mammary growth and development involves complex interactions of many hormones and growth factors. The overall hormonal requirement was originally elucidated in classical studies in rats and mice (Lyons *et al.*, 1958; Nandi, 1959). These studies showed that the hormones required for normal mammary development include the female sex hormones oestrogen and progesterone as well as growth hormone (GH) and prolactin from the pituitary. Oestrogen and GH seemed to play a major role in regulation of normal mammary development in relation to puberty. The importance of oestrogen and GH for pubertal mammary growth in heifers is well documented. The growth of the mammary ducts is almost completely abolished by ovariectomy and restored by oestrogen replacement (Wallace, 1953; Purup *et al.*, 1993). Similarly, GH stimulates mammary parenchymal growth in heifers (Sejrsen *et al.*, 1986).

Attempts to clarify the physiological mechanisms behind the negative effect of feeding level on mammary development have not come to a final conclusion. Our studies have mainly focused on the role of the GH-IGF-I axis. The rationale was that GH is required for pubertal mammary development, and GH is affected by feeding level. The conclusions of our studies are: High feeding level reduces circulating levels of GH, and exogenous administration of GH enhances pubertal mammary growth (Sejrsen *et al.*, 1983; Sejrsen *et al.*, 1986; Purup *et al.*, 1993). These findings suggest that the negative effect of feeding level on mammary development could be due to reduced GH secretion in heifers on a high feeding level.

Although the presence of GH-receptor mRNA in mammary tissue has been demonstrated, most studies have failed to show binding of GH to mammary tissue and GH does not stimulate mammary cell growth in vitro (Sejrsen and Purup, 1997). These findings therefore suggest that the in vivo effect of GH is indirect. However, using a monoclonal antibody to the GH receptor, Sinowatz *et al.* (2000), found that the receptor protein is present in the luminal layer of the mammary ducts in non-pregnant heifers, but a direct effect in vivo has still not been documented.

If, however, the action of GH on mammary growth is indirect, the effect is likely to be mediated by IGF-I according to the somatomedin hypothesis (Salmon and Daughaday, 1957). This is supported by the following facts (see Sejrsen *et al.*, 1999, 2000): the circulating level of IGF-I is increased by exogenous GH; IGF-I receptors are present in the mammary tissue; and IGF-I stimulates mammary cell proliferation in vitro. Furthermore, the mitogenic effect of serum on growth of mammary cells in vitro is closely related to the IGF-I concentration of the serum.

These findings, however, cannot explain the negative effect of feeding level on mammary growth because the IGF-I concentration in serum is increased by a high feeding level – not decreased like GH. Thus, in conflict with the in vitro data, mammary cell growth is positively correlated with serum GH and negatively correlated with IGF-I in vivo. This paradox cannot be explained by changes in blood concentrations of IGF binding proteins (Vestergaard *et al.*, 2003).

The latest possible explanation is a change in the sensitivity of the mammary tissue to endocrine effects of IGF-I. This seems to be the case. Purup *et al.* (2000) found that mammary tissue from heifers on a high feeding level responded less to IGF-I in vitro than tissue from heifers on a moderate feeding level. However, there was no reduction in the number of IGF-I receptors or the amount of IGF-I receptor mRNA in the mammary tissue. This suggests an inhibition of IGF-I action at post-receptor level. In fact, it turned out that mammary tissue from heifers on a high level of feeding had a significantly higher concentration of IGFBP-3 than heifers on a low feeding level, and addition of IGFBP-3 to cells in culture inhibited the effect of added IGF-I, serum and mammary gland extract (Weber *et al.*, 1999; Purup *et al.*, 2000). These results taken together, therefore, suggest that the negative effect of feeding level on mammary growth is caused, at least in part, by increased local production of IGFBP-3, which inhibits the effect of the increased level of IGF-I in serum. However, the question remains; what is the signal that causes the gland to produce more IGFBP-3?

The possibility that other locally-produced growth factors or binding proteins are involved cannot be ruled out. In fact, studies suggest that transforming growth factor-ß (TGF-ß) might be involved in the reduction in mammary growth (Purup *et al.*, 2000). More recently, studies have been conducted to investigate the role of leptin. Leptin is increased in the circulation of heifers on a high feeding level (Vestergaard, Boisclair and Sejrsen, unpublished), but we have not been able to demonstrate a negative effect of leptin in vitro. Mike VandeHaar and his group, on the other hand, have data that support an inhibitory role of leptin (Silva *et al.*, 2002, 2003). Another possibility might be increased local production of TNFα. TNFα is increased in adipose tissue at high feeding levels and it inhibits growth of mammary parenchymal cells in vitro (Vernon *et al.*, 2001; Purup *et al.*, unpublished).

The fact that the evidence suggests that several pathways might be involved in mediating the negative effect of high feeding level on mammary development is not surprising. In fact, it is more the rule than the exception, that several hormones with opposing actions regulate biological processes.

Growth rate during rearing and milk yield

Data from practice invariably show a positive relationship between body weight at calving and milk yield (Keown and Everett, 1986; Hoffman, 1997). This suggests that high growth rates during rearing secure the best milk yield potential of heifers after calving. This conclusion, however, ignores the fact that the relationship between body weight and milk yield depends on the biological background for the high body weight.

There are two main explanations for the observed positive relationship between body weight at calving and milk yield. Firstly, there is a positive relationship between age and body weight at calving and, over a wide range of calving ages, there is a positive relationship between age at calving and milk yield. Secondly, genetic potential for milk yield is positively related to growth rate during rearing. In a Minnesota selection experiment, growing heifers from the selection line had higher growth rate potential than heifers from the control line (Murphy, 1992; Baumgard *et al.*, 2002; Crooker, personal communication). Similarly, the results of a Danish experiment on genotype x environment interactions in dairy heifers (See Hohenboken *et al.*, 1995) revealed a positive genetic correlation between milk yield and genetic capacity for growth. The observed genetic standard deviation for growth rate was 32 g/d. This implies a 100 g/d difference in growth rate between the top and bottom 10%. Calculations based on the genetic correlations between growth and milk yield suggest that the expected yield of heifers in the top 10% for genetic growth potential is 6-7 kg higher that the yield of heifers in the bottom 10%. Interestingly, selection for growth seems to have no effect on milk yield potential (Hansen *et al.*, 1999). Thus, there is no conflict with the lower milk yield of beef breeds that are selected for growth rate not milk yield.

In spite of variation in genetic potential for growth, feeding level is the main factor determining growth rate during the rearing period (Hansson *et al.*, 1967), and it has been shown many times that high growth rates during rearing obtained by high level feeding lead to reduced milk yield of the heifers as cows (see Sejrsen *et al.*, 2000). Furthermore, the evidence suggests that it is the feeding level in the prepubertal period that is critical for subsequent milk yield, in line with the negative effect of feeding level on mammary development in that period. Although not all results are in full agreement, the data from the experiment of Hohenboken *et al.* (1995) are very convincing (Sejrsen *et al.*, 2000). The experiment compared three feeding levels in four breeds. Groups of all breeds were studied in four different years with different rations. Thus, the effect of feeding level was compared 16 times. Milk yield was reduced at the highest feeding level in all repetitions.

Nevertheless, a negative effect of feeding level is not observed in all experiments (e.g. Carson *et al.*, 2000). The reasons behind conflicting results are difficult to identify, but could be concerned with the length of treatment periods, the differences in feeding level between treatments, the pre-treatment feeding, or large variations in growth rate or milk yield making it difficult to obtain significant results. However, it is possible that the experiments where no reduction in milk yield was observed hold the key to finding a way to avoid the negative effect. It has been suggested that differences in diet composition might be the explanation, and the available data do support a role for specific nutrients (see above).

The results of the Danish experiment suggest that the full genetic potential for milk yield of Danish Friesians is achieved when heifers are raised at a feeding level resulting in an average growth rate between 600 and 700 g per day. However, average genetic potential has increased since this experiment was conducted. Considering the relationship between the genetic potential for growth and milk yield, it is likely that the optimum growth rate increases as the genetic potential for milk yield increases. Based on the genetic correlations observed and the genetic increase in milk yield, it can be calculated that the optimum growth rate in the prepubertal period is likely to increase around 50 g/d every ten years. Thus, the present optimum average growth rate is likely to approach 750 g/d (see Sejrsen *et al.*, 2000).

Although limiting growth rate in the prepubertal period potentially reduces body weight at calving, the effect is minimal. A 100 g reduction in daily gain in the prepubertal period results in only a 15 kg difference in body weight at calving, if all other factors are equal. In fact, the difference could be smaller due to compensatory growth as seen, for instance, by Van Amburgh *et al.* (1998). This difference in body weight easily disappears in random variation in growth rate outside the prepubertal period. Thus, it is not surprising that the relationship between body weight and milk yield observed in practice is positive, even if growth rate is kept at a moderate level in the prepubertal period. Therefore, there is no contradiction between the negative effect of high feeding level in the prepubertal period and the positive correlation between body weight at calving and milk yield.

Furthermore, body weight at calving is also affected to a large degree by the feeding level after the critical period. It has, however, been shown several times that higher body weight at calving achieved by high feeding during pregnancy does not influence milk yield (Foldager and Sejrsen, 1991; Lacasse *et al.*, 1993; Ingvartsen *et al.*, 1995; Carson *et al.*, 2000). In the experiment of Ingvartsen *et al.* (1995), there was 100 kg difference in body weight at calving between treatment groups.

So, the overall conclusion is, that mammary development and subsequent milk yield are unaffected by feeding level outside the prepubertal period and, in spite of conflicting results, the highest milk yield potential is best secured at a feeding level resulting in an average live-weight gain of 700 –750 g/d in the prepubertal period.

Evaluation of growth rates of heifers in practice

It is important to realise that the recommended optimal growth rate for heifers in the prepubertal period relates to the group average and not the growth rate of individual heifers.

In previous experiments, a standard deviation of approximately 10 % was found when animals were fed restricted amounts of feed individually. Since heifers rarely, if ever, are fed individually in practice, even larger variation in growth rate can be expected. Feeding system, housing system, group size, group homogeneity, animal health and quality of management are some of the factors that, in addition to feeding level and genetic growth potential, contribute to variation in growth rate in a group of healthy heifers. Therefore, to evaluate the quality of the heifer operation, it is necessary to know not only the average growth rate of the group, but also the variation in growth rate within the group.

In an attempt to develop a system to evaluate the quality of heifer operations in practice, we weighed and measured all heifers on 15 different farms; five from each of the breeds Danish Friesian, Danish Red and Danish Jersey (Fisker *et al.*, 2003). From the results, a system was developed that allows the farmer to check the average and the variation in growth rate between heifers (Fisker and Kjær, 2002). The system allows the farmer to get a quick estimate of the quality of the heifer operation from only one measurement of the weight of the animals – using a tape measure or a weigh scale. Initially, the average growth rate is compared with the planned growth rate in the various phases of the growth period. In the second stage, the body weights of individual heifers are plotted against age in a graph containing the line for the planned growth rates in the different phases of the rearing period and shaded areas reflecting the normal variation in growth rates (Figure 13.2a). The plot of individual weights gives a quick picture of individual variation in growth rate and reveals any problems in specific groups of heifers. The plot also illustrates the large variation among heifers in body weight at a given age. These evaluations can be done after a single weighing, but estimation of body weight on regular basis makes it possible to check the growth rates of individual heifers, and to see if a given intervention has had the expected effect (Figure 13.2b).

The results of the study show that average growth rate often deviates significantly from the target (Figure 13.3). In one herd, average growth rate in the prepubertal period was more than 250 g/d higher than the planned 600 g/ d. In another herd, the average was more than 200 g/d below. The variation in growth rate between the individual heifers also differed considerably among herds. In one herd, with an average growth rate far from the target, the variation

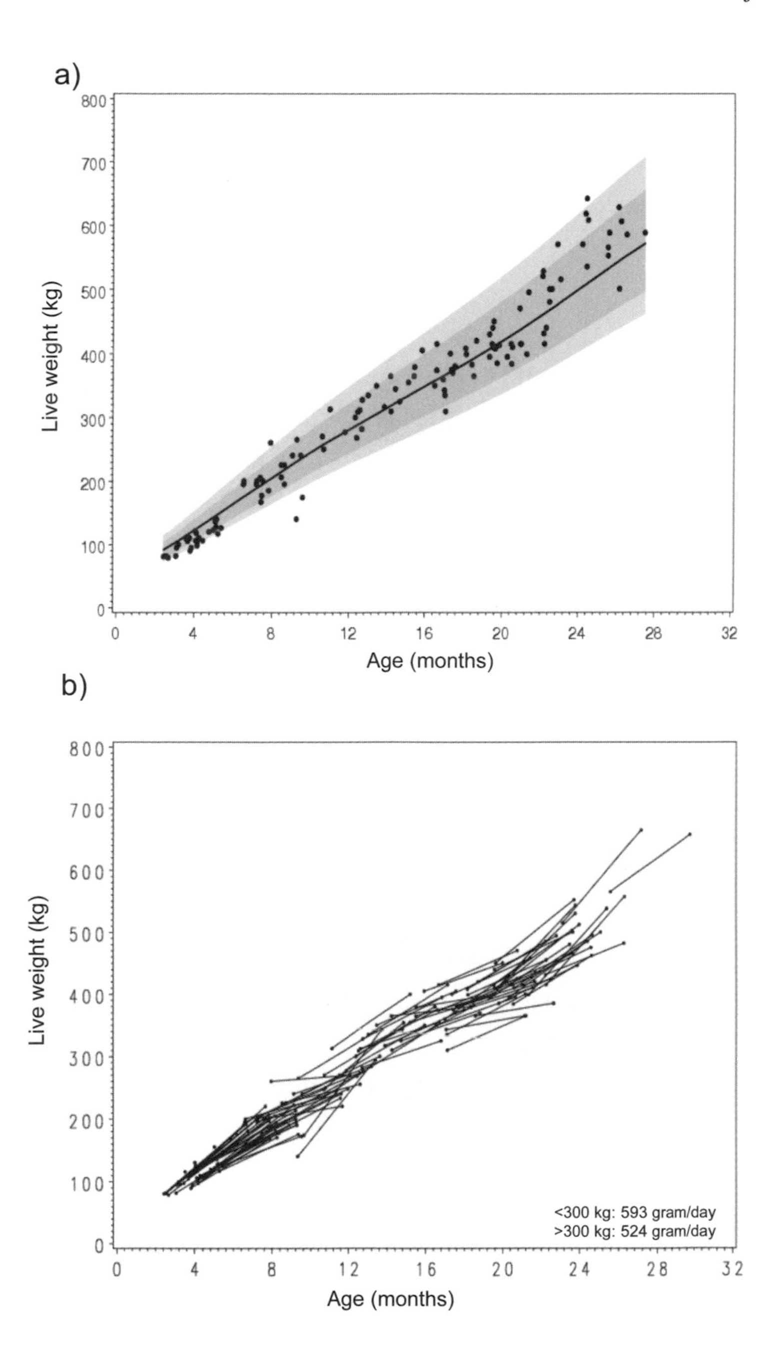

Figure 13.2. a) The weight of individual heifers from a herd is plotted against age, with shaded areas showing the average 80 and 95% confidence intervals from all herds. b) The weights of individual heifers at 2 successive weighings are connected with lines. Variations in the slopes reflect variations in growth between heifers (Fisker *et al.*, 2003).

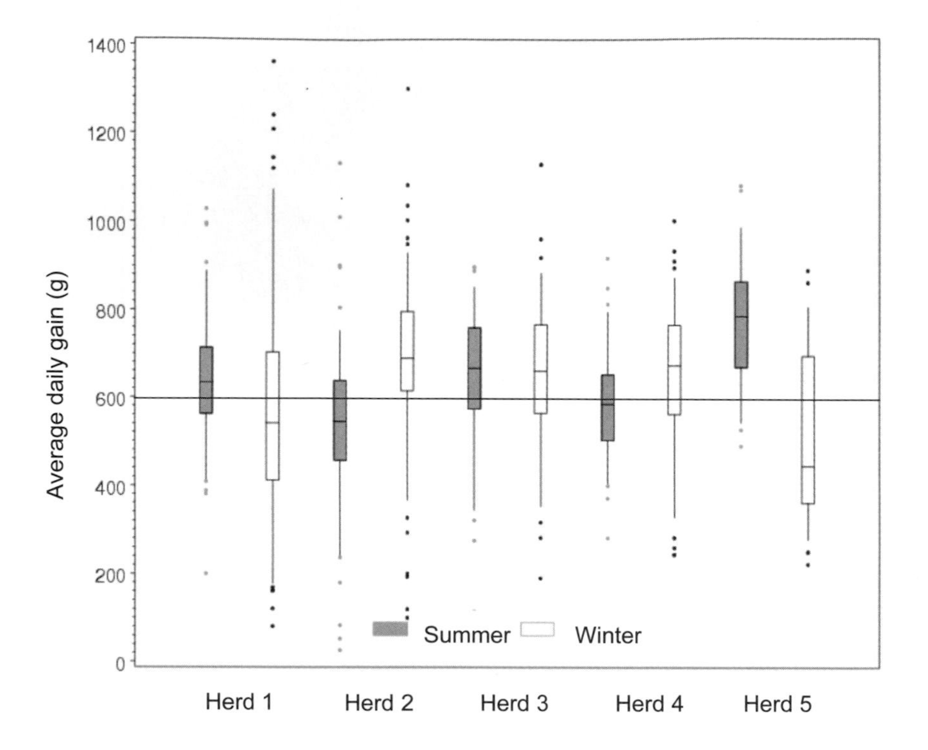

Figure 13.3 Average daily gain during the summer and winter periods from five Danish Friesian herds. The boxes show the 25/75 % fractile; the lines show the 5/95 % fractiles. (Fisker *et al.* 2003).

around the mean was as expected from the normal variation observed; the difference from the top to the bottom of the 95% fractile was 450 g/d in the summer period. In another herd, where the average growth rate was close to the target, the variation in growth rates was large; the 95 % fractile ranged from below 200 g/d to almost 1100 g/d. Figure 13.4 illustrates how growth rates can vary on different farms, and that an acceptable average growth rate can cover large variations in growth rate between different periods and between animals.

This underlines how important it is to record body weights and relate them to age of the animals, in order to get a clear picture of the status of the operation. If you ask farmers about the target growth rate for their heifers in the post pubertal period, most would probably answer that they are aiming at over 700 g/d. In this study, however, that target was achieved in only three of the ten herds with large breeds in the winter period, and in none of the herds in the summer period.

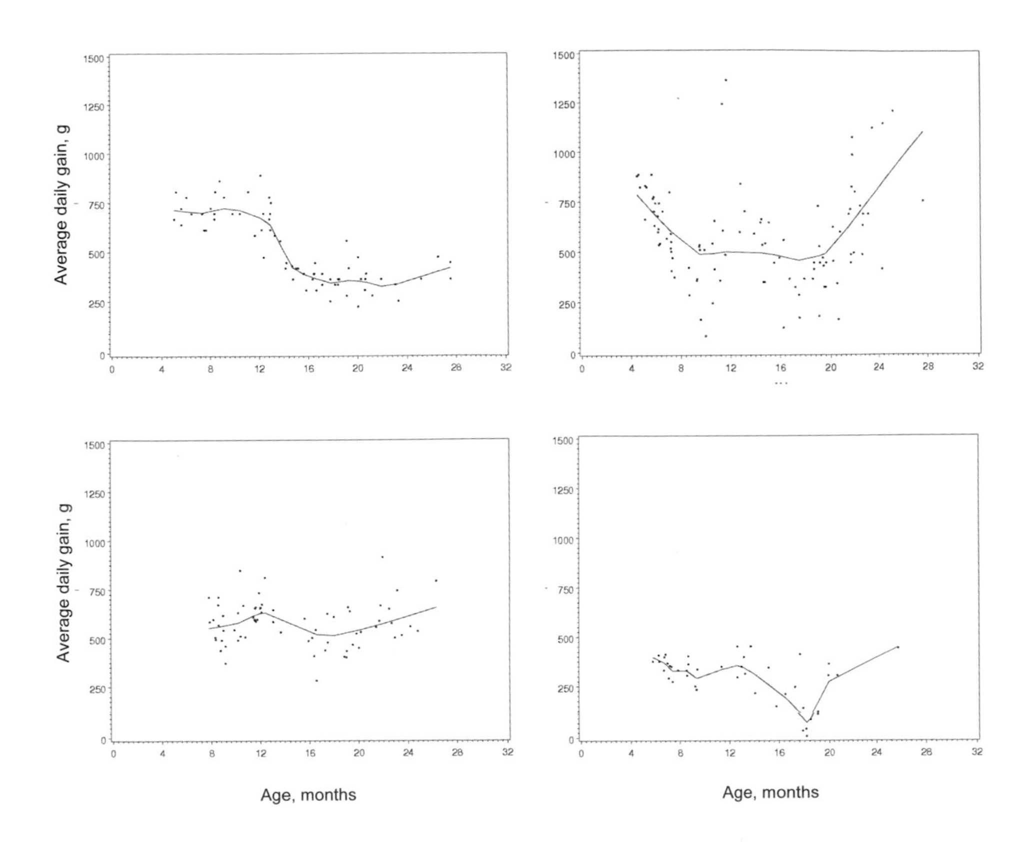

Figure 13.4 Individual and average growth rates in four different herds illustrating the large variation between animals and between herds (Fisker *et al.*, 2003)

References

Baumgard, L.H., Weber, W.J., Kazmer, G.W., Zinn, S.A., Hansen, L.B., Chester-Jones, H. and Crooker, B.A. 2002. Effects of selection for milk yield on growth hormone response to growth hormone releasing factor in growing Holstein calves. *J. Dairy Sci.* **85**, 2529-2540.

Brown, E.G., VanderHaar, M.J., Daniels, K.M., Liesman, J.S., Chapin, L.T. and Weber Nielsen, M.S. 2002. Increasing energy and protein intake of Holstein heifer calves increases mammary development. *J. Dairy Sci.* **85**, Suppl.1, 80.

Capuco, A.V., Smith, J.J., Waldo, D.R. and Rexroad, C.R. 1995. Influence of prepubertal dietary regimen on mammary growth of Holstein heifers. *J. Dairy Sci.* **78**, 2709-2725.

Carson A.F., Wylie, A.R.G., McEvoy, J.D.G., McCoy, M. and Dawson, L.E.R. 2000. The effects of plane of nutrition and diet type on metabolic hormone concentrations, growth and milk production in high genetic merit dairy herd replacements. *Animal Science*, **70**, 349-362.

Dobos, R.C., Nandra, K.S., Riley, K., Fulkerson, W. J., Lean I.J and Kellaway, R.C.. 2000. The effect of dietary protein level during the pre-pubertal period of growth on mammary gland development and subsequent milk production in Friesian heifers. *Livest. Prod. Sci.* **63**, 235-243.

Fisker, I. and Kjær K.B. 2002. Nyt værktøj til styring af kviernes vækst [New tool to control heifer growth rates]. http://www.lr.dk/kvaeg/informationsserier/lk-meddelelser/1071.htm

Fisker, I., Skjøth, F., Sejrsen, K. and Kristensen, T. 2003. *Vurdering af kviers vækst.* Rapport 107, Dansk Kvæg. 81pp.

Foldager, J. and Sejrsen, K. 1991. *Rearing intensity in dairy heifers and the effect on subsequent milk production.* In Danish with English summary and subtitles. Report 693, Statens Husdyrbrugsforsøg.

Hansen, L.B., Cole, J.B., Marx G.D. and Seykora, A.J. 1999. Productive life and reasons for disposal of Holstein cows selected for large versus small body size. *J. Dairy Sci.* **82**, 795-801.

Hansson, A., Brännäng, E. and Liljedahl, L.E. 1967. Studies on monozygous twins. XIX. The interaction of heredity and intensity of rearing with regard to growth and milk yield in dairy cattle. *Lantbr. Høgsk. Ann.* **33**, 643-693.

Harrison, R.D., Reynolds, J.P. and Little, V. 1983. A quantitative analysis of mammary glands of dairy heifers reared at different rates of live weight gain. *J. Dairy Res.* **50**, 405-412.

Herman, H.A. and Ragsdale, A.C. 1946. The influence of quantity and quality of nutrients on the growth of dairy heifers. *J. Anim. Sci.* **5**, 398.

Hoffman, P.C. 1997. Optimum body size of Holstein replacement heifers. *J. Anim. Sci.* **75**, 836-845.

Hohenboken, W. D., Foldager, J., Jensen, J., Madsen, P. and Andersen, B.B. 1995. Breed and nutritional effects and interactions on energy intake, production and efficiency of nutrient utilization in young bulls heifers and lactating cows. *Acta Agric. Scand., Sect. A, Anim. Sci.* **45**, 92.

Ingvartsen, K.L., Foldager, J., Aaes, O. and Andersen, P.H. 1995. Effekt af foderniveau i 24 uger før kælvning på foderoptagelse, produktion og stofskifte hos kvier og køer. *Intern rapport nr.* **45** fra Statens Husdyrbrugsforsøg, 60-73.

Keown J.F. and Everett, R.W. 1986. Effect of days carried calf, days dry and weight at first calving on yield. *J. Dairy Sci.* **69**, 1891-1896.

Lacasse, P., Block, E., Guilbault, L.E. and Petitclerc, D. 1993. Effect of plane of nutrition of dairy heifers before and during gestation on milk production, reproduction and health. *J. Dairy Sci.* **76**, 3420-3427.

Little, W. and Kay, R.M. 1979. The effect of rapid rearing and early calving on subsequent performance of dairy heifers. *Anim. Prod.* **29**, 131-142.

Lyons, W.R., Li, C.H. and Johnson, I.D. 1958. The hormonal control of mammary growth. *Rec. Progress in Horm. Res.* **14**, 219-254.

Mäntysaari, P., Ingvartsen, K.L., Toivonen, V. and Sejrsen, K. 1995. The effects of feeding level and nitrogen source of the diet on mammary development and plasma hormone concentrations of prepubertal heifers. *Acta Agric. Scand. Sect. A, Anim. Sci.* **45**, 236-244.

McFadden, T.B., Daniel, T.E., and Akers, R.M. 1990. Effects of plane of nutrition, growth hormone and unsaturated fat on mammary growth in prepubertal lambs. *J. Anim. Sci.* **68**, 3171-3179.

Murphy, K.D. 1992. The effects of selection for milk production on composition of growth, feed intake and growth related hormone levels in Holstein heifers. *Ph.D. Diss.*, Univ. of Minnesota, St. Paul.

Nandi, S. 1959. Hormonal control of mammogenesis and lactogenesis in the C3H/He Crgl mouse. *Univ. California Pub. Zool.* **65**, 1-128.

Petitclerc, D., Dumoulin, P., Ringuet, H., Matte, J. and Girard, C. 1999. Plane of nutrition and folic acid supplementation between birth and four months of age on mammary development of dairy heifers. *Can. J. Anim. Sci.* **79**, 227-237.

Purup, S., Sejrsen, K., Foldager, J. and Akers, R.M. 1993. Effect of exogenous bovine growth hormone and ovariectomy on prepubertal mammary growth, serum hormones, and acute in-vitro proliferative response of mammary explants from Holstein heifers. *J. Endocrinol.* **139**, 19-26.

Purup, S., Vestergaard, M., Weber, M.S., Plaut, K., Akers, R.M. and Sejrsen, K. 2000. Local regulation of pubertal mammary growth in heifers. *J. Anim. Sci.* **78**, Suppl. 3, 36-47.

Salmon, W.D. and Daughaday, W.H. 1957. A hormonally controlled serum factor which stimulates sulfate incorporation by cartilage in vitro. *J. Lab.Clin. Med.* **49**, 825-836.

Sejrsen, K. 1978. Mammary gland development and milk yield in relation to growth rate in the rearing period in dairy and dual purpose heifers. *Acta Agric. Scand.* 28, 41.

Sejrsen, K., Huber, J.T., Tucker, H.A. and Akers R.M. 1982. Influence of plane of nutrition on mammary development in pre- and postpubertal heifers. *J. Dairy Sci.* **65**, 783-790.

Sejrsen, K., Huber, J.T., and Tucker, H.A. 1983. Influence of amount fed on hormone concentrations and their relationship to mammary growth in heifers. *J. Dairy Sci.* **66**, 845-855.

Sejrsen, K., Foldager, J., Sørensen, M.T., Akers, R.M. and Bauman, D.E. 1986. Effect of exogenous bovine somatotropin on pubertal mammary development in heifers. *J. Dairy Sci.* **69**, 1528-1535.

Sejrsen, K. and Foldager, J. 1992. Mammary growth and milk production capacity of replacement heifers in relation to diet energy concentration and plasma hormone levels. *Acta Agric. Scand., Sect. A, Anim. Sci.* **42**, 99-105.

Sejrsen, K. and Purup, S. 1997. Influence of prepubertal feeding level on milk yield potential of dairy heifers: a review. *Journal of Animal Science* **75**, 828-835.

Sejrsen, K., Purup, S., Martinussen, H. and Vestergaard, M. 1998. Effect of feeding level in calves and prepubertal heifers. *J. Dairy Sci.* **81**, (Suppl. 1) 377.

Sejrsen, K., Purup, S., Vestergaard, M., Weber, M.S., and Knight, C.H. 1999. Growth hormone and mammary development. *Dom. Anim. Endocrinol.* **17**, 117-129.

Sejrsen, K., Purup, S., Vestergaard, M. and Foldager, J. 2000. High body weight gain and reduced bovine mammary growth: physiological basis and implications for milk yield potential. *Dom. Anim. Endocrinol.* **19**, 93-104.

Silva, L.F.P., VandeHaar, M.J., Weber Nielsen, M.S. and Smith, G.W. 2002. Evidence for a local effect of leptin in bovine mammary gland. *J. Dairy Sci.* **85**, 3277-3286.

Silva, L.F.P., Liesman, J.S., Weber Nielsen, M.S. and VandeHaar, M.J. 2003. Intramammary infusion of leptin decreases proliferation of mammary epithelial cells in prepubertal heifers. *J. Dairy Sci.* **85**, Suppl. 1, 680.

Sinha, Y.N. and Tucker, H.A. 1969. Mammary development and pituitary prolactin level of heifers from birth through puberty and during oestrus cycle. *J. Dairy Sci.* **52**, 507-512.

Sinowatz, F., Schams, D., Kölle, K., Plath, A., Lincoln, D. and Waters, M.J. 2000. Cellular localization of GH receptors in the bovine mammary gland during mammaogenesis, lactation and involution. *J. Endocrinol.* **166**, 503-510.

Smith, J.M., Block, S.S., Bauman, D.E. and Van Amburgh, 2002. Effect of feeding a calcium salt of conjugated linoleic acid (CLA) prior to puberty on body composition and mammary development in Holstein heifers. *J. Dairy Sci.* **85** (suppl. 1), 79.

Swanson, E.W. 1960. Effect of rapid growth with fattening of dairy heifers on their lactational ability. *J. Dairy Sci.* **43**, 377-387.

Thibault, C., Petitclerc, D., Spratt, R., Léonard, M., Sejrsen, K. and Lacasse, P. 2003. Effect of feeding prepubertal heifers with a high oil diet on mammary development and milk production. *J. Dairy Sci.* **86**, 2320-2326.

VanAmburgh, M.E., Galton, D.M., Bauman, D.E., Everett, R.W., Fox, D.G., Chase, L.E. and Erb, H.N. 1998. Effects of three prepubertal body growth rates on performance of Holstein heifers during first lactation. *J. Dairy Sci.* **81**, 527-538.

Vernon, R.G., Denis, R.G.O and Sørensen, A. 2001. Signals of adiposity. *Dom. Anim. Endocrinol.* **21**, 197-214.

Vestergaard, M., Purup, S., Frystyk, J., Løvendahl, P., Sørensen, M.T., Riis, P.M., Flint, D.J. and Sejrsen, K. 2003. Effects of growth hormone and feeding level on endocrine parameters, hormone receptors, muscle growth and performance of prepubertal heifers. *J. Anim. Sci.* **81**, 2189-2198.

Wallace, C. 1953. Observations on mammary development in calves and lambs. *J. Agric Sci.* **43**, 413-421.

Weber, M.S., Purup, S., Vestergaard, M., Ellis, S.E., Akers, R.M. and Sejrsen, K. 1999. Contribution of insulin-like growth factor-I (IGF-I) and IGF-binding protein-3 (IGFBP-3) to mitogenic activity in bovine mammary extracts and serum. *J. Endocrinol.* **161**, 365-363.

Whitlock, B.K., VandeHaar, M.J., Silva, L.A. and Tucker, H.A. 2002. Influence of dietary protein on prepubertal mammary gland development in rapidly-grown heifers. *J. Dairy Sci.* **85**, 1516-1525.

14

GRAZING SYSTEMS FOR DAIRY HERD REPLACEMENTS

L.E.R. DAWSON AND A.F. CARSON
ARINI, Large Park, Hillsborough, Co Down, BT26 6JF

Introduction

Feeding and management during the rearing period have a major impact on animal performance, reproduction, health and welfare. Most studies agree that replacement Holstein-Friesian animals should calve and enter the milking herd at approximately 24 months of age (Pirlo, *et al.* 2000). Replacement heifers should therefore be of adequate body size by 24 months of age to ensure acceptable first lactation performance and minimise dystocia (Hoffman and Funk, 1992). Research from our Institute (Carson *et al.*, 2000a; 2002b) indicates that the target weight for high genetic merit Holstein-Friesian heifers calving at two years of age should be 540 to 560 kg, at a body condition score of 2.75 to 3.0.

Recently, the value of farm output on dairy farms has been steadily decreasing with a decrease in gross margin of 26% observed in Northern Ireland between 1998 and 2003 (DARD 2003) and a 58% decrease in Net Farm Income in real terms between 1994 and 2002 (DEFRA 2003) in the United Kingdom as a whole. In order to maintain profitability, it is therefore necessary to examine strategies by which costs of production can be reduced. Increased reliance on grazed grass might be one strategy by which savings in both costs and labour can be achieved. Mayne (2000) outlined the mechanisms by which this could be achieved in the milking dairy herd. However, grazed grass also forms an important component of heifer rearing systems and should form an integral part of any rearing programme designed to minimise costs of production. A considerable volume of research has been undertaken to investigate factors that influence performance of young calves and yearlings during the grazing season (Hodgson, *et al.* 1977; Steen, 1994; Keane, 2002, Leaver 1970). Although this research has involved beef animals or Friesian type dairy herd replacements, it can be a useful tool in developing appropriate grazing regimes and can form the basis of future research programmes in heifer rearing.

The purpose of this chapter is to review target growth rates for dairy herd replacements at various stages throughout the rearing period and to identify appropriate grazing strategies to achieve these targets.

Target performance throughout the rearing period

Based on Carson *et al.* (2000a; 2002b), target growth rates throughout the rearing period for Holstein heifers calving at 550 kg live weight at 24 months of age are detailed in Table 14.1.

Table 14.1 TARGET GROWTH RATES AND LIVE WEIGHTS OF HEIFERS THROUGHOUT THE REARING PERIOD

Age	*Target live weight (kg)*	*Target live-weight gain (kg/day)*
Birth	41	
6 weeks	65	0.60
12 weeks	95	0.70
6 months	160	0.70
10 months	230	0.60
14 months	330	0.80
23 months	530	0.75
24 months	550	0.66

Within the UK, autumn-born calves are generally at grass from 6 to 13 months of age, while spring-born calves are at grass from 3 to 10 months of age. However, there is a wide variation in age within each of these groups and grazing regimes must be adapted accordingly. Both sward and animal factors affect performance at grass and these will be discussed.

Sward factors affecting performance

DIGESTIBILITY

For grazing dairy cows and growing calves, herbage intake has been shown to increase linearly up to values of 830 g/kg organic matter digestibility (OMD) (Hodgson, 1977). However, digestibility of grazed swards is related to many factors in the sward, including chemical and physical parameters, herbage mass, sward height, sward structure and the content and distribution of leaf/stem and live/dead fractions within the various horizons of the sward (Stakelum and Dillon, 1991). In addition, young calves are more sensitive to changes in herbage digestibility than older calves (Hodgson, 1968). Ultimately all these factors taken together will determine herbage intake and animal performance.

SWARD HEIGHT

An asymptotic relationship between herbage intake and sward height has been observed, with herbage intake increasing up to a 'critical' sward height and decreasing thereafter (Milne and Fisher, 1994). In view of the relationship between sward height and herbage intake, sward height can be a useful management tool to assess grass availability and to control grazing management. However, sward height can be measured in a number of ways and it is important to specify which measure is used when determining optimum sward heights for growing cattle.

(1) Sward surface height (SSH) is the vertical height of an undisturbed sward using a sward stick (Gibb, 1996).
(2) Extended tiller height (ETH) is determined as the height of an individual tiller, or the mean height of a group of tillers, extended in the vertical plane (Hodgson *et al.* 1971).
(3) Sward plate height (SPH) is determined when a plate, which usually weighs 430 g, with a surface area of 0.9 m^2 and exerts a pressure of 4.5 kg/m^2, is just compressing the sward (Hoden *et al.* 1991).
(4) Residual sward height (RSH) is the grass height after grazing in a rotational grazing system (Mayne *et al.* 1987) and is used as a management tool to determine grazing severity.

As each of these methods, i.e. SSH, ETH, SPH or RSH, will give different measurements of the same sward, the method must be specified.

The 'critical' sward height referred to previously, below which herbage intake (and ultimately performance) is reduced, varies according to whether animals are grazed continuously or in a rotational system (Sayers, 1999). It is therefore important to specify the grazing system when assessing the effect of sward height on animal performance.

Against this background, results from a range of studies undertaken with growing calves, finishing cattle, and suckler cows and their calves to assess the effect of sward height on performance are summarised in Table 14.2. Wright and Whyte (1989) showed that with continuously grazed swards, live-weight gain of young calves grazing with suckler cows increased with increasing sward height. Furthermore, live-weight gain of older calves showed a greater response to sward height, probably due to the increased reliance of the calf on herbage and decreased reliance on milk. However, in the latter study the effect of sward height on calf performance is complicated by the fact that growth rate of the calf is also influenced by milk supply from the cow. With growing

Table 14.2 EFFECT OF SWARD HEIGHT ON LIVE-WEIGHT GAIN AT PASTURE

Source	*Grazing system*	*Target sward height*	*Method of sward height measurement*	*Live-weight gain (kg/day)*	
Wright & Whyte (1989)	Continuous	4.5	Sward surface height	0.88	Charollais X calves grazing with cows May to August
		6.0		0.91	
		7.0		0.98	
		9.1		1.04	
		11.0		1.06	
	Continuous	5.4	Sward surface height	0.98	Charollais X calves grazing with cows August to September
		7.8		1.22	
		9.2		1.35	
Steen (1994)	Continuous	7	Sward surface height	0.99	Zero concentrates, British Friesian & Continental X British Friesian bulls, 5-10 months of age
		9		1.17	
		11		1.21	
Steen and Kilpatrick (1998)	Continuous	6.5	Sward surface height	0.84	Zero concentrates, Continental X Friesian bulls, 5.5-11 months of age
		10		1.21	
Carson *et al.* (unpublished)	Rotational	9	Residual sward height	0.77	Holstein heifers, 7.5 – 12.5 months, zero concentrates
Carson *et al.* (2004)	Rotational	9	Residual sward height	0.64	Holstein heifers, 7 – 12.5 months, zero concentrates
Carson *et al.* (2002a)	Rotational	5	Residual sward height	0.76	Holstein heifers 17.5 – 23.5 months, zero concentrates

bull calves, at pasture from 5 to 10 months of age, similar live-weight gains have been observed with grazing sward heights of 11 and 9 cm (Steen, 1994), but a reduction in performance was observed when sward height was reduced to 7 cm.

Similar results were reported by Steen and Kilpatrick (1998) with growing beef cattle when sward surface height was reduced from 10 to 6.5 cm. Although the effect of sward height on performance of Holstein heifer calves during their first summer or in-calf heifers in their second summer at pasture has not been extensively studied, Carson *et al.* (unpublished & 2004) demonstrated live-weight gains of 0.6 to 0.8 kg/day when autumn-born heifer calves were turned out to grass at 7 to 7.5 months of age, with a target residual sward height of 9 cm and no concentrate supplementation (Table 14.1). Live-weight gain was 0.76 kg/day when the same group of heifers was turned out to pasture at 17.5 months of age and grazed at a residual sward height of 5.0 cm without concentrate supplementation (Carson *et al.* 2002a).

SWARD DENSITY

Although sward height has been shown to affect animal performance, the density of the sward must also be taken into consideration. A close relationship between sward height and bulk density has been observed (Black and Kenny, 1984). McGilloway *et al.* (1999) observed with lactating dairy cows that a dense sward grazed to a height of 9 to 10 cm would enable the same herbage intake as an open sward (low bulk density sward) grazed to a sward height of 13 to 14 cm.

GRASS MANAGEMENT

The effects of sward height on performance cannot be considered in isolation, as management of the sward throughout the season will affect grass quality and ultimately animal performance. In the studies by Wright and Whyte (1989) and Swift *et al.* (1989), animal performance was reduced by increasing sward height from 9 to 11 cm, while Steen (1994) did not observe any detrimental affect on animal performance. The latter author attributed this to the fact that the swards were topped in late June and this prevented the build up of stemmy, low digestibility herbage which would normally occur in longer swards. The detrimental effect of lax grazing in early season on animal performance later in the season has been reported in beef cattle (Dawson, *et al.* 1981), sheep (Birrell and Bishop, 1980) and dairy cows (Mitchell and Fulkerson, 1987; Mayne *et al.* 1988), and emphasises the need for tight grazing of spring grass swards to maintain quality. Mayne *et al.* (2000) recommended that in the spring, sward surface height should be maintained at 7-8 cm to prevent the build up of stemmy, low digestibility grass. Once the risk of seedhead development has passed, then sward height can be allowed to increase to 10-11cm. However, it should also be noted that, under comparable management conditions, animals grazing autumn pastures have lower intakes than those grazing spring pastures (Jamieson and Hodgson, 1979; Marsh, 1975) and regrowths of herbage generally have lower intake characteristics than primary growths (Hodgson *et al.* 1977), although the effects on animal performance have not been quantified.

Supplementation at pasture

Supplementation at pasture can be in the form of concentrates or conserved forage. The main aim of supplementation at pasture for growing cattle is to improve daily live-weight gain, or to maintain animal growth rate if grass supply is inadequate. However, responses to concentrate supplementation are extremely variable and depend on many factors, such as growth potential of the animal, sward height, sward quality, and supplement type, composition and substitution rate. Musangi *et al.* (1965) investigated the effects of concentrate supplementation on live-weight gain with growing cattle under either restricted grazing (10 hours access to pasture) or unrestricted grazing. When herbage was freely available concentrate supplementation had no effect on animal performance, whereas under restricted grazing conditions, live-weight gain was improved by 0.49 on a proportionate basis, by provision of 3.3 kg supplement/day. Steen (1994) also noted that concentrate supplementation had no effect on live-weight gain of young growing bull calves grazing swards of 7.9 cm or greater, but increased live-weight gain of those grazing 6.7 cm swards with an average response in live-weight gain of 93 g/kg concentrates offered. Results from two studies undertaken at this Institute where autumn-born, Holstein heifer calves were supplemented with 1.5 kg concentrate/day during the first summer at grass when grazing a sward height of 8 – 10 cm are presented in Table 14.3. Responses in live-weight gain to increased supplementation were 87 g/kg concentrate in Study 1 and 73 g/kg in Study 2. For heifers in their second summer at grass a response of 60 g live weight/kg concentrate supplementation was obtained (Table 14.3).

Supplement type also influences live-weight gain. Vadiveloo and Holmes (1979) examined the effect of offering a range of supplementary feeds to young growing beef cattle on herbage intake and live-weight gain. Compared with a rolled barley supplement, supplements consisting of cobs of dried lucerne or a high protein concentrate gave slightly lower substitution rates and there was a trend towards increased live-weight gain with the high protein concentrates.

Table 14.3 LIVE-WEIGHT GAIN OF AUTUMN-BORN, HOLSTEIN HEIFER CALVES DURING THE FIRST AND SECOND SUMMER AT GRASS (GRASS SILAGE-BASED DIETS DURING WINTER PERIODS)

Live-weight gain (kg/day)	*Concentrate level (kg /day)* *0*	*1.5*
First summer at grass		
Study 1 (Carson *et al.* unpublished)		
(7.5 – 12.5 months of age)	0.77	0.90
Study 2 (Carson *et al.*, 2004)		
(7 – 12.5 months of age)	0.64	0.75
	0	2.5
Second summer at grass (17.5 – 23.5 months of age)		
(Carson *et al.* 2002a)	0.76	0.91

Animal factors affecting performance

GROWTH POTENTIAL

Steen (1994) demonstrated that breed and growth potential had a major effect on live-weight gain at pasture. This is illustrated in the results from a number of studies presented in Table 14.4, where live-weight gains at pasture are summarised for a range of animal breeds and growth potential. Live-weight gains ranged from 0.64 kg/day to 1.12 kg/day even when grazed grass was the sole feed. The high live-weight gains reported by Steen (1994) in particular indicate that good quality pasture has the potential to sustain near maximum performance, even in calves with a high growth potential. The satisfactory live-weight gains observed from grazed grass also illustrate the potential saving in costs of concentrate feed that could be achieved by maximising use of grazed grass for heifer rearing systems.

STAGE OF DEVELOPMENT

For animals of the same breed and genetic potential, stage of development also affects performance at pasture (Table 14.5). Within each study, live-weight gains were assessed for calves and yearlings in the same grazing year. The results suggest that for the breeds considered, yearlings generally had higher growth rates than calves when grazed under similar systems. These differences may be due to the fact that herbage intake is more sensitive to change in herbage digestibility in calves than adult cattle (Hodgson, 1968) and calves are more selective in their grazing habits (McKeeken, 1954; Alder & Chambers, 1958). The contrasting grazing patterns of calves and yearling/older cattle should be

Table 14.4 EFFECTS OF BREED AND ANIMAL TYPE ON LIVE-WEIGHT GAIN AT PASTURE.

Source	*Animal Breed*	*Sex*	*Age at pasture (months)*	*Live-weight gain at pasture (kg/d)*	*Additional comments*
Steen (1994)	Continental (Simmental or Limousin) X Friesian	Entire males	5-10	1.12	Zero concentrate; results averaged over a range of sward heights
Steen and Kilpatrick (1998)	Continental (Simmental or Belgian Blue) X Friesian	Entire males	5.5 – 11	0.95	Zero concentrate; results averaged over a range of sward heights
Leaver (1970)	Friesian	Female	7 - 12	0.71	Zero concentrate
Leaver (1970)	Friesian	Female	19 - 24	0.81	Zero concentrate
Wright & Whyte (1989)	Charolais cross	Entire Males & Females	2 - 8	0.97	Calves grazing with cows form May to August
Keane (2002) (study 1)	Charolais X Friesian	Entire males	2 – 8	0.78	Averaged over three seasons; concentrate (1 kg/head/day) offered from turnout until June
Keane (2002) (study 1)	Charolais X Friesian	Entire males	13 – 19	0.84	
Carson *et al.* (unpublished)	High genetic merit Holstein	Females	7.5 – 12.5	0.77	Zero concentrate supplementation
Carson *et al.* (2004)	High genetic merit Holstein	Females	7 – 12.5	0.64	
Carson *et al.* (2002a)	High genetic merit Holstein	Females	17.5 – 23.5	0.76	Zero concentrate supplementation

taken into account when deciding which grazing system to adopt and this will be discussed in more detail later in the chapter. Within calves and older cattle, differences in age and live weight might also affect live-weight gain at grass. Leaver (1974a) summarised the effect of size of animal at turnout on live-weight gain at grass (Table 14.6). Older and larger calves and heifers had greater live-weight gains than younger, smaller cattle. This can have major implications for growth rate of spring versus autumn born calves as, by virtue of their smaller size, spring-born calves have lower growth rates than autumn-born calves. Table 14.6 also illustrates the effect of gestation on live-weight gain, with pregnant heifers being heavier than non-pregnant heifers and having greater live-weight gains.

Table 14.5 EFFECT OF AGE ON LIVE-WEIGHT GAIN AT PASTURE (RESULTS AVERAGED OVER A RANGE OF GRAZING SYSTEMS)

Source	*Live-weight gain (kg/day) at pasture*		*Breed*
	Calves†	*Yearlings*†	
Keane (2002)	0.74	0.97	Charolais X Friesian bulls & heifers
Leaver (1970)	0.71	0.81	British Friesian heifers
Leaver (1974b)	0.69	0.60	British Friesian heifers
Leaver (1975)	0.77	0.80	British Friesian heifers

† calves approx 7 months at turnout, yearlings approx 18 months at turnout.

Table 14.6 SIZE OF ANIMAL AT TURNOUT AND LIVE-WEIGHT GAIN AT GRASS (LEAVER 1974a)

		Initial live weight (kg)	*Live-weight gain (kg/day)*
Calves		Over 181	0.84
		136 – 181	0.79
		91 – 136	0.75
		Under 91	0.66
Heifers	Pregnant	Over 408	0.88
	Pregnant	363 – 408	0.91
	Non pregnant	318 – 363	0.81
	Non pregnant	Under 318	0.78

COMPENSATORY GROWTH

Leaver (1970) observed that live-weight gain of heifers (average age 19 months at turnout) followed a curvilinear pattern throughout the grazing season. Live-weight gain immediately post-turnout was 1.34 kg/day and decreased to 0.49 kg/day towards the end of the season. The increased live-weight gain just after turnout was attributed to compensatory growth, as the heifers had a moderate rate of gain of 0.48 kg/day during the previous winter. More recently, Woods *et al.* (unpublished) examined the effect of live-weight gain of heifers during the winter feeding period on subsequent performance. Target live-weight gains of 0.9 kg/day or 0.6 kg/day were achieved with two groups of heifers for three months prior to turnout. At turnout, half the heifers at the target live-weight gain of 0.9 kg/day were maintained at that live-weight gain and the other half were restricted to 0.6 kg/day. Similarly, half of the group that had a target live-weight gain of 0.6 kg/day prior to turnout were maintained at that live-weight gain and the other half were increased to 0.9 kg/day. Target and actual live-weight gains are presented in Table 16.7. The heifers that were restricted prior to turnout (0.6 kg/day) had no difficulty in attaining live-weight gains greater than 0.9 kg/day after turnout. However, it was difficult to maintain live-weight gains as low as 0.6 kg/day in the same group of heifers, due to compensatory growth and these animals actually achieved a live-weight gain of 0.75 kg/day when grazed to 4 cm.

Table 14.7 EFFECT OF LIVE-WEIGHT GAIN DURING THE WINTER PERIOD ON SUBSEQUENT PERFORMANCE AT GRASS (WOODS *et al.* UNPUBLISHED)

Live-weight gains during winter period (kg/d)		*Live-weight gains during summer at grass (kg/d)*		
Target live-weight gain	*Actual live-weight gain*	*Target live-weight gain*	*Actual live-weight gain*	*Sward height (cm)*
0.9	0.89	0.9	1.00	9
0.6	0.60	0.9	1.03	9
0.9	1.00	0.6	0.53	4
0.6	0.58	0.6	0.75	4

STOCKING RATE

Stocking rate can also influence live-weight gain at grass. Leaver (1974) noted that increasing stocking rate from 6 to 8 animals/ha, where calves and heifers were grazing together, decreased live-weight gain in calves by 0.18 and heifers by 0.32 on a proportionate basis (Table 14.8). Carson *et al.* (unpublished data) reported a live-weight gain of 0.77 kg/day at stocking rates of 10 calves/ha, while Carson *et al.* (2002a) reported a live-weight gain of 0.76 kg/day when heifers (age 17.5 to 23 months) were grazed at 5.8 animals/ha (Table 14.8). However, the effect of stocking rate on performance is intrinsically linked with other parameters, many of which have been discussed previously.

For example, Leaver (1974b) observed that, at the same stocking rate, calves had lower live-weight gains late in the season compared with early in the season, due to lower herbage quality in the late-season grass.

Table 14.8 EFFECT OF STOCKING RATE ON PERFORMANCE OF CALVES AND HEIFERS

Source	*Animal type*	*Stocking rate (total animals/ha)*	*Live-weight gain (kg/day)*
Leaver (1974b)	Calves		
	(average age at turnout	6	0.75
	6 months; grazed with	7	0.70
	heifers)	8	0.62
	Heifers		
	(average age at turnout	6	0.70
	17 months non-pregnant	7	0.62
	& average age at turnout	8	0.48
	21 months pregnant;		
	grazed with calves)		
Carson *et al.* (unpublished data)	Calves (7.5-12.5 months)	10	0.77
Carson *et al.* (2002a)	Heifers (17.5 – 23.5 months)	5.8	0.76

Grazing systems for dairy heifer replacements

The development of grazing systems for dairy herd replacements must take into consideration target live-weight gains at various stages throughout development (Table 14.1) and all the sward and animal factors considered so far in this chapter. As noted by Leaver (1974), the objectives of grazing management are to achieve high live-weight gain per animal, but at the same time to efficiently utilise the grass available. Although these objectives were set over 30 years ago, they still apply today. The main problem in the development of grazing systems for dairy herd replacements stems from the fact that calves and older cattle have very different grazing habits. As calves are more selective in their grazing habits than older cattle, and graze for fewer hours per day, they are less willing to harvest a large proportion of the herbage available. As a result, live-weight gain can be reduced if they are forced to utilise more of the grass in a paddock before being moved to a new paddock (Leaver, 1974). The relative merits of a number of grazing systems will be discussed in the following paragraphs.

CONTINUOUS GRAZING

In this system, animals have access to the entire grazing area for the majority of the grazing season (Mayne *et al.* 2000). Advantages of continuous grazing over rotational grazing include lower input costs in terms of fencing, water, and labour for animal movement. A major disadvantage is the greater skill required in monitoring sward growth and managing the sward to prevent under-grazing early in the season and over-grazing later in the season or during drought. Also, some forage species e.g. Lucerne, Timothy and warm-season grasses are less tolerant of continuous grazing (Clark and Kenneganti, 1998).

ROTATIONAL GRAZING

With rotational grazing, the sward is defoliated at regular intervals following a period of regrowth (Mayne *et al.* 2000). The system involves a number of paddocks of similar size, and stock are moved from paddock to paddock throughout the grazing season; the frequency of movement depending on the individual farm. Most rotational grazing systems combine grazing and forage conservation, with paddocks set aside for silage making early in the season and then reintroduced into the grazing cycle as the season progresses. This gives a greater degree of flexibility for adjusting grass supply according to demand. However, relative to continuous grazing, rotational grazing involves more labour, particularly when the system is being set up. With rotational grazing, the ability to present tall leafy swards to animals has been shown to increase live-weight gain. From a review of 27 studies, Marsh (1976) observed a 0.053 proportionate increase in live-weight gain with rotational compared with continuous grazing in beef cattle. Similarly, Ernst *et al.* (1980) showed a proportionate benefit to rotational grazing of 0.06 for beef cattle (38 studies reviewed). However, some studies have shown improved live-weight gains with continuously grazing (0.05 proportionate improvement) relative to rotational grazing (Arnold, 1969). The instances where rotational grazing has been shown to be superior to continuous grazing have been at high stocking rates (McMeekan and Walshe, 1963), with stocking rate having a much bigger effect on performance than rotation length (Leaver, 1975).

Within rotational grazing, there are three options for grazing growing cattle – grazing calves and yearlings separately, grazing calves and yearlings together, and adopting a leader/follower system. Each system and its relative merits will be discussed in the following sub-sections

Calves and yearlings grazed separately

Until the 1970s it was recommended in the United Kingdom (UK) that calves in their first grazing season should be grazed separately from older cattle, mainly to reduce worm burdens as, even if not showing clinical signs of parasitism, older cattle can excrete worm eggs in their faeces. However, a disadvantage of this system is that the performance of younger calves depends on the area of grazing allocated to them relative to older heifers, due to their more selective grazing habits, as discussed previously.

Calves and yearlings grazed together

Grazing calves and yearlings together overcomes the problems of poor herbage utilisation by young calves, whilst enabling adequate performance of yearlings, which are able to utilise the herbage more efficiently.

Leader/follower system

This system, where calves are allowed to graze rotationally in paddocks ahead of the older cattle, was developed to overcome the problems associated with grazing calves and yearlings either together or separately. The system has been used widely in New Zealand for many years (McMeekan, 1954) and was originally investigated under UK conditions by Leaver (1974). Leaver (1974) compared the performance of calves under two systems. In the first system, calves (7 months of age) were rotationally grazed ahead of 19-month old heifers that were in their second grazing season and pregnant to calve at 24 months of age. In the second system, calves and heifers were grazed in completely separate rotations. The performance of calves and heifers in each system is summarised in Table 14.9. On average, live-weight gain of calves grazing in a leader follower system was proportionately 0.35 greater than live-weight gain of those grazing separately, while live-weight gain of the older heifers was similar in both systems. The increase in performance of calves in the leader/follower system was attributed to the fact that the calves selected high crude protein and lower crude fibre herbage whilst grazing. This implies that herbage left for the older heifers that follow is of lower quality, although it had no detrimental effects on heifer performance (Leaver, 1974).

Table 14.9 EFFECT OF LEADER/FOLLOWER VERSUS SEPARATE GRAZING SYSTEMS ON PERFORMANCE OF CALVES AND HEIFERS (LEAVER, 1970)

Live-weight gain (kg/day)	*Leader/follower*	*Separate grazing*
Calves (7 - 12 months)	0.81	0.60
Heifers (19 - 24 months)	0.78	0.84

Keane (2002, study 2) compared the performance of beef calves and yearlings under three management systems: leader/follower; calves and yearlings mixed; and calves and yearlings grazed separately (Table 14.10). In agreement with

Leaver (1970), live-weight gain of calves was greatest in the leader/follower system and poorest for calves grazing separately. Average live-weight gain was 0.41 greater for calves in the leader/follower system than for those in the mixed system, and 0.68 greater than for those in the separate grazing system. However, contrary to Leaver (1974), live-weight gains of yearlings were 0.21 greater in the separate grazing system than in the leader/follower system, and 0.14 greater than in the mixed grazing system. The lower performance of yearlings in the leader/follower system was attributed to insufficient herbage and/or reduced herbage quality.

Table 14.10 EFFECT OF LEADER/FOLLOWER, MIXED AND SEPARATE GRAZING ON PERFORMANCE OF CALVES AND YEARLINGS (KEANE 2002, STUDY 2).

Live-weight gain (kg/day)		*Grazing system*	
from turnout to housing	*Leader/follower*	*Mixed*	*Separate*
Calves	0.96	0.69	0.57
Yearling	0.89	0.95	1.08

Dairy heifer rearing without grazing

An alternative rearing system, where heifers were reared indoors throughout their life on a diet of conserved forage and concentrates without being turned out to pasture, was investigated by Troccon (1993). This research demonstrated that similar live weights and live-weight gains could be achieved in both systems throughout the rearing period (Table 14.11). However, there would be a large difference in the cost of each system because energy supply for the heifers that were housed from birth to calving was entirely from conserved forage and concentrates, compared with 0.57 for heifers that were turned out to grass.

Table 14.11 PERFORMANCE OF HEIFERS REARED INDOORS ON CONSERVED FORAGE AND CONCENTRATE DIETS IN SUMMER AND WINTER OR TURNED OUT TO GRASS IN SUMMER (TROCCON, 1993)

	Pasture during summer	*Reared indoors*
Live weight (kg)		
6 months	160	164
15 months	349	343
21 months	445	458
Live-weight gain (kg/d)		
1st winter	0.66	0.67
1st summer	0.67	0.65
2nd winter	0.66	0.67
2nd summer	0.74	0.68

Carson *et al.* (2004) observed lower live-weight gain in heifers housed during their first summer and offered a straw/concentrate-based diet relative to those turned out to grass. However, this difference had largely disappeared by 18 months of age, when both groups were of similar live weight. The increased live-weight gain from 12.5 to 18 months of age suggests that compensatory growth occurred in the heifers that had been housed during the first summer. A summary of feed requirements of the two groups during the summer period is given in Table 14.12. On average, over a 5-month grazing period, heifers at grass required 160 kg less concentrates/animal than those housed and offered straw plus concentrates.

Table 14.12. FEED REQUIREMENTS (kg/head/day) TO ACHIEVE SIMILAR TARGET LIVE WEIGHTS FOR HEIFERS HOUSED DURING THEIR FIRST SUMMER OR TURNED OUT TO PASTURE (CARSON *et al.* 2004)

	Grazed grass during first summer	*Straw/concentrate diet during first summer*
Live-weight gain during first summer (kg/day)	0.79	0.53
Grazing area required (ha)	0.095	0.00
Straw dry matter intake (kg/day)	0.00	2.58
Concentrate dry matter intake (kg/day)	1.31	2.37
Live-weight gain 12.5 – 18 months (second winter)	0.78	0.90
Live weight at 18 months	462	460

REARING PRE-WEANED CALVES AT GRASS

In terms of animal health and welfare, it is important to develop management strategies that minimise the incidence of respiratory disease in young calves. One component of such a strategy could be to turn newborn spring calves out to pasture and to feed milk replacer and concentrates as for indoor-housed calves. Research by Earley and Fallon (1999) illustrated that such a system had no effect on live-weight gain of calves reared outdoors. However, the incidence of respiratory disease in the outdoor group of calves was lower than in calves reared indoors, although the outdoor group had a greater incidence of diarrhoea. An outdoor calf-rearing system must provide adequate shelter for calves and in the study by Earley and Fallon (1999), the plots to which the calves were turned out had shelter on three sides and dry lying areas comprised of sand.

Health, milk production and grazing systems

WORM CONTROL

In any grazing system consideration has to be given to parasite control. The epidemiologies of endoparasites affecting grazing cattle (gut worms, lungworms) differ and so does their control. Control generally involves pasture management and anthelmintic treatments; vaccination is also available for lungworm control. The leader/follower system has been advocated as a useful pasture management tool to control gut worms, where young calves move on to relatively parasite-safe pasture and older, parasite-resistant cattle follow behind (Leaver,1970). However, strategic use of anthelmintics, in conjunction with grazing management is invariably required to control endoparasites in young calves (Nagle *et al.,* 1980).

INCIDENCE OF LAMENESS

Heifers turned out to grass have been observed to have a lower incidence of lesion scores in the white line and sole area of the front lateral claws compared with those housed and offered straw/concentrate diets during the summer period (Carson *et al*., 2004) (Table 14.13). McDaniel *et al.* (1982) also noted that the incidence of lameness was higher in cattle housed all year round and Whitaker *et al.* (2000) produced evidence that the incidence of lameness was reduced with grazing animals.

Table 14.13 EFFECT OF HOUSING DURING FIRST SUMMER COMPARED WITH TURNING OUT TO GRASS ON LESION SCORES OF FRONT LATERAL CLAWS (CARSON *et al.* 2004)

Lesion scores†	*Housed during first summer*	*Grazed during first summer*
White line area	0.48	0.07
Sole area	0.60	0.22
Heel erosion scores	0.78	0.70

† lesion scores recorded using the method of Livesey *et al.* (1998).

PERFORMANCE AFTER CALVING

Carson *et al.* (2004) observed that at 18 months of age, heifers reared on silage-based diets had increased fat deposition in the udder compared with heifers reared on straw-based diets during the winter period (Table 14.14). However, subsequent milk production of a similar group of heifers reared on the same diets was not significantly affected by diet during the winter period (Carson *et al.* 2002a). The only difference between the two groups of heifers was that those in the study of Carson *et al.* (2002a) were turned out to grass from 18 to 24 months of age. This led the author to conclude that the heifers reared on silage-based diets exhibited compensatory mammary development at grass, so that any differences in the mammary gland at 18 months of age had disappeared by calving.

Table 14.14 EFFECT OF GRASS-BASED DIETS ON MAMMARY GLAND DEVELOPMENT AND SUBSEQUENT MILK PRODUCTION (CARSON *et al.* 2002a & 2004)

Diet during winter *Diet during summer*	*Silage/concentrates* *Grazed grass*	*Straw/concentrates* *Grazed grass*
Udder parameters (18 months)		
Total udder weight (g)	2965	2753
Weight of fat in udder (g)	1733	1210
Weight of parenchyma in udder (g)	938	1244
Milk production (litres) per 305-day lactation(calving at 24 months)	8020	7956

The beneficial effects of grazed grass-based diets on mammary development can also be observed by comparing mammary development of heifers offered straw-based diets during the first summer relative to those turned out to grass (Carson *et al.* 2004) (Table 14.15). There was a tendency towards a greater proportion of parenchyma in the udder of heifers that had been at grass in the first summer compared with those housed and offered straw/concentrate-based diets. In beef animals, grass-based diets have been shown to decrease lipid content, increase protein content and increase concentration of conjugated linoleic acids in the carcass (Steen and Kilpatrick 1998). It has been suggested that conjugated linoleic acids stimulate mammary development (Parodi 1999), which might explain the beneficial effects of grazing pasture in the study of Carson *et al.* (2004).

Table 14.15 EFFECT OF GRASS VERSUS STRAW-BASED DIETS DURING THE FIRST SUMMER OF THE REARING PERIOD ON MAMMARY DEVELOPMENT (CARSON *et al.* 2004)

	Grazed grass during first summer	*Straw/concentrate diet during first summer*
Total udder weight (g)	2753	2409
Weight of fat in udder (g)	1210	1225
Weight of parenchyma in udder (g)	1244	833

Troccon (1993) also observed that, compared with heifers housed throughout the rearing period, heifers on rearing regimes that included grazing during the summer had lighter calves at first calving and had greater feed intakes before first calving and during early first lactation. These effects were attributed to better rumen development and lower condition score in the heifers turned out to grass. However, feeding system had no long-term effects on growth, fertility, milk yield, health or longevity (Table 14.16).

Table 14.16 PERFORMANCE OF HEIFERS HOUSED THROUGHOUT THE REARING PERIOD OR TURNED OUT TO GRASS IN THE SUMMER (TROCCON 1993)

	Pasture during summer	*Reared indoors*
Proportion of heifers in calf after two inseminations	0.76	0.74
Calf birth weight (kg)		
Males	37.0	41.8
Females	34.5	35.8
Average lifespan (years)	5.98	5.98
Lifetime milk production		
Total milk yield (kg)	21456	20842
Total fat yield (kg)	850	817
Total protein yield (kg)	703	664
Calving interval	392	401

Summary and recommendations for grazing heifers

The results reviewed in this chapter demonstrate that the performance of young calves and yearlings at grass is extremely variable and is influenced by a wide range of sward and animal characteristics. Nevertheless, the results do indicate that, with appropriate management, the target live-weight gains given in Table 14.1 are achievable on grass-based diets, often without concentrate supplementation.

GRAZING SYSTEM

From the literature reviewed, it can be concluded that rotational grazing systems offer greater potential for growing cattle in terms of live-weight gain and effective sward management. Within rotational grazing, a leader/follower system with young calves grazing ahead of yearlings, maximises performance of young calves by ensuring that calves have access to the best quality herbage. However, it is important to ensure that the performance of the older animals, i.e. the followers, is not compromised. Satisfactory performance of young calves and yearlings has also been observed when they are grazed separately in paddock systems (Keane, 2002; Carson *et al.*, 2002a, 2004), but increased levels of worm control might be necessary.

TARGET SWARD HEIGHTS

On the basis of the literature reviewed, target sward heights for autumn- and spring-born Holstein heifers in the first and second grazing season are summarised in Table 1417. The targets have been categorised according to whether continuous grazing or rotational grazing is adopted. Sward surface height in continuously grazed swards should be lower early in the season to maintain herbage digestibility and should be increased as the season progresses.

Table 14.17 TARGET RESIDUAL SWARD HEIGHTS TO ACHIEVE TARGET LIVE-WEIGHT GAINS FOR REPLACEMENT DAIRY HEIFERS THROUGHOUT THE GRAZING SEASON (ZERO CONCENTRATES) IN A ROTATIONAL OR CONTINUOUS GRAZING SYSTEM

			Rotational grazing	*Continuous grazing*	
	Age (months)	*Target live-weight gain (kg/day)*	*Residual sward height (cm)*	*Sward surface height (cm)*	
				Early season	*Late season*
First grazing season					
Autumn-born	7.5-12.5	0.70	8	7-8	10-12
Spring-born	3-7	0.70	8	7-8	10-12
Second grazing season					
Autumn-born	17.5-23.5	0.75	6	6-7	9
Spring-born	12-18	0.80	7	7-8	9

MAMMARY AND HOOF DEVELOPMENT

Including grazed grass in the diet of heifers during the rearing period has potential benefits for development of mammary parenchyma and incidence of lameness (Carson *et al.*, 2004).

SUPPLEMENTATION AT PASTURE

Supplementation at pasture might be needed if target live-weight gains are not being achieved and/or grass supplies are inadequate. The response to concentrate supplementation will depend on grass supply, with the largest responses obtained under restricted grazing conditions (sward height 7 cm or less).

Acknowledgements

The authors would like to acknowledge the Department of Agriculture and Rural Development (DARD) and AgriSearch (farmer-funded) who jointly funded the heifer research undertaken at the Agricultural Research Institute of Northern Ireland. The authors also wish to thank Dr Maurice McCoy for assistance with the parasitology section.

References

Alder, F.E. and Chambers, D.T. (1958). Studies in calf management. I. Preliminary studies of post-weaning grazing. *Journal of the British Grassland Society* **13**: 13 – 20.

Arnold, G.W. (1969). Pasture management. *Proceedings of the Australian Grassland Conference* **2**: 189 – 211.

Birrell, H.A. and Bishop, A.H. (1980). Effect of feeding strategy and area of pasture conserved on the wool production of wethers. Australian Journal of Experimental Agriculture and *Animal Husbandry* **20**: 406 – 412.

Black, J.L. and Kenny, P.A. (1984). Factors affecting diet selection by sheep. II. Height and density of pasture. *Australian Journal of Agricultural Research* **35**: 551 – 563.

Carson, A.F., Dawson, L.E.R., McCoy, M.A., Kilpatrick, D.J. and Gordon, F.J. (2002a). Effects of rearing regime on body size, reproductive performance and milk production during the first lactation in high genetic merit dairy herd replacements. *Animal Science* **74**: 553 – 565.

Carson, A.F., Dawson, L.E.R. and Gordon, F.J. (2002b). Research points the way for heifer rearing. *Proceedings of a seminar held at the Agricultural Research Institute of Northern Ireland*, March 2002, pp. 1-16.

Carson, A.F., Dawson, L.E.R., Wylie, A.R.G. and Gordon, F.J. (2004). The effect of rearing regime on the development of the mammary gland and claw abnormalities in high genetic merit Holstein-Friesian dairy herd replacement. *Animal Science*, **78**: 181-185.

Carson, A.F., Wylie, A.R.G., McEvoy, J.D.G., McCoy, M. and Dawson, L.E.R. (2000). The effects of plane of nutrition and diet type on metabolic hormone concentrations, growth and milk production in high genetic merit dairy herd replacements. *Animal Science* **70**: 349 – 362.

Clark, D.A. and Kanneganti, V.R. (1998). Grazing management systems for dairy cattle. In: J.H. Cherney and D.J.R. Cherney (eds.) *Grass for Dairy Cattle*, CAB International, pp. 311 – 334.

Dawson, K.P., Large, R.V., Jewiss, O.R., Tallowin, J.R.B. and Michell, P. (1981). Effects of management on seasonal distribution of yield. *Annual Report*, Grassland Research Institute, Hurley, pp. 55 – 56.

Department of Agriculture and Rural Development (DARD) (2003). *Statistical review of Northern Ireland Agriculture*, Economics and Statistics Division.

Department for Environment and Rural Affairs (DEFRA) (2003). *Agriculture in the United Kingdom*, Statistical publications.

Earley, B. and Fallon, R.J. (1999). *Calf Health and Immunity.* End of project report, Beef Production Series No. 17, Grange Research Centre, Dunsany, Co. Meath.

Ernst, P., Le Du, Y.L.P. and Carlier, L. (1980). Animal and sward production under rotational and continuous grazing management – a critical appraisal. *Proceedings of the International Symposium of the European Grassland Federation on the Role of Nitrogen in Intensive Grassland Production*, Wageningen pp. 119 – 126, Pudoc, Wageningen.

Gibb, M.J. (1996). Terminology for animal grazing/intake studies. *Occasional Publication No. 3.* Proceedings of a Workshop, Dublin, Ireland.

Hoden, A., Peyraud, J.L., Muller, A., Delaby, L. and Faverdin, P. (1991). Simplified and rotational grazing management of dairy cows: effects of rates of stocking and concentrate. *Journal of Agricultural Science, Cambridge* **116**: 417 – 428.

Hodgson, J. (1968). The relationship between the digestibility of a sward and the herbage consumption of grazing calves. *Journal of Agricultural Science, Cambridge* **70**: 47 – 51.

Hodgson, J. (1977). Factors limiting herbage intake by the grazing animal. In: Gilesnan, B. (ed.). *Animal Production from Temperate Grassland*, Dublin, Irish Grassland and Animal Production Association, pp. 70 – 78.

Hodgson, J., Rodriguez Capriles, J.M. and Fenlon, J.S. (1977). The influence of sward characteristics on the herbage intake of grazing calves. *Journal of Agricultural Science, Cambridge* **89**: 743 – 750.

Hodgson, J., Tayler, J.C. and Lonsdale, C.R. (1971). The relationship between intensity of grazing and herbage consumption and growth of calves. *Journal of the British Grassland Society* **26**: 231 – 239.

Hoffman, P.C. and Funk, D.A. (1992). Applied dynamics of dairy herd replacement growth and management. *Journal of Dairy Science* **75**: 2504 – 2516.

Jamieson, W.S. and Hodgson, J. (1979). The effect of daily herbage allowance and sward characteristics upon the ingestive behaviour and herbage intake of calves under strip-grazing management. *Grass and Forage Science* **34**: 261 – 271.

Keane, M.G. (2002). Development of an intensive dairy calf-to-beef system and associated grassland management. *End of Project Report*, Project No. 4582, Beef Production Series No. 41, Grange Research Institute, Teagasc.

Leaver, J.D. (1970). A comparison of grazing systems for dairy herd replacements. *Journal of Agricultural Science, Cambridge* **75**: 265 – 272.

Leaver, J.D. (1974a). Grazing management for dairy calves and heifers. *Irish Grassland and Animal Production Association Journal* **9**: 71 – 74.

Leaver, J.D. (1974b). Rearing of dairy cattle. 5. The effect of stocking rate on animal and herbage production in a grazing system for calves and heifers. *Animal Production* **18**: 273 – 284.

Leaver, J.D. (1975). Rearing of dairy cattle. 6. The effect of length of grazing rotation on animal and herbage production in a grazing system for calves and heifers. *Animal Production* **21**: 157 – 164.

Livesey, C.T., Harrington, T., Johnston, A.M., May, S.A. and Metcalf, J.A. (1998). The effect of diet and housing on the development of sole haemorrhages, while line haemorrhages and heel erosions in Holstein heifers. *Animal Science* **67**: 9 – 16.

McDaniel, B.T., Hahn, M.V. and Wilk, J.C. (1982). Floor surfaces and effect upon feet and leg soundness. In *Proceedings of a Symposium on the Management of Food Producing Animals* Vol 2, pp. 816 – 833. Purdue University, U.S.A.

McGilloway, D. A. Cushnahan, A. Laidlaw, A. S. Mayne, C. S. Kilpatrick, D. J. (1999) The relationship between level of sward height reduction in a rotationally grazed sward and short-term intake rates of dairy cows. *Grass & Forage Science* **54**: 116-126.

McKeekan, C.P. (1954). Good rearing of dairy stock. *Bulletin of the Department of Agriculture in New Zealand* No. 228.

McKeekan, C.P. and Walshe, M.J. (1963). The interrelationships of grazing method and stocking rate in the efficiency of pasture utilization by dairy cattle. *Journal of Agricultural Science, Cambridge* **61**: 147 – 166.

Marsh, R. (1975). A comparison between spring and autumn pasture for beef cattle at equal grazing pressures. *Journal of the British Grassland Society* **30**: 165 – 170.

Marsh, R. (1976). Systems of grazing management for beef cattle. *Occasional Symposium – British Grassland Society* No. 8 pp. 119 – 128.

Mayne, C.S. (2000). Getting more milk from grass. *Proceedings of a seminar held at the Agricultural Research Institute of Northern Ireland*, May 2000, pp. 77 – 89.

Mayne, C.S., Newbury, R.D., Woodcock, S.C.F. and Wilkins, R.J. (1987). Effect of grazing severity and grass utilization and milk production of rotationally grazed cows. *Grass and Forage Science* **42**: 59 – 72.

Mayne, C.S., Wright, I.A. and Fisher, G.E.J. (2000). Grassland management under grazing and animal response. In: A. Hopkins (ed). *Grass its Production and Utilization*, 3rd edition, British Grassland Society, Blackwell Science Ltd pp. 247 – 291.

Milne, J.A. and Fisher, G.E.J. (1994). Sward structure with regard to production. In: Hagger, R.J. and Peel, S. (eds.). *Grassland Management and Nature Conservation,* Occasional Symposium, No. 28, British Grassland Society, pp. 33 – 42.

Mitchell, P. and Fulkerson, W.J. (1987). Effect of grazing intensity in spring on pasture growth, composition and digestibility, and on milk production by dairy cows. *Australian Journal of Experimental Agriculture* **27**: 35 – 40.

Musangi, R.S., Holmes, W. and Jones, J.G.W. (1965). Barley supplementation and voluntary food intake of fattening beef cattle under restricted and unrestricted grazing conditions. *Proceedings of the Nutrition Society* **24**: R1.

Nagle, E.J., Brophy, P.O., Caffrey, P.J. and Nuallain, T.O. (1980). Control of ostertagiasis in young cattle under intensive grazing. *Veterinary Parasitology* **7**: 143 – 152.

Parodi, P.W. (1999). Conjugated linoleic acid and other anticarcinogenic agents of bovine milk fat. *Journal of Dairy Science* **82**: 1339 – 1349.

Pirlo, G., Miglior, F. and Speroni, M. (2000). Effect of age at first calving on production traits and on difference between milk returns and rearing costs in Italian Holsteins. *Journal of Dairy Science* **83**: 603 – 608.

Sayers, H.J. (1999). *The effect of sward characteristics and level and type of supplement on grazing behaviour, herbage intake and performance of lactating dairy cows*. PhD Thesis, Queen's University of Belfast.

Stakelum, G. and Dillon, P. (1991). Influence of sward structure and digestibility on the intake and performance of lactating and growing cattle. In: Mayne, C.S. (ed.) *Management Issues for the Grassland Farmer in the 1990's*, Occasional Symposium No. 25, British Grassland Society, pp. 30 – 42.

Steen, R.W.J. (1994). A comparison of pasture grazing and storage feeding, and the effects of sward surface height and concentrate supplementation from 5 to 10 months of age on the lifetime performance and carcass composition of bulls. *Animal Production* **58**: 209 – 219.

Steen, R.W.J. and Kilpatrick, D.J. (1998). Effects of pasture grazing or storage feeding and concentrate input between 5.5 and 11 months of age on the performance and carcass composition of bulls and on subsequent growth and carcass composition at 620 kg live weight. *Animal Science* **66**: 129 – 141.

Swift, G., Lowmand, B.G., Scott, N.A., Peebles, K., Neilson, D.R. and Hunter, E.A. (1989). Control of sward surface height and the growth of set-stocked finishing cattle. *Research and Development in Agriculture* **6**: 91 – 97.

Troccon, J.L. (1993). Dairy heifer rearing with or without grazing. *Annales de Zootechnie* **42**: 271 – 288.

Vadiveloo, J. and Holmes, W. (1979). Supplementary feeding of grazing beef cattle. *Grass and Forage Science* **34**: 173 – 179.

Whitaker, D.A., Kelly, J.M and Smith, S. (2000). Disposal and disease rates in 340 British dairy herds. *Veterinary Record* **146**: 363 – 367.

Wright, I.A. and Whyte, T.K. (1989). Effects of sward surface height on the performance of continuously stocked spring-calving beef cows and their calves. *Grass and Forage Science* **44**: 259 – 266.

15

FERTILITY IN THE MODERN DAIRY HEIFER

KEVIN D SINCLAIR AND ROBERT WEBB
Division of Agricultural and Environmental Sciences, School of Biosciences, University of Nottingham, Sutton Bonington, Leics, LE12 5RD, UK

Introduction

Sub-fertility in dairy cows is one of the most important problems currently facing the UK dairy industry, accounting for just over two-thirds of the estimated annual cost of £17,400 per 100 cows attributed to poor health and fertility (Esslemont and Kossaibati, 2002). Work at Nottingham (Royal *et al.*, 2000) confirmed earlier reports from North America (e.g. Butler, 1998) of a decline in conception rates by around 0.75-1.0% per annum over the last three decades to levels that are now less than 40%. It has been estimated that a 10% improvement in conception rate would benefit the UK industry by approximately £300 million per annum. Furthermore, because dairy cows contribute around 20% of total UK atmospheric methane emissions and 25% of total UK ammonia emissions (Misselbrook and Smith, 2002; Defra, 2003), restoring dairy cow fertility to pre-1995 levels would allow the UK to meet its 1997 Kyoto-Agreement target of a 12.5% reduction in 'greenhouse' gas emissions by 2010 (Garnsworthy, 2004). Further improvements to fertility (to levels achieved during the 1970s) could reduce methane emissions by up to 24 and ammonia emissions by up to 17%. These potential benefits arise because enhanced fertility requires smaller herds and fewer replacement heifers in order to meet milk quotas.

Whilst, understandably, much attention has been focused on the decline in fertility of the lactating dairy cow, less attention has been directed towards understanding fertility in the dairy heifer. In light of the dramatic changes that have taken place in the global dairy industry over the last 20 years, in terms genetic selection and herd size and management, the means by which these young animals are reared needs to be re-evaluated. A number of questions come to mind. For example, if selection for milk yield is genetically linked to the decline in fertility of the lactating cow has there been a similar genetically-linked decline in fertility of the heifer? If so, what genes are affected and what steps can be taken to overcome these genetic defects, i.e. is selection for enhanced fertility in the nulliparous heifer likely to enhance fertility in the multiparous cow? Also, given that there is now extensive evidence that the lifetime health and productivity of a number of mammalian species is influenced by both the level and quality of nutrition during foetal and/or pre-pubertal life, what are the implications for the management of dairy heifers with regard to optimal future milk yields and fertility? These issues are addressed in this review of our current understanding of dairy heifer rearing for optimal fertility and lifetime performance.

Genetic contributions to fertility

Although fertility among maiden heifers is superior to that of lactating cows (conception rates to first service of ~67% vs ~42%, Pryce *et al.*, 2004) there is emerging evidence to indicate that selection for milk production has led to a decline in fertility in the nulliparous animal. Pryce *et al.* (2002) analysed various fertility parameters (i.e. age at first service, conception at first service, services per conception and interval from first to last service) in two genetic lines of maiden heifer. Heifers from the line selected for enhanced yields of milk fat and protein performed relatively poorly for all of these reproductive traits. Furthermore, regression analysis revealed that, in the selected line, age at first service declined and services per conception increased with time. In spite of the fact that genetic correlations between fertility in heifers and lactating cows are low and variable (0.3 to 0.8), indicating that the genetic basis for fertility may be different between parous and non-parous animals (discussed by Pryce *et al.*, 2004), the experience of some Scandinavian countries, where selection indices include fertility of both heifers and primiparous cows, is positive. The use of such indices undoubtedly contributed to the lack of decline in daughter fertility of Swedish Red and White cattle between 1985 and 1999. This observation was in stark contrast to that for Swedish Holstein cattle, where the absence of daughter fertility data of imported Holstein semen contributed to the decline in fertility of the breed during this period.

The heritability of traditional fertility traits, such as conception rate and number of services per conception, is typically low at around 5%, making them ideal candidates for marker-assisted selection. To date, however, there have been few reports of quantitative trait loci (QTL) for fertility in dairy cattle. Ashwell *et al.* (2004) reported putative QTLs affecting pregnancy rate on bovine chromosomes 6, 14, 16, 18, 27 and 28. Other reports of QTLs on chromosomes 6 and 17 were related to interval from calving to first insemination (Schrooten *et al.*, 2000), but might not be relevant to dairy heifer fertility. An alternative or additional strategy to the use of QTL in genetic selection is to incorporate endocrine parameters of fertility with the aim of improving the accuracy of predicting breeding value for this trait. To date such attempts have been limited to incorporation of measures of luteal activity (from milk progesterone analysis) in lactating cows with the view to providing a physiological measure of post-partum anovulation and/or interval from calving to conception (Royal *et al.*, 2002a and b). This general approach, however, appears encouraging, with comparatively high estimates of heritability for endocrine fertility traits, but it remains to be applied to the non-lactating nulliparous heifer. To better understand the genetic basis of sub-fertility in dairy heifers will require a more thorough grasp of the variability in expression of genes encoding key physiological components of the reproductive axis. Although genomic mapping in ruminant species has come a long way in recent years, it still lags behind that of the mouse and human, and polymorphic differences for many of the genes that have been mapped remain to be determined (Montgomery, 2000). Furthermore, research effort in recent years has tended to focus on loci that influence ovulation rate in ruminants rather than fertility *per se*. The products of these single genes (e.g. growth and differentiation factor 9 (GDF9) and bone morphogenetic protein (BMP15), two members of the TGFß superfamily of growth and differentiation factors) play a central role in the orchestrated development of the ovarian follicle and oocyte, and are unquestionably linked to fertility (McNatty *et al.*, 2004). However, genetic variability in fertility of the modern dairy heifer is unlikely to be attributable to deficiencies in any single gene, but to polymorphisms in a number of genes that influence the reproductive axis. Furthermore, mitochondrial DNA polymorphisms have recently been linked to fertility in beef cattle, but such relationships remain to be fully investigated in the dairy heifer.

One final consideration regarding the genetic contribution to fertility in dairy cattle concerns recent efforts to identify first-lactation 'problem breeders' from metabolic and endocrine profiles in juvenile calves (Taylor *et al.*, 2004). Differences in aspects of the somatotrophic axis as juveniles in that study were correlated with irregular progesterone profiles during lactation. These relationships are thought to primarily reflect genetic differences between animals, although they may also be compounded by in utero or early post-natal environmental effects (discussed later). Nevertheless, the main findings from this study were that first lactation heifers with delayed onset to first ovulation had the lowest insulin-like growth factor 1 (IGF-I) concentrations as calves. They also exhibited impaired glucose homeostasis as calves and again during lactation. In contrast, heifers forming a persistent corpus luteum following calving exhibited high IGF-1 concentrations during the pre-pubertal period and released growth hormone from the pituitary in larger boluses during the period following feeding. Whilst more effort is required in this area to fully understand these relationships, and their possible associations with genetic selection, these findings clearly point to the fact that deviations in the somatotrophic axis of animals as juveniles can be related to impaired reproductive function during their first lactation.

Environmental contributions to fertility

PRE-NATAL PERIOD

There is evidence from sheep that undernutrition during in utero development can reduce adult ovulation rate and litter size and that this effect is programmed during the first half of pregnancy (Gunn *et al.*, 1995; Rae *et al.*, 2002). At present, although the mechanisms underlying these effects are poorly understood (Rhind, 2004), it would seem that maternal undernutrition does not alter female hypothalamic-pituitary function (Borwick *et al.*, 2003). Instead, given that maternal undernutrition appears to delay foetal ovarian follicular development (Rae *et al.*, 2001) and germ cell development (Borwick *et al.*, 1997), it would seem that these effects may be mediated by alterations to the expression and/or function of members of the TGFß superfamily of growth and differentiation factors; although this is an area of research that remains to be explored. In adult sheep, 100g of ruminally undegradable starch per day increased ovulation rate in carriers of the Booroola (*Fec*^{B}) allele (a mutation in the gene encoding BMP receptor 1B), but not in non-carriers, suggesting that specific nutrients can interact with members of the BMP system to influence the rate of folliculogenesis, at least in adults (Landau *et al.*, 1995). However, given that this specific receptor is expressed in tissues other than the ovary (e.g. pituitary gland) it remains to be established if members of this superfamily of growth and differentiation factors are involved in nutritionally-mediated development of female gonad development in utero.

There are no corresponding data for mono-ovular cattle, but the evidence of Pryce *et al.* (2002), who analysed the reproductive performance of 988 heifers from two genetic lines, suggests that there is no effect of maternal diet on daughter reproductive performance. However, maternal undernutrition might not have been great enough in that study to detect such an effect, because the nutritional management of these animals was representative of that observed in commercial practice, so the relevance of foetal programming of reproductive function in dairy cattle is questionable.

PRE-AND PERI-PUBERTAL PERIODS

In contrast to in utero effects, it is now well established that nutritional status and growth rate during the pre-pubertal period can influence both the timing of the onset of puberty and subsequent fertility in ruminants. This is most pronounced in sheep, where an extended period of underfeeding during the pre-pubertal period (from 6 weeks of age) not only delayed the onset of puberty, but also reduced ovulation and lambing rates during adulthood (Gunn, 1983). Plane of nutrition is also known to significantly influence the onset of puberty in cattle, although longer term consequences on reproductive function are not known.

Studies investigating the effects of nutrition on attainment of puberty have, to date, largely been confined to those concerned with feeding level, growth rate and body composition, and have largely been conducted in beef heifers and sheep, with comparatively few studies in dairy heifers (Rawlings *et al.*, 2003). In general, animals that are fed high planes of nutrition and grow rapidly during early life attain puberty earlier and at heavier weights than animals that are not fed so well. Some recent data from dairy heifers, however, indicated that, although the onset of puberty was attained at a younger age in well fed animals it was, nevertheless, attained at a constant body weight and body composition which was independent of dietary energy and protein concentrations (Chelikani *et al.*, 2003). Serum concentrations of leptin (a 16 kDa cytokine product of the obese gene synthesised largely by adipose tissue) are positively correlated with body fat and are known to increase in growing heifers as they approach puberty (Garcia *et al.*, 2002). However, chronically administered leptin failed to induce puberty or alter endocrine characteristics in beef heifers nearing the time of expected puberty (Maciel *et al.*, 2004), leading these authors to conclude that this metabolic hormone plays a more passive role in permitting the onset of puberty when sexual maturity is reached. Consequently, a direct causal link between body composition and the onset of puberty remains enigmatic.

Although the mechanisms remain poorly understood, nutrition is thought to operate both centrally (within the hypothalamus) and locally (within the ovary) to regulate the onset of puberty. Low planes of nutrition are known to inhibit luteinising hormone (LH) secretion and to delay the development of dominant follicles (Robinson *et al.*, 1999). In practice, however, nutritional effects are confounded by season of birth in ways that make the results of many experiments difficult to interpret. For example, autumn born beef heifers reach puberty earlier than spring born heifers, but they tend to exhibit a bimodal pattern in onset of puberty (Schillo *et al.*, 1992). Heifers born during the early autumn period, particularly those with a genetic propensity to attain puberty at early ages, will do so during the following summer, whereas those born during the late autumn and early winter period, and have a genetic propensity to attain puberty at older ages, often do not attain puberty until the following spring. Melatonin receptors have been identified in foetal tissues as early as Day 30 of gestation in both sheep (Helliwell and Williams, 1994) and deer (Williams *et al.*, 1997), and may be instrumental in mediating the prenatal photoperiodic effects on postnatal reproductive maturation reported in these two species by Helliwell *et al.* (1997) and Adam *et al.* (1994). The importance of such effects under ambient light is, however, uncertain, since they might be modified by post-natal photoperiodic changes. There are no equivalent data in cattle and so at present it is not possible to assess properly the effects of photoperiodicity in sexual maturity in this species.

PRE-PUBERTAL GROWTH, REPRODUCTIVE FUNCTION AND MILK YIELD POTENTIAL

The development of optimum feeding strategies for management of replacement heifers is complicated by the fact that the high planes of nutrition that favour the onset of puberty during the rearing period have a negative effect on mammary development and subsequent milk yield (Serjsen, 1994; Sejrsen and Purup, 1997; Serjsen, 2005). Knowledge of the timing and extent of compensatory growth during the pre-pubertal period can, however, be used to overcome problems of impaired mammary growth and development. Compensatory growth is normally expressed over a relatively short period of time (usually 4 to 8 weeks on high planes of nutrition typical of spring pastures), and arises as a consequence of increased voluntary food intake, reduced maintenance energy requirements and increased efficiency of energy and protein utilisation (O'Donovan, 1984). Early body-weight gains, following restriction, comprise increased proportions of protein and water and a reduced proportion of fat, which is then followed by a period in which gains consist of an increased proportion of fat such that, eventually, body weight and composition of restricted cattle returns to that of non-restricted cattle (Wright and Russel, 1991). Protein turnover increases during compensatory growth and is characterised by an increase in protein synthesis relative to degradation, and increased nitrogen balance. The endocrinology associated with compensatory growth is only partially understood. Basal metabolism is reduced during feed restriction, mainly as a consequence of reduced mass and metabolic activity of the viscera. Consequently, plasma insulin, triiodothyronine (T_3), thyroxine (T_4) and IGF-I concentrations decrease, while cortisol and growth hormone (GH) concentrations increase (Hornick *et al.*, 2000). During compensatory growth insulin concentrations increase sharply and plasma GH concentrations remain high (at least initially) or decrease slightly, and there is a marked decrease in the GH:insulin ratio. Effects on IGF-I are less clear; it would seem that circulating levels of this metabolic hormone take longer to recover following a period of feed restriction. Similarly, circulating concentrations of T_3 and T_4 take several weeks to recover following dietary re-alignment.

STAIR-STEP FEEDING, MAMMARY GLAND DEVELOPMENT AND FERTILITY

A number of groups have now attempted to exploit knowledge of the timing and extent of compensatory growth during the peri-pubertal period in order to overcome problems of impaired mammary gland development which arise as a consequence of rapidly growing heifers striving to attain puberty at an early age. The so-called 'stair-step' pattern of feeding and growth subjects heifers to short periods (3 to 4 months) of restricted growth interspersed with short (2 month) periods of compensatory growth (Park *et al.*, 1987). In these studies feed restriction and reduced growth were associated with higher plasma concentrations of GH and lower concentrations of IGF-I, T_3, T_4 and insulin (Yambayamba *et al.*, 1996). During the early phase of re-alimentation (first 2 to 3 weeks), plasma T_3 and T_4 concentrations remained low, but plasma concentrations of insulin and IGF-I had increased by Day 10 to be in line with those of unrestricted heifers, and plasma GH concentrations remained high. Little is known about the effects of dietary re-alimentation in this system on IGF binding protein (IGFBP) concentrations either at tissue level or in peripheral circulation. This is unfortunate, as these binding proteins are thought to be central in mediating the effects of IGF-I within the mammary gland (Serjsen, 2004). IGFBP-3 is the most abundant binding protein in peripheral circulation and its concentration decreases during feed restriction (Hornick *et al.*, 2000), but concentrations of this and other binding proteins during the early period of re-alimentation in heifers are not known. Nevertheless, from the limited evidence available, it would seem that the 'stair-step' system of feeding can allow high peri-pubertal growth rates in heifers with no detrimental effects on mammary gland development or fertility (Choi *et al.*, 1997).

Reproductive function in heifers

HYPOTHALAMIC-PITUITARY AXIS

In heifers the maturational processes that result in puberty are initiated before birth, continue post-natally and are completed early after puberty, with the hypothalamus being the primary site of maturation during this period. (see Kinder *et al.*, 1995). It is now well established that LH pulses, which can be detected soon after calving (Schams *et al.*, 1981; Nakada, *et al.*, 2002), occur due to the stimulated release of GnRH from the hypothalamus (see Clarke, 2002). For example, there is an increase in the release of GnRH pulses into the hypothalamic pituitary-portal vessels during the peri-pubertal period (Rodriguez and Wise, 1989). These changes are also accompanied by changes in the pattern of secretion of the pituitary gonadotrophins, LH and FSH, the primary drivers of follicular development (Webb *et al.*, 2003; 2004). An increase, both in the frequency of episodic LH pulses and LH response to GnRH have been demonstrated (Schams *et al.*, 1981; Nakada *et al.*, 2002), with a gradual increase from 4 months of age, with a peak in LH output occurring at about 50 days prior to puberty (Schams *et al.*, 1981; Day *et al.*, 1984; Day *et al.*, 1987). In heifers an increase in peripheral concentrations of LH also accompanies this increase in episodic LH release (Day *et al,* 1984; Dodson *et al.*, 1988; Schams *et al.*, 1981; Kinder, Day and Kittok, 1987).

FSH is also elevated during the first 4 months after calving in heifers, peaking at around 2 months of age (Williams, Campbell and Webb, unpublished observations). In support of this observation, FSH concentrations in the first 4 months of life have been reported to be higher than in cycling heifers (Nakada *et al.*, 2002). There is then a reduction of FSH with increasing age that probably reflects increased numbers of oestrogenic follicles, and elevated peripheral inhibin concentrations, which have been shown to inhibit FSH concentrations from 60 days of age onwards in cattle (Kaneko *et al.*, 1993; 1995). In support of these changes, the FSH response to a GnRH challenge decreases with increasing age up to 8 months in cattle (Nakada *et al.*, 2002).

There is a marked increase in frequency of release of LH pulses around the time of puberty, which is the primary driver inducing the onset of oestrous cycles. Treatment of heifers with oestrogen will induce preovulatory surges of both LH and FSH, thus the preovulatory surge mechanisms are in place even prior to puberty (Kinder *et al.*, 1995). Based on ultrasound scanning data, most heifers attain puberty by 18 months of age (Williams, Campbell and Webb, unpublished observations), and the average age for the Holstein breed is approximately 11 months (Evans *et al.*, 1994). Variation in age of puberty is affected by a number of environmental variables, in particular changes in nutrition and metabolic status, which can influence the secretory pattern of LH. There is a complex, and not completely understood, relationship linking body mass index and/or adiposity to the control of the reproductive axis (see Williams *et al.*, 2002).

OVARIAN FOLLICULOGENESIS AND OOCYTE DEVELOPMENT

Reproductive tract development is a gradual process and although oestrus can be observed between 8 and 13 months of age, maturation of the bovine female reproductive tract can continue until at least 3 years of age. Follicles begin to grow as soon as the primordial follicle store has been established, usually during foetal development in ruminants (Juengel *et al.*, 2002). Initiation of follicle growth is defined as the transition of primordial follicles from the quiescent phase to the growth phase. The mechanisms that are involved in either initiation of follicle growth or the number of primordial follicles that start to grow each day are still not known, but a range of local growth factors appear to be involved (see Webb *et al.*, 2003; 2004). Braw-Tal (2002) speculated that regulation of follicle growth initiation in cattle is a two-phase process, the first phase involving local inhibitory and stimulatory factors and the second phase involving oocyte and granulosa derived factors, such as GDF9, BMP15 and kit ligand. For example, sheep with a BMP15 gene mutation are infertile with follicle growth arrested at the primary stage of growth (Galloway *et al.*, 2000). Once initiated, follicle growth continues without halting until the follicle becomes either atretic or proceeds to ovulation. Gonadotrophins are probably not involved in the initiation of follicle growth (Wandji *et al.*, 1992; McNatty *et al.*, 1999; Campbell *et al.*, 2000; Fortune *et al.*, 2000), although FSH receptor (FSHr) mRNA can be detected in follicles with only one or two layers of granulosa cells (Bao and Garverick, 1998). Ovarian autograft studies have confirmed that it takes several months for a primordial follicle to reach the preovulatory stage in ruminants (see Campbell *et al.*, 2000). This is a relatively long period when various perturbations, including environmental factors such as nutrition, can influence follicular development, oocyte quality and hence fertility (Garnsworthy and Webb, 1999; Webb *et al.*, 1999a; 1999b), as discussed in the later sections of this review. In monovulatory species, such as cattle, only a few follicles from the original store of primordial follicles (<0.1%) will proceed through all stages of development and go on to ovulate.

Changes in number of follicles have been examined in heifers from birth to 12 months of age (Desjardin and Hafs; 1969) and in this study no antral follicles were visible, macroscopically, on the surface of the ovary at birth. However, more recent microscopy studies have detected antral follicles pre-natally (Tanaka *et al.*, 2001, Fouldadi-Nashta *et al.*, 2005a). Follicle numbers, observed on the surface of ovaries (Desjardin and Hafs; 1969) in calves increased to a maximum at 4 to 6 months of age, decrease at eight months of age and appeared to remain constant thereafter. However, using transrectal ultrasound, Hopper *et al.* (1993) and Adams *et al.* (1994) observed that, in pre-pubertal heifers, follicular development occurs in waves similar to those in post-pubertal heifers. Bergfield *et al.* (1994) evaluated pre-pubertal ovarian development during the later pre-pubertal period and reported that dominant ovarian follicles were larger as first ovulation approached, compared with earlier in the pre-pubertal period. Although generally smaller than dominant follicles in mature adults, pre-pubertal heifers do have some >8mm follicles (see Driancourt, 2001). This increase in follicle size was correlated with increased17ß-oestradiol secretion, necessary to induce oestrous behaviour and for induction of the preovulatory gonadotrophin surge and hence the first ovulation. Bovine oestrous cycles are therefore characterized by two or three waves of follicular growth (Adams, 1999), with each wave being characterized by recruitment of 3 to 5 follicles, that grow to >4 mm in diameter. This group of recruited follicles continues to grow until follicles reach approximately 6-8 mm in diameter, when one follicle is selected for continued growth and becomes dominant over the others. This pattern of follicular waves can be found prior to puberty, after puberty, throughout most of pregnancy and starting again a few weeks after calving (see Adams, 1999; Ireland *et al.*, 2000).

As discussed, the very early stages of folliculogenesis can occur without gonadotrophins, but FSH along with other extra-ovarian and locally produced growth factors may affect the rate of preantral follicle growth (Hulshof *et al.*, 1995; Gutierrez *et al.*, 2000; Webb *et al.*, 2004). Antral follicle development from 1-4 mm in heifers is completely gonadotrophin dependent. These recruited follicles express a range of mRNA encoding steroidogenic enzymes, gonadotrophin receptors and local regulatory factors and their receptors. For example, FSH infusion of mature heifers, in which pituitary gonadotrophin secretion had been significantly reduced by either GnRH agonist treatment or GnRH immunisation, stimulated follicle growth up to 8.5 mm in diameter (Crowe *et al.*, 2001; Garverick *et al.*, 2002). Furthermore, this infusion of FSH for 48 h, as well as inducing follicular growth, increased expression of steriodogenic enzyme mRNAs (P450scc and P450arom in granulosa cells and P450c17 in theca) compared with recruited follicles of similar-size in animals with normal oestrous cycles (Garverick *et al.*, 2002). As follicles continue to mature beyond 8.5mm in diameter, there is a transfer of dependency from FSH to LH (Webb *et al.*, 2003; 2004).

Locally produced growth factors, such as IGFs and members of the TGFß super-family, such as BMPs, work in concert with gonadotrophins throughout the follicular growth continuum and can have significant effects on follicle selection. Indeed, both we and others have demonstrated that by the preantral stage of development bovine follicles have acquired IGF and BMP systems within the follicle. These include the presence of both the ligands and the receptors, in addition to binding proteins and proteases in the case of the IGF system (see Webb *et al.*, 2004; Armstrong *et al.*, 1998; 2000; 2002; Dugan *et al.*, 2004; Glister *et al.*, 2004; Fouladi-Nashta *et al.*, 2005a).

As stated, follicular development is a lengthy process. During preantral stages of development the oocyte increases in volume and it is during this phase that the greatest increase in oocyte mass is achieved (see Telfer *et al.*, 1999). This early period is critical for oocyte development as many of the factors persist right through to pre-implantation development (Howe and Solter, 1979). At the antral stage of development oocytes of most species have acquired the ability to resume meiosis. In heifers the acquisition of meiotic competence and completion of metaphase I to metaphase II are sequentially acquired during late antral development. Therefore, the ability of the oocyte to complete all stages of development is related to its volume and hence follicular size (see Telfer *et al.*, 1999). A recent allometric study of the relationship between growth of follicles and oocytes found that bovine oocytes reach their full size when follicles attain a diameter of 4-6 mm (Izumi *et al.*, 2003), the stage when follicles are responsive to FSH (Webb *et al.*, 1999a). Throughout development, the majority of follicles undergo atresia and regress, with subsequent loss of oocyte competence through reduction of contact with granulosa cells, via gap junctions. We have recently demonstrated that IGFBP expression profiles of follicular fluid, along with changes in caspase-3 activity, can be used to better predict oocyte developmental competence (Nicholas *et al.*, 2005). Progress is therefore being made in identification of better markers of oocyte and embryo (Fouladi-Nashta, 2005b) quality and hence the development of better strategies to improve fertility. The quality of the oocyte and hence the ability to develop further, with a subsequent impact on fertility in heifers, can be affected by a number of extra-ovarian factors and this is discussed in later sections.

POST-FERTILISATION DEVELOPMENT

Establishment of pregnancy in heifers depends on a balance between development of the luteolytic signal in the mother (Pate, 2003) and production of an antiluteolytic protein, interferon tau, by the growing embryo (see Mann *et al.*, 1999; Pate, 2003). Embryo development can be influenced by a whole range of factors, including those that affect oocyte quality and influence embryo development after fertilization, such as GH (Thatcher *et al.*, 2003, Gutierrez *et al.*, 2005). A further example was an enhancement in early embryo development observed in heifers fed a diet generating an increased insulin:glucagon ratio. In dairy cattle, inadequate luteal function, affecting the ability of the uterine environment to support normal embryo development, has also been established as a major reproductive problem (see Wathes *et al.*, 2003). However, in beef heifers, progesterone concentrations seem to be less important (Mann *et al.*, 2003), although progesterone concentrations during the period of the post-ovulatory rise (from Days 4-5) were somewhat lower in heifers with less well-developed embryos. There is increasing evidence that the time at which progesterone concentrations start to increase is the critical determinant of the outcome of mating (Mann and Lamming, 1999). There appears also to be a strong correlation between progesterone, coming from the developing corpus luteum, on days 4 and 5 after mating and uterine concentrations of interferon tau, produced by the bovine conceptus (Wathes *et al.*, 2003). Interferon tau is synthesized by the trophectoderm, demonstrating a direct relationship with embryo size (Mann *et al.*, 1999; Robinson *et al.*, 2001) and hence potential to develop into a foetus.

Nutrition and reproductive function in the dairy heifer

Under good systems of husbandry, nutritional anoestrus in dairy heifers seldom occurs. Indeed, studies with post-pubertal crossbred beef heifers indicate that such animals would require to be chronically underfed for several months and to have lost more than 20% of their initial body weight in order to induce anovulation (Diskin *et al.*, 2003). In contrast, periods of acute food shortage, and/or deficiencies or excesses of certain dietary nutrients can have a profound effect on reproductive function, and these may be common in modern systems of dairy heifer management. Few studies, however, have assessed such effects and much of our understanding of the basic mechanisms underlying nutritional regulation of reproductive function (Figure 15.1) have been conducted either in beef heifers or in other ruminant species.

HYPOTHALAMIC-PITUITARY (HP) AXIS

Short-term periods of under nutrition which may occur, for example, when heifers are moved, housed, turned out or mixed with other heifers can dramatically influence central mechanisms controlling gonadotrophin release from the anterior pituitary gland. An acute (< 14 days) period of underfeeding (0.4 x maintenance) resulted in 60% of beef heifers becoming anovular and

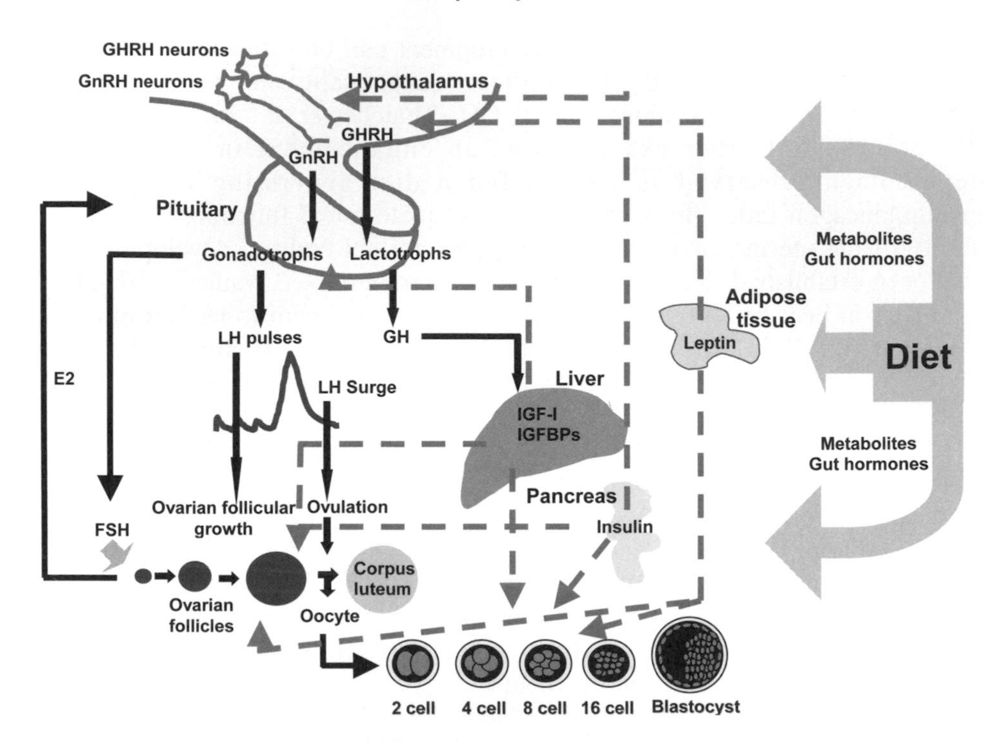

Figure 15.1 Sites of action within the hypothalamic-pituitary-ovarian axis for dietary derived nutrients and metabolic hormones. See text for details. GnRH, gonadotrophin releasing hormone; GHRH, growth hormone releasing hormone; LH, luteinizing hormone; FSH, follicle stimulating hormone; GH, growth hormone; IGF-1, insulin-like growth factor-1; IGFBPs, IGF binding proteins; E2, oestradiol. (Modified from Lucy *et al.*, 2003).

this was associated with an absence of a preovulatory surge in LH and FSH (Mackey *et al.*, 1999). Furthermore, this was often, but not always, associated with the absence of a pro-oestrus increase in plasma oestradiol concentrations, suggesting that the acute effects of undernutrition may be acting directly both within the hypothalamus and adenohypophysis to inhibit LH release. Fasting for 72 hours significantly reduced LH pulse frequency in heifers and this was associated with reduced plasma concentrations of both insulin and IGF-I (Maciel *et al.*, 2004). However, this effect on LH was prevented when leptin was injected subcutaneously at 12-hour intervals during a 3-day period. Under normal conditions of fasting in rodents and sheep, circulating concentrations of leptin decrease rapidly (within 24 hours in sheep), whilst expression of mRNA for the long form of the receptor increases rapidly within the arcuate nucleus and ventromedial hypothalamus (Adam *et al.*, 2003). This would have rendered the hypothalamus hypersensitive to injected leptin in the study of Maciel *et al.* (2004). Other metabolic cues that may be implicated in short-term regulation of gonadotrophin release during feed restriction include the gut hormones (e.g. cholecystokinin (CCK), glucagon-like peptides 1 and 2, and the recently discovered hormone ghrelin). Although receptors for these gut-derived peptides are situated at key locations within the hypothalamus (Holmberg and Malven, 1997; Horvath *et al.*, 2001; Burrin and Guan, 2003), their role in nutritional regulation of gonadotrophin release remains unexplored.

In contrast, there is a growing body of data that describe the effects of IGF-I and insulin on the HP axis. For example, IGF-I is known to interact with oestradiol to increase the GnRH stimulated release of LH in cultured bovine anterior pituitary cells (Hashizume *et al.*, 2002). Predominant IGFBPs detected in bovine anterior pituitary tissue include IGFBP-2, -3 and -5, and activities of these are known to fluctuate with the oestrous cycle, although their moderating effects on IGF-I mediated gonadotrophin release remain to be fully elucidated (Roberts *et al.*, 2001). Less is known about the direct effects of insulin within the HP axis in cattle although, in mice, insulin is known to stimulate production and secretion of GnRH from hypothalamic neurons (Burcelin *et al.*, 2003). Cerebroventricular infusion of physiological concentrations of insulin or insulin plus glucose also led to a significant increase in serum LH concentrations in underfed ewes within two days (Daniel *et al.*, 2000). The action of many of these systemic metabolic hormones is likely to be mediated by a number of locally produced neuropeptides within the hypothalamus, including γ-aminobutyric acid (which in mice has been shown to mediate the effects of leptin on GnRH release; Sullivan *et al.*, 2003), neuropeptide Y (which in cattle can suppress LH release either in the presence or absence of oestradiol; Thomas *et al.*, 1999) and opioid peptides (which in beef cows and heifers are known to mediate the effects of negative energy balance and excess dietary protein on the inhibition of LH release (Sinclair *et al.*, 1995; Dawuda *et al.*, 2002).

Effects of excess dietary nitrogen on central mechanisms controlling anterior pituitary function are of some interest, but remain relatively unexplored. In a series of experiments to assess effects of excess rumen degradable nitrogen on oocyte quality, Sinclair *et al.* (2000a) monitored the diurnal pattern of voluntary food intake and observed that it was inversely correlated with plasma ammonia concentrations, so that animals in these studies altered their pattern of intake to minimise the post-prandial increase in plasma ammonia concentrations. Although the mechanisms underlying this modified pattern of intake are not known, they are likely to include those operating within the central nervous system. Consequently, one possible explanation for this effect on intake and the depression in LH release observed following the feeding of high nitrogen diets in the study of Sinclair *et al.* (1995) might lie with the excitatory amino acids L-glutamate and L-aspartate, which are recognised key intermediates of ammonia metabolism within the brain but, in their capacity as neurotransmitters, also act as powerful appetite stimulants within the lateral hypothalamus and nucleus accumbens (discussed by Sinclair *et al.*, 2000a); they are also known to be powerful stimulators of GnRH and LH release (Dhandapani and Brann, 2000). It might be that excess dietary nitrogen, leading to elevated plasma concentrations of ammonia crossing the blood-brain barrier, simultaneously depresses appetite and gonadotrophin release by depleting the central nervous system of these excitatory amino acids.

Other products of metabolism that have a direct effect on pituitary function and gonadotrophin release include glucose. A single subcutaneous injection of 2-deoxy-D-glucose (2DG; an inhibitor of glucose oxidation) significantly reduced LH release, but had no effect on GH or prolactin release in mature, ovariectomised crossbred ewes (Funston *et al.*, 1995). The effects of 2DG in this study appeared to be mediated entirely within the hypothalamus or higher centres within the brain because the responsiveness of the pituitary gland to GnRH was unaltered. In a recent study with goats, effects of both glucose and free fatty acids on the regulation of GnRH were assessed (Ohkura *et al.*, 2004). Glucoprivation was induced either by intravenous administration of 2DG or insulin at various doses, and lipoprivation induced by peripheral infusion of sodium mercaptoacetate (MA; an inhibitor of fatty acid oxidation). GnRH release was measured electrophysiologically within the arcuate nucleus and median eminence. Infusion of 2DG or insulin reduced GnRH pulse frequency, and the co-infusion of glucose with insulin returned this to normal. In contrast, infusion of MA failed to influence GnRH pulse frequency. The results from this study confirmed the importance of glucose as a metabolic regulator of pulsatile GnRH release in ruminants. Free fatty acids, however, do not seem to serve as a metabolic signal, but the authors of this report went on to comment that other metabolites of fatty acid catabolism (e.g. ketone bodies) and short-chain volatile fatty acids (the major energy-yielding substrates in ruminants) may have some role as metabolic signals. In support of this latter claim, Boukhliq and Martin (1997) demonstrated that LH secretion could be increased when sheep were offered a maintenance diet supplemented with volatile fatty acids.

OVARIAN FOLLICULOGENESIS AND OOCYTE DEVELOPMENT

Environmental influences such as chronic and acute changes in dietary intake have an impact on ovarian function in heifers. These changes can occur without significant variation in circulating gonadotrophin concentrations and are correlated with changes in circulating concentrations of various metabolites and metabolic hormones including insulin, IGF-I, GH and leptin (Webb *et al.*, 2004). Growth rate and eventual size of the dominant follicle (DF) of chronically underfed (0.7 x maintenance) beef heifers decrease, but these animals continue to ovulate until they have lost around 20 to 25% of their initial live weight (Diskin *et al.*, 2003). Furthermore, during the period of feed restriction, up to but preceding the point of anovulation, neither oestrous cycle length nor the proportion of animals with two or three follicular waves per oestrous cycle change. In contrast, the acute and severe level of undernutrition (0.4 x maintenance) in the study of Mackey *et al.* (1999) described earlier, caused a significant and immediate decrease in both growth rate and maximum diameter of the DF of the first wave of follicles to emerge following dietary restriction. Using data from this study, Diskin *et al.* (2003) predicted that a DF of < 9 mm had a probability of ovulating of < 20%. Failure of ovulation in such circumstances arises as a consequence of reduced pulsatile LH release from the pituitary gland in conjunction with a follicle that is probably less responsive to LH; for very soon after DF selection, at around 8 - 9 mm diameter, receptors for LH begin to be expressed in granulosa cells (Webb *et al.*, 2003).

Nutritional status is known to affect circulating concentrations of a number of growth factors that promote follicular development (Armstrong *et al.*, 2003). Results from a series of studies highlight the importance of insulin which, when infused into underfed beef heifers, can increase both the diameter of the DF and ovulation rate (Harrison and Randel, 1986; Simpson *et al.*, 1994). Indeed, granulosa cell studies have demonstrated that physiological concentrations of insulin are necessary for normal cell growth (Gutierrez *et al.*, 1997) and steriodogenesis (Armstrong *et al.*, 2002). Circulating concentrations of both insulin and IGF-I fluctuate during the oestrous cycle, with peak concentrations coinciding with oestrus and ovulation. Circulating concentrations of insulin in particular are known to fluctuate with each follicular wave (Armstrong *et al.*, 2001). Differences between well fed and undernourished heifers in mean plasma concentrations of these two metabolic hormones peak during oestrus but, although oestradiol is known to stimulate the expression of mRNAs encoding both insulin and IGF-I in pancreatic and liver cells respectively (Johnson *et al.*, 1998; Morimoto *et al.*, 2001), there is no evidence to suggest that oestradiol might be responsible for the increased differences in circulating concentrations of these two growth factors during oestrus.

Like insulin, IGF-I is known to interact synergistically with gonadotrophins to promote follicular growth and steriodogenesis within the follicular compartment, and dietary induced increases in circulating concentrations of IGF-I promote oestradiol production in cultured granulosa cells (Armstrong *et al.*, 2002). IGF-I is known to increase LH binding sites and LH-stimulated androstenedione and progesterone production by thecal cells. Collectively, these results indicate that this growth factor has a critical role in development of the pre-ovulatory follicle. The bioavailability of this growth factor, however, is controlled by IGFBPs and circulating concentrations of these binding proteins are also regulated by nutritional status. For example, IGFBP-3 is positively correlated with dietary intake (Rausch *et al.*, 2002), which is thought to increase the half-life of IGF-1 in circulation. In contrast, high energy diets in the study of Armstrong *et al.* (2001) significantly reduced steady-state concentrations of mRNA encoding IGFBP-2 and -4 locally within small (< 4mm) bovine antral follicles. This is thought to increase the bioavailability of intra-follicular IGFs, both locally produced IGF-II and systemically derived IGF-I.

The most significant observation to emerge from this latter study, however, was that, in contrast to the beneficial effects on ovarian follicular development, high intakes of dietary energy and protein had a detrimental effect on oocyte development, reflected by the significant decrease in blastocyst yields in vitro following harvest of oocytes from heifers offered these diets. The ovarian IGF system has the potential to interact directly with the oocyte through the type I IGF receptor and the reduced levels of IGFBP-2 and -4, together with increased concentrations of IGF-I, might have led to the oocyte being 'over-stimulated'. More recent data from our laboratory suggest that such effects might be cumulative and dependent on body-condition score of the heifer. For example, Adamiak *et al.* (2003) reported that post-fertilisation developmental potential of oocytes from thin heifers offered a high plane of nutrition (2 x maintenance) improved over a 9-week period (coinciding with 3 oestrous cycles) whereas post-fertilisation development of oocytes from fat heifers offered the same level of feeding deteriorated over the same time period. Collectively, these observations have important implications for heifer rearing during the period leading up to first insemination. Level of feeding during this period needs to be adjusted for heifer body condition, and formulation of diets designed to optimise heifer fertility needs to consider the possibility of divergent actions of nutrient supply on follicular growth and oocyte quality.

The detrimental effects of high levels of dietary nitrogen on post-fertilisation developmental potential of oocytes discussed earlier might operate by mechanisms other than those that act through the insulin and/or IGF system. For example, Sinclair *et al.*, (2000b) reported that high nitrogen diets, leading to elevated concentrations of ammonia and urea in peripheral circulation, significantly reduced blastocyst yields from oocytes recovered from crossbred beef heifers and matured, fertilised and cultured in vitro. Although circulating insulin concentrations were reduced when high ammonia-generating diets were offered in that study, the mechanisms of action of ammonia were thought to be direct and non-specific on the growing follicle and/or follicle-enclosed oocyte. A subsequent study by the same group (Rooke *et al.*, 2004) confirmed that elevated concentrations of ammonium chloride in vitro reduces granulosa cell growth, alters steroidogenesis and reduces the ability of granulosa cells to support the maturation of bovine oocytes in vitro. However, an earlier study by Sinclair *et al.* (2000c) presented evidence to suggest that these detrimental effects on oocyte development could be 'programmed' during the earlier, pre-antral stages of follicular development, two to three months prior to the expected date of ovulation. These observations have important implications for nutritional management of heifers during the rearing period.

POST-FERTILISATION DEVELOPMENT

Although the nutrient demands of the pre-attachment bovine embryo are quantitatively small they are, nevertheless, qualitatively specific, and reflect the changing needs of the developing embryo during its migration from the oviduct to the uterine lumen. These needs are satisfied almost entirely from exogenous nutrients available from the female reproductive tract. One might expect that the precise nature and quantity of these nutrients within the uterine environment, and how they might be affected by maternal nutrition, is well characterised in the ruminant, but this is certainly not the case. What is known was comprehensively reviewed by Sinclair *et al.* (2003) and is briefly summarised below and in Table 15.1.

Table 15.1 CHANGING ENVIRONMENTAL CONDITIONS AND NUTRIENT SUPPLY IN THE BOVINE OVIDUCT AND UTERUS (DERIVED FROM SINCLAIR *et al.*, 2003)

Day of oestrous cycle *Location*	*1-3* *Isthmus*	*4-7* *Uterus*
Oxygen tension (mm Hg)	55	20
Osmolarity	290	310
pH	~7.1	
Protein content	10% that of serum	
Albumin	60 to 80% of total protein	
Principal essential amino acids (mM)		
Threonine	0.8	1.7
Leucine	0.5	1.8
Arginine	0.3	1.4
Principal nonessential amino acids (mM)		
Glycine	14.0	12.0
Glutamate	5.5	4.2
Alanine	3.7	3.1
Glucose and carboxylic acids (mM)		
Glucose	1.6	2.2
Pyruvate	1.1	1.7
Lactate	8.3	11.7
Lipids		
Cholesterol (mg)	143	175
Phospholipid (nM)	324	90
Phospholipid:cholesterol ratio	2.3	0.5

Oviductal secretions are hormonally regulated, increasing during pro-oestrus and decreasing during the luteal phase of the cycle. There comprises a subtle mixture of compounds found in the serum and manufactured by the oviductal epithelium. The relative contribution of each is not well understood. However, the protein content of oviductal fluid is low, around 10% of that in serum, and albumin is the main protein present. The most comprehensive analysis of the amino acid content of bovine oviductal fluid published to date was by Elhassan *et al.* (2001). So called non-essential amino acids, including glycine, glutamate and alanine, account for around 80% of total amino acids in the oviduct, a value that drops to around 60% in uterine fluids. Glycine concentrations are particularly high where, in addition to protein synthesis, glycine is thought to play an important role in osmoregulation. However, amino acids also function as important energy sources for the early embryo and threonine and glutamine are readily metabolised in this way. One of the products of such metabolism is alanine, significant amounts of which are secreted from the pre-elongation embryo. The bovine oviduct can also synthesise and release a variety of lipids (particularly cholesterol and phospholipids). Again concentrations are generally lower than that of blood serum and vary with the stage of cycle. The ratio of phospholipids to cholesterol is particularly high in the ampulla and isthmus in the first few days following ovulation; conditions thought favourable for the acrosome reaction in spermatozoa. Finally, glucose concentrations in oviductal fluid are also lower than in blood serum, whereas lactate concentrations are much higher. This is due principally to the metabolism of glucose to lactate by the oviductal epithelium. Again, reliable estimates of concentrations of these carbohydrates in ruminants are limited, and effects of specific dietary treatments are not well understood.

Feeding of high nitrogen diets, particularly those high in ruminally degradable nitrogen, has been shown to increase urea and ammonia concentrations in oviductal fluid in beef heifers (Kenny *et al.*, 2002) and to reduce uterine pH around Day 7 after oestrus in dairy heifers (Elrod and Butler, 1993). First-service conception rates in the latter study were also significantly reduced (61 vs 82 %) for heifers on the high compared with the low nitrogen diet. However, given the technical difficulties of studying in vivo embryo development in cattle, there have been few studies in this area, and so our understanding of how these nutrients, together with the various growth factors and metabolic hormones that are produced and/or secreted into the reproductive tract, affect bovine pre-attachment embryo development is limited mainly to *in vitro* studies. In one of the few studies that were able to assess the effects of nutrition on *in vivo* embryo development in cattle, beef heifers were superovulated and, following artificial insemination, Day 7 embryos flushed from the reproductive tracts and cultured for 24 hours in vitro (Nolan *et al.*, 1998). Blastocyst yields and blastocyst cell numbers were significantly greater (73 vs 42%) for heifers on the Low (40 MJ metabolisable energy (ME) per day) diet compared with heifers on the High (120 MJ ME per day) diet. A subsequent study from the same group (Yaakub *et al.*, 1999), in collaboration with researchers from Germany and North America (Wrenzycki *et al.*, 2000), showed that ad libitum feeding of high starch diets significantly reduced blastocyst yields following superovulation in beef heifers. This occurred as a consequence of deranged embryo metabolism, with increased transcripts for the antioxidant enzyme Cu/Zn-superoxide dismutase detected in blastocysts.

Clearly, therefore, high planes of nutrition are detrimental for early embryo development, just as they are for oocyte development. Indeed, feeding of high starch diets is likely to result in sub-clinical acidosis, a consequence of excessive production of lactate from rapidly fermented carbohydrates within the rumen. The resulting decrease in blood pH might correspond to reduced uterine pH and result in impaired embryo development and the deranged patterns of gene expression described by Wrenzycki *et al.* (2000). In a separate study with beef heifers, however, Mann *et al.* (2003) reported that feeding a high starch diet to heifers of moderate body condition led to a modest increase in plasma insulin:glucagon ratio following feeding, an increase in mean interferon tau (IFN-τ) concentration in uterine fluid, and a significant increase in the proportion of well-elongated embryos on Day 16 following mating (85 vs 39% respectively). Interferon tau secreted from the embryo around this time is known to suppress oxytocin-induced release of prostaglandin $F_{2\alpha}$ ($PGF_{2\alpha}$) and in so doing helps to maintain the corpus luteum of pregnancy (Thatcher *et al.*, 2001). However, although the glucogenic diet of Mann *et al.* (2003) enhanced embryo development, it remains to be determined if it would subsequently have resulted in enhanced embryo survival rate beyond Day 16 of gestation.

Dietary components that have been shown to enhance embryo survival in lactating dairy cows by attenuating the release of $PGF_{2\alpha}$ include fat sources rich in linoleic acid and/or eicosapentaenoic (EPA; C20:5) and docosahexanaenoic (DHA; C22:6) acids (Binelli *et al.*, 2001). The physiological basis for these effects lies in their ability to reduce conversion of arachidonic acid (C20:4) to $PGF_{2\alpha}$, by competing with arachidonic acid for binding on the key enzyme prostaglandin endoperoxide synthase (Thatcher *et al.*, 2001). The benefits of such dietary ingredients in heifer diets, however, remain to be established. Furthermore, from the foregoing discussion it could be assumed that feeding of high starch diets supplemented with fish oils (rich in EPA and DHA) might be beneficial for fertility as in combination they would respectively promote embryo development (and hence IFN-τ production) whilst independently inhibiting $PGF_{2\alpha}$ synthesis. However, as eluded to earlier, high starch diets can lead to suboptimal uterine conditions and inclusion of long chain polyunsaturated fatty acids in the diet might, if incorporated within embryonic membranes, render ova vulnerable to free radical damage, although this remains to be established from dietary studies with heifers.

Conclusions

Sub-fertility in dairy cows is one of the most important problems currently facing the UK dairy industry, accounting for just over two-thirds of the estimated annual cost attributed to poor health and fertility; it also has important implications for environmental pollution and 'greenhouse' gas emissions. There is evidence that selection for milk yield has reduced fertility in the maiden heifer. Future improvements in this trait will therefore necessitate the inclusion of selection for fertility in broader breeding goals (selection based on physiological parameters and the use of modern molecular-genetic techniques may assist), but consideration must also be given to adjustments in current heifer rearing practices. Attention should be directed towards developing a better understanding of the underlying physiology controlling pre- and peri-pubertal growth and, in particular, the process of mammogenesis and the development of the hypothalamic-pituitary-ovarian axis (the ontogeny of the latter may also be influenced by in utero events). In order to optimise fertility during post-pubertal development, consideration should be given to the effects of body composition in concert with dietary formulations that promote ovarian follicular development whilst safeguarding oocyte quality and maintaining a receptive uterine environment.

Acknowledgements

Much of the work of the authors was supported by Defra and SEERAD.

References

Adam, C. L., Archer, Z. A. & Miller, D. W. (2003) Leptin actions on the reproductive neuroendocrine axis in sheep. *Reprod. Suppl* , **61,** 283-297.

Adam, C. L. & Robinson, J. J. (1994) The role of nutrition and photoperiod in the timing of puberty. *Proc.Nutr. Soc.*, **53,** 89-102.

Adamiak, S. J., Mackie, K., Powell, K. A., Watt, R. G., Dolman, D. F., Webb, R. & Sinclair , K. D. (2003) Body composition and dietary energy intake affect folliculogenesis, oocyte quality and early embryo development. *Reproduction, Abstract Series No.* ***30,*** 62-63.

Adams, G. P. (1999) Comparative patterns of follicle development and selection in ruminants *J. Reprod. Fert Supplement,* **54**, 17-32.

Adams, G. P., Evans, A. C. O. & Rawlings N. C. (1994). Follicular waves and circulating gonadotrophins in 8 month-old prepubertal heifers. *J. Reprod. Fert.*, **100**, 27-33.

Armstrong, D. G., Baxter, G., Gutierrez, C. G., Hogg, C. O., Glazyrin, A. L., Campbell, B. K., Bramley, T. A. & Webb, R. (1998) Insulin-like growth factor binding protein -2 and -4 mRNA expression in bovine ovarian follicles: effect of gonadotropins and developmental status *Endocrinol.,* **139**, 2146-2154.

Armstrong, D. G., Baxter, G., Hogg, C. O. & Woad, K. J. (2002) Insulin-like growth factor (IGF) system in the oocyte and somatic cells of bovine preantral follicles *Reprod.*,**12**, 789-797.

Armstrong, D. G., Gong, J. G., Gardner, J. O., Baxter, G., Hogg, C. O. & Webb, R. (2002) Steroidogenesis in bovine granulosa cells: the effect of short-term changes in dietary intake. *Reprod.*, **123**, 371-378.

Armstrong, D. G., Gong, J. G. & Webb, R. (2003) Interactions between nutrition and ovarian activity in cattle: physiological, cellular and molecular mechanisms. *Reprod. Suppl.*, **61**, 403-414.

Armstrong, D. G., Gutierrez, C. G., Baxter, G., Glazyrin, A. L., Mann, G.E., Woad, K. J., Hogg, C. O., & Webb, R. (2000) Expression of mRNA encoding IGF-I, IGF-II and type 1 IGF receptor in bovine ovarian follicles *J. Endocrinol.,* **165**, 101-113.

Armstrong, D. G., McEvoy, T. G., Baxter, G., Robinson, J. J., Hogg, C. O., Woad, K. J., Webb, R. & Sinclair, K. D. (2001) Effect of dietary energy and protein on bovine follicular dynamics and embryo production in vitro: associations with the ovarian insulin-like growth factor system. *Biol.Reprod.*, **64**, 1624-1632.

Ashwell, M. S., Heyen, D. W., Sonstegard, T. S., Van Tassell, C. P., Da, Y., VanRaden, P. M., Ron, M., Weller, J. I. & Lewin, H. A. (2004) Detection of quantitative trait loci affecting milk production, health, and reproductive traits in Holstein cattle. *J. Dairy Sci.*, **87**, 468-475.

Bao B & Garverick HA (1998) Expression of steroidogenic enzyme and gonadotropin receptor genes in bovine follicles during ovarian follicular waves: a review *J. Anim. Sci.,* **76**, 1903-1921.

Bergfeld, E. G. M., Kojima, F. N., Cupp, A. S., Wehrman, M. E., Peters, K. E., Garcia-Winder, M. & Kinder, J. E. (1994). Ovarian follicular development in prepubertal heifers is influenced by level of dietary intake. *Biol. Reprod.*, **51**, 1051-1057.

Borwick, S. C., Rae, M. T., Brooks, J., McNeilly, A. S., Racey, P. A. & Rhind, S. M. (2003) Undernutrition of ewe lambs in utero and in early post-natal life does not affect hypothalamic-pituitary function in adulthood. *Anim Reprod. Sci.*, **77**, 61-70.

Borwick, S. C., Rhind. S. M., McMillen, S. R. & Racey, P. A. (1997) Effect of undernutrition of ewes from the time of mating on fetal ovarian development in mid gestation. *Reprod. Fert. Dev.,* **9**, 711-715.

Boukhliq, R. & Martin, G. B. (1997) Administration of fatty acids and gonadotrophin secretion in the mature ram. *Anim Reprod. Sci.*, **49**, 143-159.

Braw-Tal, R. (2002) The initiation of follicle growth: the oocyte or the somatic cells? *Domestic Anim. Endo.*, **187**, 11-18.

Burcelin, R., Thorens, B., Glauser, M., Gaillard, R. C. & Pralong, F. P. (2003) Gonadotropin-releasing hormone secretion from hypothalamic neurons: stimulation by insulin and potentiation by leptin. *Endocrinol.*, **144**, 4484-4491.

Burrin, D. G., Stoll, B. & Guan, X. (2003) Glucagon-like peptide 2 function in domestic animals. *Domest. Anim Endocrinol.*, **24**, 103-122.

Butler, W. R. (1998) Review: effect of protein nutrition on ovarian and uterine physiology in dairy cattle. *J. Dairy Sci.*, **81**, 2533-2539.

Campbell, B. K., Telfer, E. E., Webb, R. & Baird, D. T. (2000) Ovarian autografts in sheep as a model for studying folliculogenesis *Mol. Cell Endocrinol.*, **163**, 137-139.

Chelikani, P. K., Ambrose, J. D. & Kennelly, J. J. (2003) Effect of dietary energy and protein density on body composition, attainment of puberty, and ovarian follicular dynamics in dairy heifers. *Theriogenology*, **60**, 707-725.

Choi, Y. J., Han, I. K., Woo, J. H., Lee, H. J., Jang, K., Myung, K. H. & Kim, Y. S. (1997) Compensatory growth in dairy heifers: the effect of a compensatory growth pattern on growth rate and lactation performance. *J. Dairy Sci.*, **80**, 519-524.

Clarke, I. J., (2002). Two decades of measuring GnRH secretion. *Reprod. Supplement*, **59**, 1-13.

Crowe, M. A., Kelly, P., Driancourt, M. A., Boland, M. P. & Roche, J. F. (2001) Effects of follicle-stimulating hormone with and without luteinizing hormone on serum hormone concentrations, follicle growth, and intrafollicular estradiol and aromatase activity in gonadotropin-releasing hormone-immunised heifers. *Biol. Reprod.*, **64**, 368-374.

Daniel, J. A., Thomas, M. G., Hale, C. S., Simmons, J. M. & Keisler, D. H. (2000) Effect of cerebroventricular infusion of insulin and (or) glucose on hypothalamic expression of leptin receptor and pituitary secretion of LH in diet-restricted ewes. *Domest. Anim Endocrinol.*, **18**, 177-185.

Dawuda, P. M., Scaife, J. R., Hutchinson, J. S. & Sinclair, K. D. (2002) Mechanisms linking under-nutrition and ovarian function in beef heifers. *Anim Reprod. Sci.*, **74**, 11-26.

Day, M. L., Imakawa, K., Garcia-Winder, M., Zalesky, D. D., Schanbacher, B. D., Kittok, R. J. & Kinder, J. E. (1984). Endocrine mechanisms of puberty in heifers: estradiol negative feedback regulation of luteinizing hormone secretion. *Biol. Reprod.*, **31**, 332-341.

Day, M. L., Imakawa, K., Wolf, P. L., Kittok, R. J. & Kinder, J. E. (1987). Endocrine mechanisms of puberty in heifers: role of hypothalamo-pituitary estradiol receptors in the negative feedback of estradiol on luteinising hormone secretions. *Biol. Reprod.*, **37**, 1054-1065.

Defra (2003) http://www.defra.gov.uk

Desjardins, C. & Hafs, H. D. (1969). Maturation of bovine female genetalia from birth through puberty. *J. Anim. Sci.*, **28**, 502-507.

Dhandapani, K. M. & Brann, D. W. (2000) The role of glutamate and nitric oxide in the reproductive neuroendocrine system. *Biochem. Cell Biol.*, **78**, 165-179.

Diskin, M. G., Mackey, D. R., Roche, J. F. & Sreenan, J. M. (2003) Effects of nutrition and metabolic status on circulating hormones and ovarian follicle development in cattle. *Anim Reprod. Sci.*, **78**, 345-370.

Dodson, S. E., McLeod, B. J., Haresign, W., Peters, A. R. & Lamming, G. E. (1988). Endocrine changes from birth to puberty in the heifer. *J. Reprod. Fert.*, **82**, 527-538.

Driancourt, M. A. (2001). Regulation of ovarian follicular dynamics in farm animals. Implications for manipulation of reproduction. *Theriogenology*, **55**, 1211-1239.

Dugan, K. J., Campbell, B. K., Skinner, A., Armstrong, D. G. and Webb, R. (2004) Expression of bone morphogenetic protein-6 (BMP-6) in the bovine foetal ovary at different stages of gestation. *Reprod. Abstract Series*, **31**, O46.

Elhassan, Y. M., Wu, G., Leanez, A. C., Tasca, R. J., Watson, A. J. & Westhusin, M. E. (2001) Amino acid concentrations in fluids from the bovine oviduct and uterus and in KSOM-based culture media. *Theriogenology*, **55**, 1907-1918.

Elrod, C. C. & Butler, W. R. (1993) Reduction of fertility and alteration of uterine pH in heifers fed excess ruminally degradable protein. *J. Anim. Sci.*, **71**, 694-701.

Esslemont, R. J. & Kossaibati, M. A. (2002) *Daisy Research Report No. 5*, Intervet UK. Ltd.

Evans, A. C. O., Adams, G. P. & Rawlings, N. C. (1994). Endocrine and ovarian follicular changes leading up to the first ovulation in prepubertal heifers. *J. Reprod. Fertil.*, **100**, 187-194.

Fortune, J. E., Cushman, R. A., Wahl, C. M., Kito, W. S. (2000) The primordial to primary follicle transition *Mol. Cell. Endocrinol.*, **163**, 53-60.

Fouladi-Nashta A. A., Alberio R., Kafi M., Nicholas B., Campbell K. H. S. & Webb, R. (2005b) Differential staining combined with TUNEL labelling (DST) to detect apoptosis in preimplantation embryos. *Reprod. BioMed. Online* (In press).

Fouladi-Nashta A. A., Dugan, K. J. & Webb R. (2005a) Immunohistochemical localisation of BMP receptors in bovine foetal ovaries. *Reprod. Abstract Series*, **32**, (in press).

Funston, R. N., Roberts, A. J., Hixon, D. L., Hallford, D. M., Sanson, D. W. & Moss, G. E. (1995) Effect of acute glucose antagonism on hypophyseal hormones and concentrations of insulin-like growth factor (IGF)-I and IGF-binding proteins in serum, anterior pituitary, and hypothalamus of ewes. *Biol. Reprod.*, **52**, 1179-1186.

Galloway, S. M., McNatty, K. P., Cambridge, L. M., Laitinen, M. P. E., Juengel, J. L., Jokiranta, T. S., McLaren, R. S., Luiro, K., Dodds, K. G., Montgomery, G. W., Beattie, A.E., Davis, G. H. & Rivito, O. (2000) Mutations in an oocyte–derived growth factor gene (BMP 15) cause increased ovulation rate and infertility in a dosage sensitive manner. *Nat. Genet.*, **25**, 279-283.

Garcia, M. R., Amstalden, M., Williams, S. W., Stanko, R. L., Morrison, C. D., Keisler, D. H., Nizielski, S. E. & Williams, G. L. (2002) Serum leptin and its adipose gene expression during pubertal development, the estrous cycle, and different seasons in cattle. *J. Anim Sci.*, **80**, 2158-2167.

Garnsworthy, P. C. (2004) The environmental impact of fertility in dairy cows: a modelling approach to predict methane and ammonia emissions. *Anim. Feed Sci.Tech.*,**112**, 211-223.

Garnsworthy, P. C. and Webb, R. (1999) The influence of nutrition on fertility in dairy cows. In *Recent Advances in Animal Nutrition - 1999* (Eds P.C. Garnsworthy and J. Wiseman), 39-57, Nottingham University Press, Nottingham.

Garverick, H. A., Baxter, G., Gong, J., Armstrong, D. G., Campbell, B. K., Gutierrez, C. G. & Webb, R. (2002) Regulation of expression of ovarian mRNA encoding steroidogenic enzymes and gonadotrophin receptors by FSH and GH in hypogonadotrophic cattle *Reprod.,* **123**, 651-661.

Glister, C., Kemp, C.F. & Knight, G. (2004) Bone morphogenetic protein (BMP) ligands and receptors in bovine ovarian follicle cells: actions of BMP-4, -6 and –7 on granulosa cells and differential modulation of Smad-1 phosphorylation by follistatin. *Reprod.*, **127**, 239-254.

Gunn, R. G. (1983). The influence of nutrition on the reproductive performance of ewes. In: *Sheep Production* (Ed W. Haresign), 99-212, Butterworths, London.

Gunn, R. G., Sim, D. A. & Hunter, A. (1995) Effects of nutrition *in utero* and in early life on the subsequent lifetime of reproductive performance of Scottish Blackface ewes in two management systems. *Anim. Sci.*, **60**, 223-230.

Gutierrez, C. G., Aguilera, I., Leon, H., Rodriguez, A., & Hernandez-Ceron, J. (2005) The metabolic challenge of milk production and the toll it takes on fertility. *Cattle Practice* (in press).

Gutierrez, C. G., Campbell, B. K. & Webb, R. (1997) Development of a long-term bovine granulosa cell culture system: induction and maintenance of estradiol production, response to follicle-stimulating hormone, and morphological characteristics. *Biol.Reprod.*, **56**, 608-616.

Gutierrez, C. G., Ralph, J. H., Telfer, E. E., Wilmut, I. & Webb, R. (2000) Growth and antrum formation of bovine antral follicles in long-term culture *in vitro. Biol. Reprod.*, **62**, 1322-1328.

Harrison, L. M. & Randel, R. D. (1986) Influence of insulin and energy intake on ovulation rate, luteinizing hormone and progesterone in beef heifers. *J. Anim Sci.*, **63**, 1228-1235.

Hashizume, T., Kumahara, A., Fujino, M. & Okada, K. (2002) Insulin-like growth factor I enhances gonadotropin-releasing hormone-stimulated luteinizing hormone release from bovine anterior pituitary cells. *Anim Reprod. Sci.*, **70**, 13-21.

Helliwell, R. J. A., Wallace, J. M., Aitken, R. P., Racey, P. A. & Robinson, J. J.

(1997) The effect of prenatal photoperiodic history on the postnatal endocrine status of female lambs. *Anim Reprod. Sci.,* **47**, 303-314.

Helliwell, R. J. & Williams, L. M. (1994) The development of melatonin-binding sites in the ovine fetus. *J. Endocrinol.*, **142**, 475-484.

Holmberg, B. J. & Malven, P. V. (1997) Neural levels of cholecystokinin peptide and mRNA during dietary stimulation of growth and LH secretion in growth-retarded lambs. *Domest. Anim. Endocrinol.*, **14**, 304-315.

Hopper, H. W., Silcox, R. W., Byerley, D. J. & Kiser, T. E. (1993). Follicular development in prepubertal heifers. *Anim. Reprod. Sci.*, **31**, 7-12.

Hornick, J. L., Van Eenaeme, C., Gerard, O., Dufrasne, I. & Istasse, L. (2000) Mechanisms of reduced and compensatory growth. *Domest. Anim Endocrinol.*, **19**, 121-132.

Horvath, T. L., Diano, S., Sotonyi, P., Heiman, M. & Tschop, M. (2001) Minireview: ghrelin and the regulation of energy balance—a hypothalamic perspective. *Endocrinol.*, **142**, 4163-4169.

Howe, C. C. & Solter, D. (1979) Cytoplasmic and nuclear protein synthesis in preimplantation mouse embryos. *J. Embryol. Exp. Morph.*, **52**, 209-225.

Hulshof, S. C. J., Figueiredo, J. R., Becker, J. F., Severs, M. M., van der Donk, J. A. & van den Hurk, R. (1995) Effects of fetal bovine serum, FSH and 17ß-estradiol on the culture of bovine preantral follicles. *Theriogenology*, **44**, 217-226.

Ireland J. J., Mihm, M., Austin, E., Diskin, M.G. & Roche J. F. (2000) Historical perspective of turnover of dominant follicles during the bovine estrous cycle: key concepts, studies, advancements, and terms. *J. Dairy Sci.*, **83**,1648-1658.

Izumi, T., Sakakida, S., Nagai, T. & Miyamoto, H. (2003) Allometric study on the relation between the growth of preantral and antral follicles and that of oocytes in bovine ovaries. *J. Reprod. Dev.*, **49**, 361-368.

Johnson, B. J., White, M. E., Hathaway, M. R., Christians, C. J. & Dayton, W. R. (1998) Effect of a combined trenbolone acetate and estradiol implant on steady-state IGF-I mRNA concentrations in the liver of wethers and the longissimus muscle of steers. *J. Anim. Sci.*, **76**, 491-497.

Juengel, J. L., Hudson, N. L., Heath, D. A. , Smith, P. , Reader, K. L., Lawrence, S. B., O'Connell, A. R. , Laitinen, M. P. E. Cranfield, M., Groome, N. P., Ritvos, O. & McNatty, K. P. (2002) Growth differentiation factor 9 and bone morphogenetic protein 15 are essential for ovarian follicular development in sheep. *Biol. Reprod.*, **67**,1777-1789.

Kaneko, H., Kishi, H., Watanabe, G. Taya, K., Sasamoto, S. & Hasegawa, Y. (1995). Changes in plasma concentrations in immunoreactive inhibin, oestradiol and FSH associated with follicular waves during the estrous cycle of the cow. *J. Reprod. Dev.*, **41**, 311-320.

Kaneko, H., Nakanishi, Y., Taya, K., Kishi, H., Watanabe, G., Sasamoto, S. & Hasegawa, Y. (1993). Evidence that inhibin is an important factor in the regulation of FSH secretion during the mid-luteal phase in cows. *J.*

Endocrinol., **136**, 35-41.

Kenny, D. A., Humpherson, P. G., Leese, H. J., Morris, D. G., Tomos, A. D., Diskin, M. G. & Sreenan, J. M. (2002) Effect of elevated systemic concentrations of ammonia and urea on the metabolite and ionic composition of oviductal fluid in cattle. *Biol. Reprod.*, **66**, 1797-1804.

Kinder, J. E., Bergfeld, E. G. M., Wehrman, M. E., Peters, K. E. & Kojima, F. N. (1995). Endocrine basis for puberty in heifers and ewes. *J. Reprod. Fert., Supplement*, **49**, 393-407.

Kinder, J. E., Day, M. L. & Kittok, R. J. (1987). Endocrine regulation of puberty in cows and ewes. *J. Reprod. Fert., Supplement*, **34**, 167-189.

Landau, S., Bor, A., Leibovich, H., Zoref, Z., Nitsan, Z. & Madar, Z. (1995). The effect of ruminal starch degradability in the diet of Booroola crossbred ewes on induced ovulation rate and prolificacy. *Anim. Reprod. Sci.,* **38**, 97-108.

Lucy, M. C. Mechanisms linking nutrition and reproduction in postpartum cows. *Reprod., Suppl.* **61**, 415-427.

Maciel, M. N., Zieba, D. A., Amstalden, M., Keisler, D. H., Neves, J. P. & Williams, G. L. (2004) Chronic administration of recombinant ovine leptin in growing beef heifers: effects on secretion of LH, metabolic hormones, and timing of puberty. *J. Anim. Sci.*, **82**, 2930-2936.

Mackey, D. R., Sreenan, J. M., Roche, J. F. & Diskin, M. G. (1999) Effect of acute nutritional restriction on incidence of anovulation and periovulatory estradiol and gonadotropin concentrations in beef heifers. *Biol. Reprod.*, **61**, 1601-1607.

Mann, G. E., Green, M. P., Sinclair, K. D., Demmers, K. J., Fray, M. D., Gutierrez, C. G., Garnsworthy, P. C. & Webb, R. (2003) Effects of circulating progesterone and insulin on early embryo development in beef heifers. *Anim. Reprod. Sci.*, **79**, 71-79.

Mann, G. E. & Lamming, G. E. (1999) The influence of progesterone during early pregnancy in cattle. *Reprod. Dom. Anim.*, **34**, 269-274.

Mann, G. E., Lamming, G. E., Robinson, R. S. & Wathes D. C. (1999) The regulation of interferon-ô production and uterine hormone receptors during early pregnancy. *J. Reprod. Fert. Supplement* **54**, 317-328.

McNatty, K. P., Heath, D. A., Lindy, F., Fidler, A. E., Quirke, L., O'Connell, A., Smith, P., Groome, N, & Tisdall, D. J. (1999) Control of early ovarian follicular development *J. Reprod. Fert., Supplement*, **54**, 3-16.

McNatty, K. P., Moore, L. G., Hudson, N. L., Quirke, L. D., Lawrence, S. B., Reader, K., Hanrahan, J. P., Smith, P., Groome, N. P., Laitinen, M., Ritvos, O. & Juengel, J. L. (2004) The oocyte and its role in regulating ovulation rate: a new paradigm in reproductive biology. *Reprod.*, **128**, 379-386.

Misselbrook, T. and Smith, K. (2002) In: *Ammonia in the UK.* pp 40-47. Defra Publications, London.

Montgomery, G. W. (2000) Genome mapping in ruminants and map locations for genes influencing reproduction. *Rev. Reprod.*, **5,** 25-37.

Morimoto, S., Fernandez-Mejia, C., Romero-Navarro, G., Morales-Peza, N. & Diaz-Sanchez, V. (2001) Testosterone effect on insulin content, messenger ribonucleic acid levels, promoter activity, and secretion in the rat. *Endocrinol.*, **142**, 1442-1447.

Nakada, K., Ishikawa, Y., Nakao, T. & Sawamukai, Y. (2002). Changes in responses to GnRH on LH and FSH secretion in prepubertal heifers. *J. Reprod. Dev.*, **48**, (6), 545-551.

Nicholas, B, Alberio R., Fouladi-Nashta, A. A. & Webb, R. (2005) Relationship between low molecular weight insulin-like growth factor binding proteins, caspase-3 activity and oocyte quality. *Biol. Reprod.* (in press).

Nolan, R., O'Callaghan, D., Duby, R. T., Lonergan, P. & Boland, M. P. (1998) The influence of short-term nutrient changes on follicle growth and embryo production following superovulation in beef heifers. *Theriogenology*, **50**, 1263-1274.

O'Donovan, P. B. (1984) Compensatory gain in cattle and sheep. *Nutr. Abstr. Rev. - Series B*,, **54,** 389-410.

Ohkura, S., Ichimaru, T., Itoh, F., Matsuyama, S. & Okamura, H. (2004) Further evidence for the role of glucose as a metabolic regulator of hypothalamic gonadotropin-releasing hormone pulse generator activity in goats. *Endocrinol.*, **145**, 3239-3246.

Park, C. S., Erickson, G. M., Choi, Y. J. & Marx, G. D. (1987) Effect of compensatory growth on regulation of growth and lactation: response of dairy heifers to a stair-step growth pattern. *J. Anim Sci.*, **64**, 1751-1758.

Pate, J. L. (2003) Lives in the balance: responsiveness of the corpus luteum to uterine and embryonic signals. *Reprod. Supplement*, **61**, 207-217.

Pryce, J. E., Royal, M. D., Garnsworthy, P. C. & Mao, I. L. (2004). Fertility in the high yielding dairy cow. *Live. Prod. Sci.,* **86**, 125-135.

Pryce, J. E., Simm, G. & Robinson, J. J. (2002). Effects of selection for production and maternal diet on maiden dairy heifer fertility. *Anim. Sci.,* **74,** 415-421.

Rae, M. T., Kyle, C. E., Miller, D. W., Hammond, A. J., Brooks, A. N. & Rhind, S. M. (2002) The effects of undernutrition, in utero, on reproductive function in adult male and female sheep. *Anim Reprod. Sci.*, **72**, 63-71.

Rae, M. T., Palassio, S., Kyle, C. E., Brooks, A. N., Lea, R. G., Miller, D. W. & Rhind, S. M. (2001) Effect of maternal undernutrition during pregnancy on early ovarian development and subsequent follicular development in sheep fetuses. *Reprod.*, **122**, 915-922.

Rausch, M. I., Tripp, M. W., Govoni, K. E., Zang, W., Webert, W. J., Crooker, B. A., Hoagland, T. A. & Zinn, S. A. (2002) The influence of level of feeding on growth and serum insulin-like growth factor I and insulin-like growth factor-binding proteins in growing beef cattle supplemented with somatotropin. *J. Anim Sci.*, **80**, 94-100.

Rawlings, N. C., Evans, A. C., Honaramooz, A. & Bartlewski, P. M. (2003)

Antral follicle growth and endocrine changes in prepubertal cattle, sheep and goats. *Anim Reprod. Sci.*, **78**, 259-270.

Rhind, S. M. (2004) Effects of maternal nutrition on fetal and neonatal reproductive development and function. *Anim. Reprod. Sci.*, **82-83**, 169-181.

Roberts, A. J., Funston, R. N. & Moss, G. E. (2001) Insulin-like growth factor binding proteins in the bovine anterior pituitary. *Endocrinol.*, **14**, 399-406.

Robinson, J. J., Sinclair, K. D., Randel, R. D. & Sykes, A. R. (1999). Nutritional management of the female ruminant: mechanistic approaches and predictive models. In: H. -J. G. Jung and G. C. Fahey, Jr. (Ed). *Vth International Symposium on the Nutrition of Herbivores.* pp 550-608 American Society of Animal Science, Savoy, IL.

Robinson, R. S., Mann G. E., Lamming G. E. & Wathes, D. C. (2001) Embryonic-endometrial interactions during early pregnancy in cows. *Brit. Soc. Anim. Sci. Occasional Publication,* **26**, 289-295.

Rodriguez, R. E. & Wise, M. E. (1989). Ontogeny of pulsatile secretion of gonadotropin-releasing hormone in the bull calf during infantile and pubertal development. *Endocrinol.*, **124**, 248-256.

Royal, M. D., Darwash, A. O., Flint, A. P. F., Wooliams, J. A. & Lamming, G. E. (2000) Declining fertility in dairy cattle: changes in traditional and endocrine parameters of fertility. *Anim. Sci.*, **70**, 487-501.

Schams, D., Schallenberger, E., Gombe, S. & Karg, H. (1981). Endocrine patterns associated with puberty in male and female cattle. *J. Reprod. Fert., Supplement*, **30**, 103-110.

Schillo, K. K., Hall, J. B. & Hileman, S. M. (1992) Effects of nutrition and season on the onset of puberty in the beef heifer. *J. Anim Sci.*, **70**, 3994-4005.

Schrooten, C., Bovenhuis, H., Coppieters, W. & Van Arendonk, J. A. (2000) Whole genome scan to detect quantitative trait loci for conformation and functional traits in dairy cattle. *J. Dairy Sci.*, **83**, 795-806.

Sejrsen, K. (1994) Relationships between nutrition, puberty and mammary development in cattle. *Proc. Nutr. Soc.*, **53**, 103-111.

Sejrsen, K. (2005) Mammary development and milk yield potential .In *Calf and Heifer Rearing* (Ed P.C. Garnsworthy) pp 237-252. Nottingham University Press, Nottingham.

Sejrsen, K. & Purup, S. (1997) Influence of prepubertal feeding level on milk yield potential of dairy heifers: a review. *J. Anim Sci.*, **75**, 828-835.

Simpson, R. B., Chase, C. C., Jr., Spicer, L. J., Vernon, R. K., Hammond, A. C. & Rae, D. O. (1994) Effect of exogenous insulin on plasma and follicular insulin-like growth factor I, insulin-like growth factor binding protein activity, follicular oestradiol and progesterone, and follicular growth in superovulated Angus and Brahman cows. *J. Reprod. Fertil.*, **102**, 483-492.

Sinclair, K. D., Broadbent, P. J. & Hutchinson, J. S. M. (1995) Naloxone evokes a nutritionally dependent LH response in post partum beef cows but not in mid luteal phase maiden heifers. *Anim. Sci.,* **61**, 219-230.

Sinclair, K. D., Sinclair, L. A. & Robinson, J. J. (2000a) Nitrogen metabolism and fertility in cattle: I. Adaptive changes in intake and metabolism to diets differing in their rate of energy and nitrogen release in the rumen. *J. Anim Sci.*, **78**, 2659-2669.

Sinclair, K. D., Kuran, M., Gebbie, F. E., Webb, R. & McEvoy, T. G. (2000b) Nitrogen metabolism and fertility in cattle: II. Development of oocytes recovered from heifers offered diets differing in their rate of nitrogen release in the rumen. *J. Anim Sci.*, **78**, 2670-2680.

Sinclair, K. D., Kuran, M., Staines, M. E., Aubailly, S., Mackie, K., Robinson, J. J., Webb, R. & McEvoy, T. G. (2000c). In vitro blastocyst production following exposure of heifers to excess rumen degradable nitrogen during either the pre-antral or antral stages of follicular growth. *J. Reprod. Fert., Abstract Series*, **25**, 41.

Sinclair, K. D., Rooke, J. A. & McEvoy, T. G. (2003) Regulation of nutrient uptake and metabolism in pre-elongation ruminant embryos. *Reprod. Supplement*, **61**, 371-385.

Sullivan, S. D., DeFazio, R. A. & Moenter, S. M. (2003) Metabolic regulation of fertility through presynaptic and postsynaptic signaling to gonadotropin-releasing hormone neurons. *J. Neurosci.*, **23**, 8578-8585.

Tanaka, Y., Nakada, K., Moriyoshi, M. & Sawamukai, Y. (2001) Appearance and number of follicles and change in the concentration of serum FSH in female bovine fetuses. *Reprod.*, **121**, 777-782.

Taylor, V. J., Beever, D. E., Bryant, M. J. & Wathes, D. C. (2004). First lactation ovarian function in dairy heifers in relation to prepubertal metabolic profiles. *J. Endocrinol.,* **180**, 63-75.

Telfer, E. E., Webb, R., Moor, R. M. & Gosden, R. G. (1999) New Approaches to increasing oocyte yield from ruminants. *Anim. Sci.*, **68**, 285-298.

Thatcher, W. W., Guzeloglu, A., Mattos, R., Binelli, M., Hansen, T. R. & Pru, J. K. (2001) Uterine-conceptus interactions and reproductive failure in cattle. *Theriogenology*, **56**, 1435-1450.

Thatcher, W. W., Guzeloglu, A., Meikle, A., Kamimur, S., Bilby, T., Kowalski, A. A., Badinga, L., Pershing, R., Bartolome, J. & Santos, J. E. P. (2003). Regulation of embryo survival in cattle. *Reprod. Supplement*, **61**, 253-266.

Thomas, M. G., Gazal, O. S., Williams, G. L., Stanko, R. L. & Keisler, D. H. (1999) Injection of neuropeptide Y into the third cerebroventricle differentially influences pituitary secretion of luteinizing hormone and growth hormone in ovariectomized cows. *Domest. Anim. Endo.*, **16**, 159-169.

Wandji, S-A, Pelletier, G. & Sirard, M-A. (1992) Ontogeny and cellular localization of ^{125}I-labelled insulin-like growth factor-1, ^{125}I-labelled follicle-stimulating hormone, and ^{125}I-labelled human chorionic gonadotropin binding sites in ovaries from bovine fetuses and neonatal calves. *Biol. Reprod.*, **47**, 814-822.

Wathes, D. C., Taylor, V. J., Cheng, Z. & Mann, G. E. (2003) Follicle growth, corpus luteum function and their effects on embryo development in postpartum dairy cows. *Reprod. Supplement*, **61**, 219-237.

Webb, R., Campbell, B. K., Garverick, H. A., Gong, J. G., Gutierrez, C. G. & Armstrong, D. G. (1999a) Molecular mechanisms regulating follicular recruitment and selection *J. Reprod. Fert. Supplement*, **54,** 33-48

Webb, R., Garnsworthy, P. C., Gong, J. G. & Armstrong, D. G. (2004) Control of follicular growth: Local interactions and nutritional influences. *J. Anim Sci.*, **82**, E63-E74.

Webb, R., Gosden, R. G., Telfer, E. E. & Moor, R. M. (1999b) Factors affecting folliculogenesis in ruminants *Anim. Sci.*, **68**, 257-284.

Webb, R., Nicholas, B., Gong, J. G., Campbell, B. K., Gutierrez, C. G., Garverick, H. A. & Armstrong, D. G. (2003) Mechanisms regulating follicular development and selection of the dominant follicle. *Reprod. Supplement*, **61,** 71-90.

Williams, G. L., Amstalden, M., Garcia, M. R., Stanko, R. L., Nizielski, S. E., Morrison, C. D. & Keisler, D. H. (2002) Leptin and its role in the central regulation of reproduction in cattle. *Dom. Anim. Endo*. **23**, 339-349.

Williams, L. M., Hannah, L. T., Adam, C. L. & Bourke, D. A. (1997) Melatonin receptors in red deer fetuses (Cervus elaphus). *J. Reprod. Fertil.*, **110**, 145-151.

Wrenzycki, C., De Sousa, P., Overstrom, E. W., Duby, R. T., Herrmann, D., Watson, A. J., Niemann, H., O'Callaghan, D. & Boland, M. P. (2000) Effects of superovulated heifer diet type and quantity on relative mRNA abundances and pyruvate metabolism in recovered embryos. *J. Reprod. Fertil.*, **118**, 69-78.

Wright, I. A. & Russel, A. J. F. (1991). Changes in the body composition of beef cattle during compensatory growth. *Anim. Prod.*, **52**, 105-113.

Yaakub, H., O'Callaghan, D. & Boland, M. P. (1999) Effect of roughage type and concentrate supplementation on follicle numbers and in vitro fertilisation and development of oocytes recovered from beef heifers. *Anim Reprod. Sci.*, **55**, 1-12.

Yambayamba, E. S., Price, M. A. & Foxcroft, G. R. (1996) Hormonal status, metabolic changes, and resting metabolic rate in beef heifers undergoing compensatory growth. *J. Anim Sci.*, **74**, 57-69.

16

GUIDELINES FOR OPTIMAL DAIRY HEIFER REARING AND HERD PERFORMANCE

J. MARGERISON[1] AND N. DOWNEY[2]
[1]University of Plymouth, School of Biological Sciences, Drakes Circus, Plymouth PL4 8AA; [2]Provimi Ltd, SCA Mill, Dalton Airfield Industrial Estate, Dalton, Thirsk Y07 3HE

Introduction

This chapter has been written to provide a set of guidelines which have been formulated using the reviews from the previous chapters and as a consequence it contains summarisation and extracts from these chapters.

OBJECTIVES OF DAIRY HEIFER REARING

The objectives of dairy heifer rearing must be to maximize profitability by minimizing calf disease and mortality, optimizing early growth, optimizing nutrition, and minimizing fat deposition in the body and mammary tissues, thus optimizing lifetime performance while optimizing rearing costs. A heifer rearing programme will have a specified age at first calving, but must also have a target live weight or proportion of mature weight. As a consequence, the heifer rearing programme will have set targets for calving age, and live weight at parturition and thus will have target growth rates to be achieved at each part of the programme that will optimize the future life time performance of the dairy heifer in the milking herd.

TARGETS TO BE ACHIEVED IN A DAIRY HEIFER REARING SYSTEM

The targets that need to be achieved are for the heifer to be served between 13 and 15 months of age, at 55 to 60 % of mature weight. This should result in parturition between 22 and 24 months of age at 85 to 90 % of mature weight. This would maximize heifer feed intake, health and fertility and, with careful attention to housing and management, particularly around parturition, this will minimize heifer and cow culling rates, thus achieving optimum lifetime performance. The subsequent paragraphs contain guidelines that have been formulated from the main areas involved in the successful rearing of dairy heifers to achieve optimal lifetime performance.

OPTIMISING REARING AND DAIRY REPLACEMENT COSTS

Rearing a dairy heifer up until calving, including fixed costs, has been found to cost from £1037 calving at 2 years of age, to £1340 at 2.5 years and £1638 at 3 years (Esslemont and Kassaibati, 1998). In dairy cattle, greater age at first calving (Esslemont and Kassaibati, 1998) and higher culling rate (Jagannatha *et al.*, 1999) increases the number of heifers being reared at any one time, increases rearing costs and subsequently reduces farm profitability. Unfortunately, in practice heifer rearing costs are not always transparent and all too frequently heifer rearing has a low priority in the farm business. However, timely and efficient provision of dairy replacements for the dairy herd is key to long term economic success of a dairy enterprise. As a consequence, a planned approach to rearing dairy heifer replacements is required and the lack of this type of an approach will have a significantly negative impact on farm profitability. **As a guideline**, greatest net dairy farm incomes have been associated with a herd life greater than 5 years (Jagannatha *et al.*, 1999), which would be synonymous with a replacement rate of 20%.

Pre weaning period

MINIMISING CALF MORTALITY

Colostrum intake

The majority of mortality occurs in young calves and reported levels of mortality in the UK are 2 to 5% (Blowey, 2005), and 7.8 % (Esslemont and Kossaibati, 1996); mortality levels in the US are around 8 % (Van Amburgh and Drackley, 2005), 8.2 (Tyler *et al.*, 1999), and greater than 10% (NAHMS, 1996).

In a national survey of US dairy operations high calf mortality levels were associated with a number of management factors, which included higher levels of milk production (>7710 kg), housing pre-weaned heifers in groups of seven or more, male staff having primary responsibility for care and feeding of preweaned heifers, calves not receiving hay or other roughages until > 20 d of age, calves being offered solely mastitic or antibiotic containing milk after colostrum, and calves not given whole milk after colostrum (Losinger and Heinrichs, 1997). However, the single most important factor associated with between 39 % (Tyler *et al.*, 1999) and 50 % (NAHMS, 1996) of cases of calf mortality was inadequate transfer of passive immunity.

Acquisition of adequate passive immunity is imperative as the structure of the ruminant placenta does not allow the transfer of maternal antibodies (Tizard, 2000) and at birth the calf is dependent on supply of antibodies from colostrum, which needs to be achieved during the first 24h of life, while the intestine is permeable to macromolecules and antibodies can be absorbed and transported intact into the bloodstream (Sangild, Fowden and Trahair, 2000). The efficiency of passive transfer decreases to half at 12 h after birth (Drackley, 2000) and, as a consequence, the sooner colostrum is consumed following birth the greater will be the possibility of effective immunoglobulin transfer. Total colostrum intake, time of first feeding and concentration of IgG in colostrum have a great impact on efficiency of immune passive transfer (Quigley and Drewry, 1998). The amount of IgG absorbed by the calf is determined by many factors, including concentration of IgG in colostrum, feeding practices and metabolic state of the animal (Quigley *et al.*, 2005). Colostrum IgG concentration ranges from 10 to >100 g/l and variation is a result of the animals' disease history, age, volume of colostrum produced, season of year, breed and other factors. Unfortunately, estimation of colostrum quality is difficult on the farm (Quigley *et al.*, 2005). However, intake of colostrum (4 l) with a high IgG (60 g/l) content has been found to transfer high levels of IgG into calf plasma (20.8 g/l) (Morin, McCoy and Hurley, 1997) and it has been recommended that calves should receive colostrum from multiparous cows, as primiparous dairy cows have been found to have low milk IgG concentrations (Quigley, Martin, Dowlen, Wallis and Lamar, 1994).

More recently, researchers have reported that 36 to 82% of colostrum samples contain high bacterial counts (>100,000 cells/ml) and colostrum has been recognized as a vector for transmission of a number of disease causing organisms, including *Mycobacterium paratuberculosis*, and farms with significant Johne's infestation often have inadequate supplies of colostrum to feed to their newborn calves. In these cases, farmers have to rely on feeding colostrum or milk replacers or antibiotics until the calf has developed an active immune system (Quigley *et al.*, 2005). Colostrum supplements or replacers are used by dairy producers and studies suggest that most supplements derived from lacteal secretions are ineffective in improving circulating IgG concentration. Animal sources, such as blood, eggs, milk, and colostrum, are currently the only available sources of IgG; as a consequence, the safety of these proteins sources is critical. However, in some countries regulations preclude the feeding of ruminant derived blood proteins, with the exception of milk, in ruminant animal diets including calves, and blood-derived products are unavailable in other locations. Therefore, under these circumstances, IgG derived from lacteal secretions would have the greatest potential. Responses to IgG in milk or colostrum based supplements do not exceed 4 g/l,, although IgG in bovine plasma are well absorbed and can increase plasma IgG concentrations by approximately 7 g/L. In the US, one colostrum replacer, derived from fractionated bovine plasma, has shown that feeding one or two doses were sufficient to increase plasma IgG concentrations at 24 hours of age by 12.2 g/l.

The survival, health and growth of calves fed the colostrum replacer were similar to calves fed maternal colostrum. Unfortunately, addition of colostrum supplements can reduce the efficiency of absorption of IgG in maternal colostrum (Quigley *et al.*, 2005).

As a guideline it is recommended that suckling of colostrum should be ensured as soon as possible following birth and the effect on mortality of timing of colostrum intake is presented in Figure 16.1. Suckling should always be supervised and assisted where necessary, to achieve colostrum intake levels of 2.5 l within the first 6 hours of life and 4 l within the first 12 hours of life. Where the calf is not able to suckle successfully then calves should be offered milked or stored colostrum from a pail or by stomach tube and in these cases all feeding utensils should be kept scrupulously clean. Finally, to ensure adequate colostrum 'quality', colostrum consumed by the calf should be from multiparous cows, as heifers tend to have lower immunoglobulin levels in their colostrum. The amount of colostrum required to be consumed by the calf according to age is presented in Table 16.1.

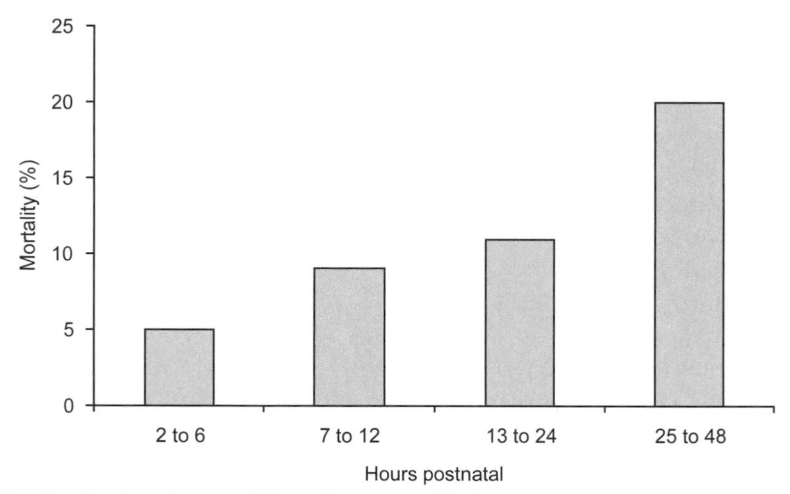

Figure 16.1 Mortality in calves receiving colostrum at differing numbers of hours postnatal.

Table 16.1 RECOMMENDED COLOSTRUM ALLOWANCES ACCORDING TO POSTNATAL PERIOD.

Days postnatal	*Period*	*Allowance (l)*
1	< 6.0 h	2.5
1	< 12.0 h	4.0
2 to 3	Daily	3 to 4
Continued ‡	Continued	5 % of total milk offered

‡ - Continued feeding of 5 % is recommended for the control of enteric gastrointestinal disease

Farms with significant *Mycobacterium paratuberculosis* and Johne's infestation may need to use colostrum replacers or milk replacers or antibiotics where these are permitted as animal feeds. See Quigley *et al.* (2005) for further details.

Enteric disease

Enteric problems are frequently seen in milk-fed calves and can be subdivided into infectious and management-based diseases. During the first three days of life, the main infectious problems are bacterial conditions such as E. coli, salmonella and clostridial infections, while at older ages, rotavirus, coronavirus and cryptosporidia become important (Blowey, 2005).

The initial source of enteric disease infection is typically either the dam and or other calves which previously used the same or adjacent housing facilities (Blowey, 2005). As a consequence, control of enteric disease is based on good hygiene standards, preferably with the use of an all in - all out housing management policy.

Calves that have achieved an adequate level of immunity can withstand low levels of challenge. However, in all of these conditions, good management provides the opportunity to reduce the level of infection and calves with good passive immune responses will have lower levels of sloughing of villi epithelial cells in the GIT due to rotavirus and coronavirus and suffer less from the loss of fluid absorption and digestion capacity which leads to dehydration (Blowey, 2005). As a consequence, treatment of enteric conditions includes feeding electrolyte solutions, that positively promote uptake of fluid through the damaged intestinal wall (Scott *et al.*, 2004) for approximately 2 days maximum and housing under warm clean conditions, preferably without contact with other potentially infectious calves (Blowey, 2005). Where enteric disease is a problem, it is recommended that colostrum feeding is continued for the first three weeks of life, at approximately 5% of total milk intake (Blowey, 2005) (Table 16.1). Cryptosporidia, a protozoan parasite, infections can be managed in a similar way to enteric diseases and treated using *halofuginome* as a prophylactic (Blowey, 2005).

Guidelines for the control of enteric diseases, including Cryptosporidia, prevent the disease by adequate colostrum intake and good hygiene standards, preferably with the use of an all in - all out housing management policy. Where enteric diseases are a problem, continued feeding of colostrum at 5% of total milk intake for up to 3 weeks of age is recommended. Withholding milk for up to a maximum of 3 days with the use of electrolytes is recommended for treatment of animals affected by enteric disease. *Halofuginome* can been used as a prophylactic where Cryptosporidia protozoan parasite infections are encountered regularly. See Blowey (2005) for further details.

NUTRITION OF THE GROWING DAIRY HEIFER

Nutritional value of colostrum

Colostrum provides a source of energy, protein, minerals and vitamins (Baumrucker *et al.*, 1994; Bühler *et al.*, 1998; Hammon *et al.*, 2000; Hammon and Blum, 1999; Hammon and Blum, 1998; Rauprich *et al.* 2000a; Rauprich, Hammon and Blum, 2000b; Kühne *et al.*, 2000). Colostrum contains three times more gross energy than milk on an as-fed basis (Kühne *et al.*, 2000) and neonatal animals utilize this energy to support glucose levels and maintain critical body temperature (Girard, 1986). Feeding 10 to 13 l of colostrum during the first three days, followed by intake of whole milk, has been found to result in growth rates of 250 to 500 g/d during the first week of life (Kühne *et al.*, 2000). In Table 16.2 the composition of colostrum and whole milk are compared with requirements of the calf from NRC (2001).

Table 16.2 COMPOSITION OF COLOSTRUM AND WHOLE MILK, AND CALF NUTRITIONAL REQUIREMENTS FOR CRUDE PROTEIN, FAT, MINERALS AND VITAMINS.

	Colostrum	*Whole milk*	*Calf requirements †*
Component			
Crude protein (g/kg)	58	26	20 to 26
Fat (g/kg)	28	30	16 to 18
Minerals (mg/kg)			
Iron	10	3	75 to 100
Copper	2.5	1	10
Selenium	0.1	0.002	0.3
Vitamins (iu/kg)			
A	12,400	11,500	11,000
D^3	500	307	600
E	15	8	50

† - NRC (2001).

Effect of colostrum on gastro intestinal development

Colostrum has been found to contain hormones, bioactive peptides and enzymes that play a major role in the development of the digestive tract, thus increasing growth and development of the of the calf (Kühne, Hammon, Bruckmaier, Morel, Zbinden and Blum, 2000; Rauprich, Hammon and Blum, 2000a; Quigley, Martin, and Dowlen, 1995; Guilloteau, Le Huërou-Luron, Chayvialle, Toullec, Zabielski and Blum 1997; Hammon and Blum 1998; Rauprich *et al.*

2000a). IGF-1 and 2 are in higher concentrations in colostrum compared with whole milk (Kühne *et al.*, 2000; Baumrucker *et al.*, 1994;Skarr *et al.*, 1994), and specific receptors for both IGF-1 and IGF-2 exist in the intestinal mucosa (Baumrucker *et al.*, 1994). Levels of IGF-1 and 2 are presented in Table 16.3. Colostrum (750 ng/ml of IGF-1) was found to increase villi surface area by 1.5 fold compared with milk substitute (Bühler *et al.*, 1998) and similar results have been found with milk substitute offered for one week were intestinal DNA synthesis was significantly increased in the duodenal, jujenal and ileal tissues (Baumrucker *et al.*, 1994). However, not only IGFs but also other non-nutritional factors present in colostrum may contribute to development of the digestive tract (Bühler *et al.* 1998). This has been found to increase growth rate (Table 16.4).

Table 16.3 GROWTH FACTOR (IGF-1 AND 2) CONCENTRATIONS IN COLOSTRUM AND WHOLE MILK (ng/ml).

	Colostrum	*Whole milk*	*Source*
IGF-1	512 to 1537	3	Baumrucker *et al.* (1994)
IGF-1	313	2 to 3	Kuhne *et al.* (2000)
IGF2	393	3 to 4	Skarr *et al.* (1994)

Table 16.4 EFFECT OF FEEDING LEVEL OF COLOSTRUM OR MILK REPLACER ON GROWTH RATE BETWEEN 0 AND 7 DAYS POSTNATAL.

		Liquid feed offered			
Age (d)	*Feeding level*	*Colostrum*		*Milk replacer*	
0 to 3	(l/d)	3.2	4.4	3.2	4.4
4 to 7	(% LW in kg)	0.52	0.64	0.52	0.65
Birth weight	(kg)	46.2	43.8	48.0	46.5
Growth rate	(kg, 0 to 7 d)	1.78 [a, b]	3.47 [a]	-0.67 [c]	0.72 [a, b]

[a, b] - Data in rows followed by differing letters differ significantly ($P<0.05$)
Amended from: Kuhne *et al.* (2000).

MILK AND MILK REPLACERS

As a guide, the recommended amounts of colostrum or milk replacer for twice daily feeding from a pail or bucket are presented in Table 16.5.

To avoid reduced growth rate following colostrum feeding, offer a good quality calf milk replacer. Highest levels of performance are achieved if the product is highly digestible and consists of fractionated milk proteins containing natural immunoglobulins, which increase disease resistance in the GIT and provide protein to enable the calf to gain 0.8 to 0.9 kg live weight per day during the first 3 months of life.

Table 16.5 RECOMMEND AMOUNTS OF COLOSTRUM OR MILK OFFERED FOR TWICE DAILY FEEDING FROM PAIL OR BUCKET

Age (days)	*a.m.*	*p.m.*	*Daily total (l)*
		Colostrum (l)	
1 to 3 or 4	1.5 to 2	1.5 to 2	3 to 4
		Milk replacer (l at 125 g/l milk replacer concentration)	
4 to 7	1.5	1.5	3
8 to 10	2	2	4
11 to weaning	2.5	2.5	5

Traditional calf milk replacers contain between 18 and 20% fat and 22 to 24% crude protein, and are fed at 500 grams per head per day contributing around 2.5 MJ of ME per litre of milk fed. A milk replacer of this specification can only support growth rates of 400 – 450 grams per head per day. Increasing the feeding rate of milk replacer will increase growth and feed conversion rates (Diaz *et al*; 1998, 2001). A Milk replacer powder contained 200g/kg fat, 420 g/kg lactose and a high concentration of CP (300g/kg) to ensure that protein was not limiting fed at 3 different levels (actual levels 900, 1530 and 1690g/d) increased significantly weight gain (600, 1040, 1170 g/d) and improved feed conversion efficiency (0.65, 0.60, 0.76) and the fat content of empty body weight (EBW) increased as feeding rate increased (5.7, 9.8 and 9.1g fat/kg EBW). However, there was no concomitant reduction in protein content of EBW, which was similar across treatments (21.3, 20.5 and 21.1g CP/kg EBW) (Diaz *et al*; 1998, 2001).

As a guide it is presently agreed that the crude protein content of the diet should be approximately 280g/kg DM, which is similar to the crude protein content of whole milk (Thorbek, 1977; Diaz et al., 2001), and at least 260 to 280g/kg DM for a live-weight gain of 700 g/d (Davis and Drackley, 1998; Donnelly and Hutton, 1976 a,b). The effect of milk replacer crude protein level on feeding rate, feed conversion efficiency, live-weight gain and fat content of the body are presented in Table 16.6. For a more detailed description of energy and protein requirements see Van Amburgh and Drackley (2005).

Higher growth rates can be achieved when calves are fed high levels of milk replacer containing high concentrations of protein (Table 16.7). Results from Drackley (2001) showed that body weight gain (BWG) was driven mainly

Table 16.6 FEED INTAKE, GROWTH RATE AND BODY COMPOSITION OF CALVES OFFERED WHOLE MILK OR MILK REPLACER AT 21/20 OR 20/18 FAT (g/kg FAT/% CP RESPECTIVELY)

		Milk replacer	
	Whole milk	*21/20*	*20/18*
DMI (% live weight)	1.44	2.45	2.67
ME (MJ/kg FM)	2.81	2.52	2.46
Contribution to ME (%)			
CP	25	25	23
Lactose	26	40	45
Fat	49	35	32
Growth rate (g/d)	**600**	**1040**	**1170**
Body composition (g/kg)			
CP	213 [a]	205 [c]	211 [a]
Fat	57 [c]	98 [a]	91 [b]

[a, b, c] - Data in rows followed by differing superscript letters differ significantly (P < 0.05)
Source: Diaz *et al.* (2001).

Table 16.7 EFFECTS OF FEEDING RATE AND CP CONTENT OF MILK REPLACER ON FEED:GAIN EFFICIENCY, AVERAGE DAILY GAIN, ENERGY RETENTION AND FAT CONTENT OF WHOLE BODY GAIN.

		Feed rate	*CP in milk replacer (g/kg)*			
		DM (g/kg LW)	*140*	*180*	*220*	*260*
DMI intake	g/d	13	625	646	672	655
		18	981	949	958	986
Feed:gain ratio (a, c)		13	0.40	0.48	0.61	0.55
		18	0.52	0.59	0.72	0.71
Daily live-weight gain (a, b)	g/d	13	250	310	410	360
		18	510	560	690	700
Whole body:						
Energy retained (a)	MJ/d	13	2.49	2.13	2.47	2.03
		18	4.89	4.98	5.28	4.99
Fat content (a, c)	g/kg	13	68	58	56	51
		18	88	81	71	66

(a) - Effect of feeding rate (P<0.05)
(b) - Linear effect of increasing dietary CP (P<0.05)
(c) - Quadratic effect of increasing dietary CP (P<0.05)
Amended from: Drackley (2001).

by energy intake, however protein intake can influence both BWG and composition of body weight gain. Live-weight gain (g/d) of calves offered diets of differing levels of crude protein is presented in Figure 16.2. Increasing level of milk replacer and crude protein increased LWG, with LWG being optimised when the milk replacer contained 22% crude protein. However, within each level of feeding, increased milk replacer CP concentration decreased the fat content of whole body, without any effect on the quantity of energy retained. Body fat (%) and energy (MJ) retained from calves offered milk replacer with varying crude protein composition are presented in Figure 16.3.

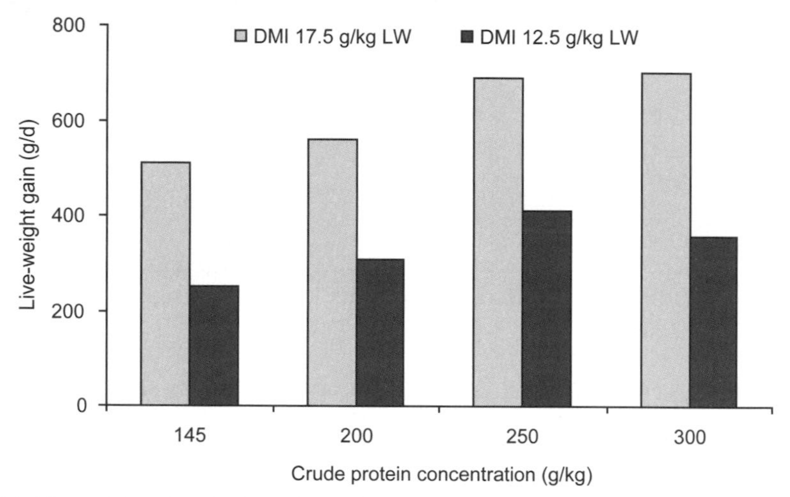

Figure 16.2 Live-weight gain of calves offered diets with differing concentrations of crude protein at DMI of 12.5 and 17.5 g/kg live weight (Amended from: Drackley, 2001).

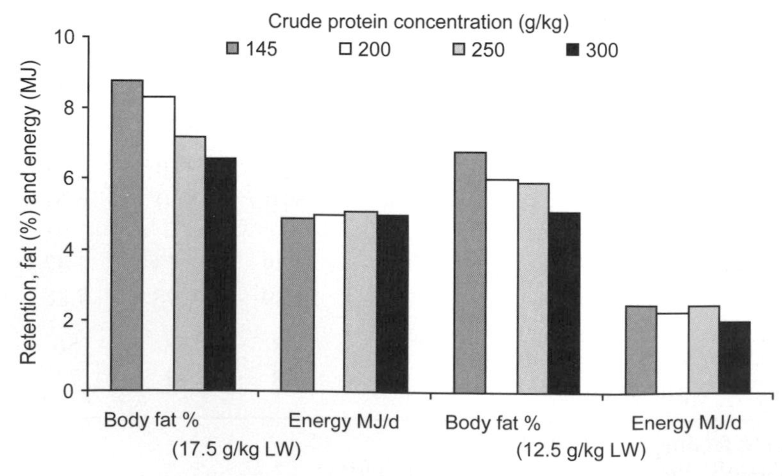

Figure 16.3 Body fat and energy retained by calves offered milk replacer with crude protein concentrations of 145, 200, 250 and 300 g/kg at DMI of 12.5 and 17.5 g/kg live weight (Amended from Drackley, 2001).

Body composition is affected by energy source, particularly the fat content of the diet (Tikofsky *et al.*, 2001), and high lactose low fat concentrations in milk replacer favour lower levels of fat deposition. Live-weight gain and body fat content of calves offered low, medium and high fat to lactose ratio diets are presented in Figure 16.4. Selecting a milk replacer with the correct combination of protein, oil and lactose will minimise fat deposition and achieve body weight gains of over 650 g per head per day. The effect of feeding rate and CP content of milk replacer on feed conversion and energy and fat content of the body is presented in Table 16.7.

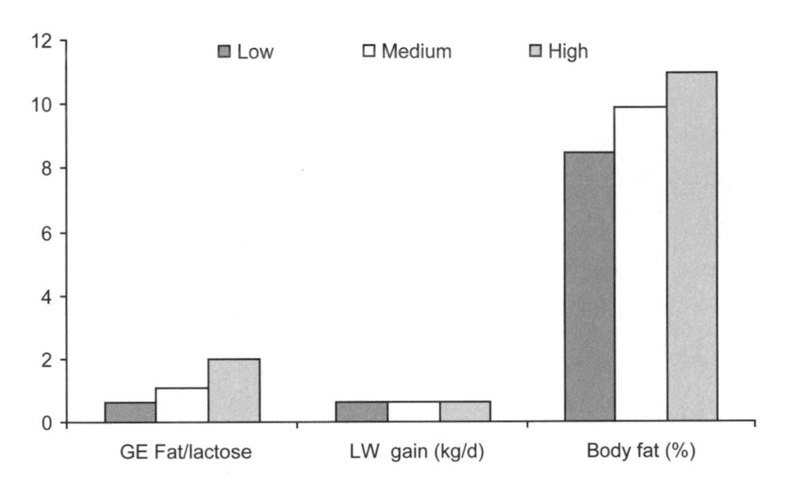

Figure 16.4 Live-weight gain and body fat content of calves offered diets with low, medium and high fat to lactose ratio.

Nutrient sources for milk powders

Potential nutrient sources for inclusion in milk replacers and the effect of heat on the quality of milk powders have been reviewed by Tanan (2005). Milk replacers typically used in calf rearing are based on either skim milk or whey powder. Skim milk powders, like whole milk, contain caseins that coagulate in the abomasum increasing the retention time (Le Huerou-Luron *et al.*, 1992), but whey based milk powders do not form clots under the action of chymosin in the abomasum. Lammers *et al.* (1998) compared skim milk and whey protein concentrate and concluded that calves fed on milk replacer based on skim milk do not seem to perform better than those receiving milk replacer based on whey protein concentrate.

Quality of milk replacers is affected by their composition, nutrient sources, particularly their digestibility and biological value, and heat treatment. Heat damaged skim milk powders do not clot properly (Emmons and Lister, 1976), have lower digestibility, can cause diarrhoea (Roy, 1970) and are not suitable for a young calf of less than one month old (Longenbach and Heinrichs, 1998).

According to Tanan (2005), lysine content of the powder can be a good indicator of previous thermal treatment and potential damage or denaturing of protein.

According to Tannan (2005), the vegetable sources of crude protein (CP) most commonly used in milk replacers are hydrolised wheat gluten (800 g/kg CP) and soya protein concentrate (665 g/kg CP). However, calf growth rates have been variable when offered soya bean concentrate (SBC), which had low digestibilites (0.77 to 0.87) (Toullec, Lalles and Bouchez, 1994) possibly due to the variability in the quality of soya (Tannan, 2005). Soya bean isolate has a higher CP content (900 g/kg). However, products such as soya bean flour contain antinutritional factors (Lalles, 1993 Davis and Drackley, 1998), so they can be used only following processing and have CP digestibilities between 0.84 (Caugant *et al.*, 1993a) and 0.91 (Montage, Toullec and Lalles, 1998). Crude protein content does not distinguish between true protein and non-protein nitrogen (NPN) and gives no estimation of the amino acid profile and subsequent potential biological value of protein sources. Soluble wheat gluten and SBC have differing amino acid profiles and have lower levels of essential amino acids (0.453 and 0.348 respectively) compared with skim or whey based milk replacers (Tannan, 2005). However, synthetic lysine and threonine have been added to wheat gluten based milk replacers to give similar growth rates (Toullec and Grongnet, 1990).

The source of carbohydrate in milk is lactose and scouring can occur, probably due to the limitations of lactose absorption and subsequent reduction in pH of the gastro intestinal tract, when milk replacer is offered at approximately 1.1. kg DM/d (Tannan, 2005). Lactose is difficult to replace due to the limitations of the development and subsequent ability of the digestive tract to digest other carbohydrate sources. Consequently, only limited levels of processed starch should be included in milk replacers (Toullec, 1989). In terms of energy, fat provides around 0.5 of the gross energy in whole milk and 0.35 in milk replacer (Tannan, 2005). Tallow has been replaced by vegetable fat sources in the EU, due to EU regulations regarding animal feeds, and coconut, palm and soya oils are the main sources of energy. Oils, typically 0.2 coconut and 0.8 palm, should be offered in combination to prevent liver infiltration found when coconut oil (Graulet, Gruffat-Mouty, Durand and Bauchart, 2000; Jenkins and Kramer, 1986) or soya oil (Leplaix-Charlat, Durand and Bauchart, 1996) are offered to young calves. Finally, the true digestibility of fat is greatly affected by particle size, so the technology used in homogenization and incorporation of fat are of considerable importance. Spray drying techniques not only reduce the potential for heat damage of protein, but also give good levels of emulsification and stabilisation of fat particles thus increasing fat digestibility.

Dietary calcium and phosphorus levels are important for skeletal development in growing dairy heifers. Whey based products and vegetable sources of nutrients are low in calcium and dicalcium phosphate, a highly digestible and retentive source of these minerals (Gronnet *et al.*, 1981), or calcium formate (Tannan 2005) should be included. It is well known, however, that high levels of dietary calcium can reduce fat digestion causing diarrhoea, so calcium should not be offered in excess of maximum requirements. Finally, macro minerals can affect osmotic pressure of the intestine and increase the water content of faeces, resulting in a watery diarrhoea. In particular, high levels of ash (Ullerich and Kampheus, 1998; 2000), sulphate (< 0.6 g/l or 0.1 inclusion rate offered at 125 g of milk powder / kg) (Kampheus, Stole and Rust, 2002) and Na, K and Cl will increase the water content of faeces. Whey products can vary considerable in their sulphate content (0.3 to 0.43 g/kg DM; Kampheus, 2000) The osmotic effect of minerals included in the diet can be calculated (Tannan, 2005) from the Na, K and Cl content of the milk powder and compared with Na, K and Cl requirements, which provide 59 mOsm/l (NRC, 2001) to 103 mOsm/l for whole milk (Tannan, 2005). However, osmotic levels can be up to 250 mOsm/l because no changes in dry matter of faeces was found at this level, with an ash concentration of 135 g/kg (Ullerich and Kampheus (1998).

According to NRC (2001) whole milk is deficient in vitamins D and E and trace elements, Fe, Mn, Zn, Cu, I, Co and Se. As a consequence, supplementation of these and additional vitamin A is necessary. Skim and whey powders have only traces of fat soluble vitamins (Mahaud *et al.*, 2000b) and vitamins such as thiamine (B1), Cobalamin (B12) and C are heat sensitive. As a consequence, supplementation of these is necessary. Similarly, liquid whey requires supplementation with Fe and Cu.

As a guide, Calves fed on milk replacer based on skim milk do not seem to perform better than calves receiving milk replacer based on whey protein concentrate. Tannan (2005) also describes the merits of protein sources from skim milk powder and whey powder. In the production of milk replacers the vegetable protein sources most commonly used are hydrolised wheat gluten and soya protein concentrate. However, calf growth rates have been variable when offered soya bean concentrate, probably due to the variability of soya bean concentrate quality. Lactose and fat are the main sources of energy and only low levels of starch may be included in calf diets. Coconut (0.2) and palm oils (0.8) should be used in combination to prevent liver infiltration. The digestibility of fat and the quality of protein in milk replacers are greatly affected by the technology used in their production; homogenization and incorporation give good levels of emulsification and stabilisation of the fat particles thus increasing fat digestibility and spray drying techniques reduce the potential for heat damage of protein. Heat damaged skim milk powders do not clot properly, are less digestible and can cause diarrhoea and are not suitable for a young calf of less than one month old. It is important to remember that whey powders do not form clots in the abomasum. According to Tanan (2005), lysine content of the powder can be a good indicator of previous thermal treatment and potential damage or denaturing of protein. High levels of lactose, calcium and/or minerals affect the osmotic pressure of the digestive system and fat digestion causing diarrhoea in dairy calves. However, all milk products, including whole milk, require supplementation with minerals and vitamins in order to meet dietary requirements of growing dairy heifers.

NUTRIENT SOURCES FOR SOLID FEEDS

Relative masses and volume capacities of forestomachs, small intestine and colon are greatly modified by nutrition, especially dependent on whether calves are fed liquid or solid diets. Forestomachs markedly develop, more than the abomasum (that relatively decreases in size), when calves start to ingest solid foods (Church, 1971; Davis and Drackley, 1998). Development of specific metabolic functions (for instance ketogenesis) of ruminal keratinocytes is initiated and modified by microbial fermentation products and is influenced by hormones and growth factors (Galfi, Neogrady and Sakata, 1991; Baldwin, 1999, 2000; McLeod and Baldwin, 2000; Lane and Jesse, 1997; Lane, Baldwin and Jesse, 2000 and 2002; Zanming, Seyfert, Löhrke, Schneider, Zitnan, Chudy, Kuhla, Hammon, Blum, Martens, Hagemeister and Voigt, 2004). See Blum (2005) for details.

Food ingestion also modulates the activity of nerves, especially of the autonomous nervous system, that release acetycholine, noradrenaline, serotonine, nitric oxide and other neurotransmitters (Ekblad and Sundler, 2002). The effects on the GIT of ingested hormones and growth factors depend on localization and availability of receptors from the luminal and (or) basolateral sites, on regional versus overall control, and on coordination of endocrine and neuronal regulation, as for example shown for cholecystokinin in calves (Guilloteau *et al.*, 2002). Increasing solid food intake increases the development of the rumen, but high levels of liquid feed will reduce this rate.

Overall, solid feeds for calves should be; highly digestible, palatable and contain a limited but defined degree of bulk (Hill, Aldrich and Schlotterbeck, 2005).

The young calf has limited rumen fermentation and has been found to respond to volatile fatty acids (VFA) produced from rumen fermentation (Hill *et al.*, 2005). As a consequence, it is recommended that the solid feeds for calves are; palatable, to encourage high consumption levels, and highly digestible, to facilitate rapid development of the rumen and the transition from liquid to solid diets (Hill *et al.*, 2005). Some farms are using one solid feed from birth to approximately four months of age and using significant amounts of fibrous ingredients in the diet reduces the appearance of scours and acidosis in the weaned calf.

Feed source materials

Predominantly in Europe, more than in the US, grains are used as a source of energy and maize is steam-rolled or flaked to make it more digestible, rather than dry rolled or other dry processing. In both the EU and US fibre levels are increased by use of co-products such as wheat middlings and soyabean hulls to reduce costs; in other areas of the world they can be one of the main components (Hill *et al.,* 2005).

Forages

In pre-weaned calf diets, abrasive fibre of adequate particle size may be required to ensure normal development of rumen papillae, to strengthen the musculature if the rumen and omasal lining and to prevent keratinization (Hill *et al.,* 2005). Forages are, however, considered to be excessively bulky, and potentially limit DM intake; when long hay has been offered intake has been found to be variable (Thomas and Hinks, 1968 from Beharka, Nagaraja, and Morrill, 1991). Hay intakes have been estimated to be up to 150 mg/kg (Jasper and Weary, 2002; Chua, Coenan, van Delen, and Weary, 2002) and straw and hay intakes have been as low as 20 and 50 mg/kg respectively, of total solid feed intake (706 and 741 g daily) (Hill *et al.,* 2005). Moreover, calves wasted 3 times more forage than they consumed (Hill *et al.,* 2005). As a consequence, it may be better to delay feeding forage until after weaning (Davis and Drackley, 1998) and, in this case, production of volatile fatty acids from concentrate fermentation would be the key to development of the rumen epithelium. However, calves offered finely ground diets have been found to consume bedding to achieve rumen development when forage has not been provided (Greenwood, Morrill, Titgemeyer, and Kennedy, 1997b; Beharka, Nagaraja, Morrill, Kennedy, and Klemm, 1998) and calves seem to have an innate requirement to consume some form of large particle fibre.

At 2 to 4 weeks post-weaning, when concentrate intake increases rapidly, hay intake levels remained relatively low at 50 mg/kg (Funaba, Kagiyama, Inki, and Abe, 1994; Jasper and Weary, 2002; Chua *et al.*, 2002) 18 mg/kg when offered restricted concentrate (4.5 kg DM) or 50 to 100 mg/kg when hay was restricted to 250 g of total solid feed intake (Quigley, Steen and Boehms, 1992). There seemed to be little work on the effect of forage quality, particularly hay, but Quigley *et al.* (1992) had 160 mg/kg CP and 620 mg/kg NDF. However, if straw or hay is offered this should be good quality, clean and relatively dust free to avoid potential respiratory disease.

The use of ground forage or coarse grain materials in calf diets has had limited research attention. However, processing of the feeds affects digestibility and as a consequence feed intake levels may be affected. However, solid feed intake and live-weight gain were greatest for calves offered a textured feed, intermediate for ground feed and lowest for a pelleted feed (Franklin, Amaral-Phillips, Jackson and Campell, 2003); unfortunately, the ingredient composition and nutrient profiles of the diets differed. Greater calf solid feed intake and gains were found with a meal feed having 850 mg/kg of the particles retained on an 1190 micron sieve vs. a pellet having 200 mg/kg of the particles retained on an 1190 micron sieve (no calf bedding material) (Porter cited by Warner, 1991). The textured feed mentioned above had approximately 500 mg/kg remaining on a 1,190 micron sieve, not counting the pellets in the diet. There have been some suggestions regarding grinding and incorporation of 100 to 150 mg/kg of hay in complete solid feed mixtures (Morrill, personal communication), but this would pose problems for manufacturing and increase the heterogeneity of feeds. However, there was limited information regarding the provision and or type of bedding material used. Clearly, the provision of and thus potential consumption of bedding by the calf is a critical issue when considering the offering of coarse materials in the form of either forage or textured diet. For a more detailed review see Hill, *et al.*, 2005).

As a guideline, rumen development can take place either due to consumption of forage or fibrous feed intake, or due to the presence of volatile fatty acids from fermentation of concentrated feed sources. Intake of 'starter' diets will be increased by palatability of the components of the diet and where available steam-rolled or flaked maize will make these feeds more digestible. Fibrous feeds should be offered in line with current welfare guidelines for each country. It is clear that calves offered finer feed resources should be offered functional fibre or forages or they will consume bedding material. As a consequence, adequate levels of bedding to allow consumption of clean dust free straw and other bedding materials such as sawdust would not be suitable should forage not be offered. Straw or hay offered as forage should be good quality, clean and relatively dust free to avoid potential respiratory disease. Similarly, clean water should be on offer at all times, in line with the welfare regulations of each country, unless otherwise advised due to ill health by a veterinary surgeon.

As a guideline, calves should be weaned when they are consuming between 1 and 1.5 kg of dry feed for 2 consecutive days. This usually occurs around 6 weeks of age and calves should be removed from milk, unless there are restrictions within a particular assurance scheme or other circumstances preventing this, and offered a good quality 18% crude protein calf 'starter' diet and clean water. The recommended daily feed, energy and protein levels according to growth rate are presented in Table 16.8.

Table 16.8 RECOMMENDED DAILY INTAKES OF DRY MATTER (DMI), METABOLISABLE ENERGY (ME), AND PROTEIN (MP AND CP), PROTEIN:ENERGY RATIO (MP:ME) AND DRY MATTER INTAKE PER KG LIVE WEIGHT (LW) ACCORDING TO GROWTH RATE (g/d).

Growth rate g/d	DMI ‡ kg/d	ME MJ/d	MP g/d	CP g/kg DMI	MP:ME g/MJ	DMI g/kg LW
0	0.36	7.3	28	84	3.9	8.0
200	0.46	9.2	75	177	8.2	10.0
400	0.58	11.8	123	226	10.4	13.0
600	0.72	14.6	170	252	11.6	16.0
800	0.87	17.7	217	267	12.3	19.0
1000	1.03	20.8	265	276	12.7	23.0

‡ - DMI: calculated from ME of milk replacer containing 20.2 MJ/kg DM
Amended from NRC (2001).

Post weaning period

In terms of guidelines, for heifers to calve for the first time at 24 months of age they need to be served at 13 to 15 months of age with a target body weight of 390 kg live weight or 55 to 60 % mature weight. Research indicates that growing heifers too quickly from 3 to 10 months (Pre-puberty period) will reduce mammary gland development and hence milk production potential (Sejrsen, 1978). High levels of feeding during this period, promoting growth rates in excess of 0.85 kg per day, have been shown to increase fat deposition in the udder and to reduce development of milk secretary tissue. According to Sejrsen (2005), mammary development in the rearing period can influence milk yield capacity of the heifers as cows. At 2 to 3 months of age the mammary glands start to grow at a faster rate than the rest of the body; the growth becomes allometric (Sinha and Tucker, 1969) and in this phase there is rapid growth of the mammary fat pad and of the mammary ducts, but no alveoli are formed. The allometric growth phase of the mammary glands is closely linked to reproductive development and most studies suggest that the allometric growth phase ends at onset of puberty or shortly thereafter. However, in large dairy breeds, the average age and body weight at onset of puberty is 9 to 11 months and approximately 275 kg live weight. Variation in age and body weight at onset of puberty between animals is wide, but onset occurs at approximately the same live weight independent of feeding level (Sejrsen and Purup, 1997). Sejrsen et al. (1982) found that the amount mammary parenchyma was reduced by high feeding level in the prepubertal period, but not in the post-pubertal period of virgin heifers. Since then the effect of feeding level in the prepubertal period has been investigated in many experiments (Sejrsen et al., 2000). The initial finding was confirmed in most of the subsequent experiments. Sejrsen (2005) clearly states that mammary development and subsequent milk yield can be seriously reduced when feeding level results in average growth rates above 700 to 750 g per day during the prepubertal period and this is considered to be a critical period (see Figure 5). The effect of nutrition and management of mammary development has been reviewed in detail by Sejrsen (2005).

Growing heifers need to be supplemented with the correct mineral and vitamin package. Calcium and Phosphorus are vital for development of large framed animals with strong healthy bones. According to Tanan (2005), whey based products and vegetable ingredients are low in calcium and dicalcium phosphate (Grognet *et al.*, 1981), but calcium formate can be used as a supplement. However, if the calcium level is too high there is a risk a reduced fat digestibility. Phosphorus also promotes feed intake. Minerals are important nutrients that can also affect fertility and hence the ability of the heifer to calve within 24 months. Mineral deficiencies are most likely to occur in older heifers when there is a greater reliance on home grown forages and cereals. The minerals that are most often associated with infertility include copper, zinc, selenium, iodine and phosphorus.

Guidelines for growth rate and mature weight development according to age of dairy heifers are presented in Figure 16.5 and **guidelines** for weight gain, wither height and dietary energy and protein intake of dairy heifers according to age are presented in Table 16.9.

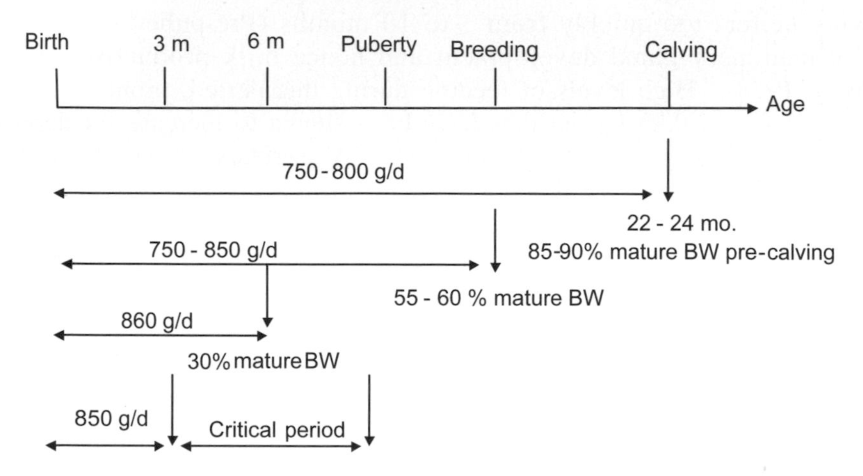

Figure 16.5 Growth rate and mature weight development according to age of dairy heifers (Amended from: Troccon, 1993; Penno, 1997; Van Amburgh et al., 1998; Mourrits, 2000).

Table 16.9 GUIDELINES FOR AGE, LIVE-WEIGHT GAIN, LIVE WEIGHT, WITHER HEIGHT, DIETARY METABOLIZABLE ENERGY AND CRUDE PROTEIN CONCENTRATIONS FOR GROWING DAIRY HEIFERS

Age (months)	*Weight gain (g/d)*	*Live weight (kg)*	*Wither height (cm)*	*ME (MJ/kg)*	*CP (g/kg)*
2	900	95	86	12	180
3	900	123	90	12	180
6	850	200	104	11	160
14	800	392	128	10	140
20	750	527	135	10	140
22	750	572	139	11	150
23	750	595	140	11	150

Source: NRC (2001)

DISEASE MANAGEMENT

Pneumonia

In calf rearing pneumonia can be a key problem, with a range of infections involved (Blowey, 2005). Prevention of high air humidity (> 70 %) should be achieved through a combination of adequate building ventilation, appropriate stocking density and frequent replenishment and full replacement of bedding material. Avoid mixing calves of differing ages and source, as they will inevitably have differing levels and types of infection and subsequent immunity to disease. As a consequence, disease can be avoided by not mixing together calves of differing age or environmental origin within the same housing or more correctly shared air space.

On rearing units where, despite these management conditions, pneumonia remains a problem the use of vaccines and antibiotics may be required. There is, however, no one vaccine that is effective against all pathogens. However, vaccination against bacterial components, especially Pasteurella is considered a logical approach (Blowey, 2005). Similarly, administration of antibiotics has generally been found to be effective in treatment of pneumonia.

Clostridial disease

Vaccines can be used to control a range of clostridial diseases such as tetanus, blackleg and enterotoxaemia.

NUTRITION FROM SERVICE TO CALVING

Heifers can be rapidly grown from service until calving provided the ration is correctly balanced for energy and protein. However, it is advisable to avoid letting the heifer get over fat as this will result in calving difficulties and increase the risk of ketosis, poor feed intake and lower milk production. To minimise metabolic stress during early lactation, steps need to be taken to avoid negative energy balance prior to calving. This means conditioning the rumen to the lactating diet. Feeding a concentrate containing a good quality source of Digestible Undegradable Protein (DUP) will help to improve post calving body condition, by reducing the mobilisation of body protein reserves to meet foetal and maternal growth requirements in late gestation and also help to increase dry matter intake.

Periparturient heifer

Management of the periparturient heifer is of major importance in relation to survival and productivity in her first lactation (Margerison, 2003) and reducing longer term effects of heifer management on health and productivity and, thus the longevity of dairy cattle

MASTITIS

Calving itself is a major risk for many diseases. Heifers are at risk of mastitis, lameness and infertility. During the peripartum period, the immune system becomes depressed and the natural keratin teat seal at the teat end begins to dissolve making the teat becomes susceptible to the entry of new infections the risk of disease and infection such as peracute toxic mastitis. Clean dry housing pre and during parturition by increasing bedding frequency is essential in reducing the incidence of mastitis and somatic cell count.

LAMENESS

Relaxation of ligaments within the suspensor system of the pedal bone within the hoof capsule, which leads to the corium becoming trapped, bruising and subsequent lameness (Tarlton *et al.*, 2004). Levels of sole bruising can be reduced by housing lactating heifers in a loose housed straw bedded system (Whay, 1999; Webster, 1998), using high dry matter diets (Offer, Leach, Brocklehurst and Logue, 2003) and housing of heifers separately from cows (Whay, 1998) to reduce competition for feed and aggressive social interaction. This results in higher food intake and reduced levels of a range of post-partum disorders such as ketosis and left displacement of the abomasum (Blowey, 2005). Increased feed intake and reduced live-weight loss result in increased milk yield and fertility.

In terms of guidelines, heifers are better housed separately from cows and abrupt changes of diet and ‘wet’ diets should be avoided. Abrupt changes of housing particularly onto hard floor types should also be avoided. As a consequence, heifers would be better housed in loose bedded yards in a ‘solely heifer group’ and bedding should be added as frequently as possible. To prevent mastitis, the area were feed is offered should be scraped clean to reduce soiling of bedding, and animals that are on heat should be removed to prevent excessive disturbance of the bedding material.

FERTILITY

Although conception rates to first service have been found to be higher in heifers (67%) compared with cows (42%) (Pryce *et al.*, 2004) there is evidence to suggest that selection for milk yield has reduced fertility in the maiden heifer (Pryce *et al.*, 2002).

The effect of nutrition and management on heifer fertility has been reviewed by Sinclair and Webb (2005). Plane of nutrition is known to significantly influence the onset of puberty in dairy cattle; lower levels of nutrition delay the onset of puberty and puberty is reached at a constant body weight and body composition, independently of dietary energy and protein concentrations (Chelikani et al., 2003). **In terms of guidelines**, it is recommended to avoid under nutrition as this has been found to reduce luteinising hormone secretion and to delay development of dominant follicles (Robinson *et al.*, 1999). This effect is affected by season as autumn born beef heifers tend to reach puberty earlier than spring born heifers, but they tend to exhibit a bimodal pattern in onset of puberty (Schillo *et al.*, 1992). There is a direct relationship between embryo size (Mann *et al.*, 1999; Robinson *et al.*, 2001) and the potential to develop into a foetus, which influences subsequent fertility.

In terms of guidelines, sudden reductions in feed intake and subsequent (20 to 25%) loss of initial live weight (Diskin et al., 2003) or undernutrition, 0.7 x maintenance (Webb et al., 2004) 0.4 x maintenance (Mackey et al., 1999), should be avoided as they could cause anoestrus, or reduce dominant follicle size and viability, leading to poor conception and embryo survival.

In terms of diet composition, high nitrogen, particularly rumen degradable nitrogen, should be avoided as this is known to increase urea and ammonia concentrations (Kenny *et al.*, 2002), reduce uterine pH around Day 7 after oestrus and reduce conception rate to first service (Elrod and Butler, 1993). Similarly, high starch diets can reduce blood pH and potentially uterine pH. High planes of nutrition (120 MJ ME compared with 40 MJ ME) (Yaakub *et al.*, 1999) and starch diets offered ad libitum have been found to reduced blastocyst yield (Wrenzycki *et al.*, 2000), and are detrimental to oocyte and early embryo development in heifers.

In order to improve fertility, dietary fat sources rich in linoleic acid and/or eicosapentaenoic (EPA; C20:5) and docosahexanaenoic (DHA; C22:6) acids, such as fish oil, have been found to increase embryo survival in lactating dairy cows by increased PGF_2a release (Binelli *et al.*, 2001). However, this effect has not been assessed in heifers.

GRAZING

During the rearing period, including grass as part of the diet has been found to have potential benefits to mammary parenchyma development and to reduce the incidence of lameness (Carson *et al.*, 2004). Grazed heifers have lighter calves at first calving and greater feed intakes before first calving and during early first lactation compared with heifers housed throughout the rearing period (Troccon, 1993). However, there was no long-term effects on growth, fertility, milk yield, health or longevity.

Performance at grass is influenced by a wide range of factors including sward height, forage quality, supplementation levels, grazing system, age of the animal, animal health and management (Dawson and Carson, 2005; Troccon, 1993; Steen, 1994).

Sward height and grazing system

Sward height is a useful tool to assess grass availability. Higher sward heights

(SH), up until forage quality declines, increase live-weight gain. (Hodgson, 1968) As a consequence, good grassland management, mediated through the effects on grass digestibility, increase grass quality due to high stocking rates during high growth periods typical to the early grazing season, result in higher growth rates during the grazing season. At turn out, Carson *et al.* (unpublished, 2000a and 2004) using a sward height of 9 cm found live-weight gains of 0.6 to 0.8 kg/h/d at 7 to 12.5 months of age and at a SH of 5 cm found a growth rate of 0.76 kg/h/d in 17.5 month old heifers.

Possibly as a consequence of forage quality, rotation systems which give maximum control over grazing height have been found to result in high growth rates, potentially due to a reduction in worm infestation, especially during the late grazing season. Similarly, within rotational grazing, the leader/follower system offers the greatest potential, although performance of the follower animals should be considered. Alternatively, heifer rearing systems which do not involve grazing for the first and/or second summer period have the potential to achieve similar live-weight gains (Troccon 1993; Carson *et al.* 2002a, 2004), however heifers at grass required 160 kg less concentrates/animal to achieve similar live-weight gains to the housed heifers (Carson *et al.* 2004).

Supplementary feeding at pasture is important in achieving planned growth rates in dairy heifers. Growth rate response to supplementary feeding at pasture depends on herbage availability. Supplementary feeding will be required to maintain adequate growth rates particularly when the pasture available is of low quality or restricted levels (Steen, 1994). Carson *et al.* (unpublished and 2004) found growth rates, in response to supplementation at pasture, of 87 g/kg and 73 g/kg in heifers during the first summer at grass and 60 g/kg during the second summer at grass. It has been found in younger animals (12 months of age), that herbage intake was more sensitive to changes in herbage digestibility than in adult cattle (Hodgson, 1968). As a consequence, young heifers during the first grazing season require particular attention to be paid to pasture availability and quality and prompt allocation of supplementation when required.

Heifers have been found to have a good propensity for compensatory growth at pasture following a winter period of restricted growth. Under these circumstances, heifers have been found to achieve live-weight gains of 0.9 kg/day after turnout (Woods *et al.* unpublished), however it was found to be extremely difficult to maintain a live-weight gain of 0.6 kg/day in these animals.

GROWTH RATE, CONDITION SCORE, LIVE WEIGHT AND AGE AT CALVING

Holstein-Friesian heifers calving at two years of age should weigh 540 to 560 kg, but actual weight would clearly be dependant on mature weight, with a body condition score of 2.75 to 3.0 (Carson *et al.* 2000a, 2000b).

Conclusions

In dairy heifer rearing optimum profitability will be achieved by the use of a planned and monitored approach to live-weight gain and feeding programmes. The aim should be to first calve heifers at 22 to 24 months of age and optimal production levels and heifer longevity are achieved by heifers being between 85 and 90 % of mature weight at parturition. To achieve this heifers should be served at 13 to 15 months of age at 55 to 60 % mature weight. Live weight is a highly heritable and therefore predictable trait, making it a good tool for measurement of heifer maturity. However, where it is not possible to measure live weight, wither height and belly band measurements can be used. The body condition score of a heifer at calving is as important as that of cows and should be 2.75 to 3.0, but this will vary according to genetic merit, with lower genetic merit or dual purpose breeds tending to have higher body condition scores. Animals that are too fat or too thin at parturition should be avoided to minimise calving difficulties, metabolic disorders and potential subsequent culling.

To minimise mortality and to maximise intestinal development and growth rate, dairy heifers need to be provided with adequate levels of good 'quality' colostrum as soon as possible following birth. Suckling should always be supervised to ensure intake of 2.5 l of colostrum within 6 h of birth and 4 l within the first 12 h of life. Multiparous cows provide better 'quality' colostrum than first calving heifers. Enteric diseases and pneumonia should be prevented wherever possible, and control and treatment of these diseases should be applied rapidly. Continuing to feed lower levels of colostrum (5 % of total milk intake) after the first few days of life can reduce the impact of endemic enteric diseases.

Target growth rates required in each growth stage and age of the dairy heifer are best described in Tables and Figures, as set out in this chapter of guidelines and in other chapters of this book. The rate of growth around puberty has been highlighted as critical, and a slightly lower growth rate during this phase is recommended to reduce deposition of fat in the developing mammary parenchyma which would reduce subsequent milk yield. Onset of puberty occurs between 3 and 9 months of age, but is determined more by live weight or maturity than by age. As a consequence, puberty occurs earlier in faster growing or higher genetic merit animals compared with lower merit or dual purpose cattle.

Dietary composition and components have been shown to affect body tissue composition. Increasing dietary protein concentration allows increased growth rates and deposition of lean tissue while reducing fat deposition. Similarly, the source of energy affects deposition of lean tissue and diets with lower fat concentrations and a greater proportion of energy from lactose increase growth rates while minimising deposition of fat. As a consequence, higher growth rates and feeding levels of dairy heifers need to be achieved using milk powders or whole milk with specifically designed additives and solid feeds that are specifically formulated to achieved high growth rates and lean tissue deposition without encouraging deposition of fat in body tissues or the mammary parenchyma.

Milk replacers are based on either skim milk or whey protein concentrate and although they behave differently in the digestive system, with whey powders not forming discernable clots in the abomasum, calf performance is similar when fed either type milk replacer. The vegetable protein sources most commonly used in milk replacers are hydrolysed wheat gluten and soya protein concentrate. However, calf growth rates have been variable when offered soya bean concentrate, probably due to variability of soya bean concentrate quality.

Lactose and fat are the main sources of energy and only low levels of starch may be included in calf diets. Coconut (0.2) and palm oils (0.8) are used in combination to prevent liver infiltration found when coconut oil (Graulet, Gruffat-Mouty, Durand and Bauchart, 2000; Jenkins and Kramer, 1986) or soya oil (Leplaix-Charlat, Durand and Bauchart, 1996) are offered individually to young calves. Digestibility of fat and quality of protein in milk replacers are greatly affected by the technology used in their production; homogenization and incorporation give good levels of emulsification and stabilisation of fat particles thus increasing fat digestibility, and spray drying techniques reduce potential for heat damage of protein and produce small digestible fat particles. Lysine content of the powder can be a good indicator of previous thermal treatment and potential damage or denaturing of protein. Heat damaged skim milk powders do not clot properly, are less digestible and can cause diarrhoea, so they are not suitable for a young calf of less than one month of age. However, it is important to remember that whey powders do not form clots in the abomasum. High levels of lactose, calcium and/or minerals that affect osmotic pressure of the digestive system and fat digestion cause diarrhoea in dairy calves. However, all milk products, including whole milk, require supplementation with minerals and vitamins in order to meet dietary requirements of growing dairy heifers.

Intake of solid feeds will facilitate rumen development and this might be due to consumption of either forage or fibrous feed intake, or due to the presence of volatile fatty acids from fermentation of concentrated feed sources such as calf 'starter' diets. Intake of 'starter' diets is affected by palatability of diet components and steam-rolled or flaked maize will make these feeds more digestible. Where calves are offered finer ground feeds, they should be offered functional fibre or forages and or bedding with clean dust free straw and other bedding materials. In these cases, sawdust is not be suitable as calves will consume this bedding material. Weaning calves from milk consumption should take place according to starter feed intake, and intake levels of 1 to 1.5 kg of dry feed for two consecutive days is generally used as a target. This usually corresponds with and an age of 6 to 7 weeks. In general, fibrous feeds, forages and clean water should all be offered in accordance with the welfare regulations of each country, unless otherwise advised due to ill health by a veterinary surgeon.

At pasture, provision of adequate levels of good quality grass and clean pastures will minimise parasite burden and maximise growth rates of dairy heifers. Regular estimation of pasture availability will be required to achieve the required growth rate targets for dairy heifers. Possibly as a consequence of this, the highest heifer growth rates are achieved using rotational grazing systems, however these grazing systems are not commonly adopted for dairy heifer rearing. In any grazing and housing system, prevention and control of worm burden and subsequent damage of the lung, stomach and liver should be applied appropriately according to the individual conditions of each heifer rearing site. Similarly, timely spring turn out and housing should be managed to prevent reductions in growth rate. In dairy heifer rearing, short term body condition changes should not be used as indicators to alter feeding levels. As a consequence, a planned approach with long term feeding, growth rates, maturity and health management is required for successful and economic dairy heifer rearing.

In recent years selection of dairy cattle for increased milk yield seems to have reduced fertility and conception rates in both heifers and cows. In terms of nutrition, under nutrition and sudden reductions in live weight, high levels of dietary protein, particularly rumen degradable protein and high dietary starch levels have been found to reduce fertility and/or embryo survival. Conversely, dietary fats rich in linoleic acid and/or eicosapentaenoic (EPA; C20:5) and docosahexanaenoic (DHA; C22:6) acids, such as fish oil have been found to increase embryo survival and subsequent fertility.

Culling rates of dairy heifers can be minimised, and longevity can be increased, by reducing disease and maximising feed intakes. Ideally, newly calved heifers should be kept with other heifers on a loose bedded system when housed. Housing newly calved heifers with cows on concrete flooring has been found to increase the incidence of sole bruising, ulceration and lameness and to reduce forage intake. In loose bedded systems, frequent bedding and prevention of mastitis in dairy heifers must be given high priority to minimise clinical mastitis and maintain low somatic cell counts.

References

Baldwin, R.L. (1999) The proliferative actions of insulin, insulin-like growth factor-I, epidermal growth factor, butyrate and propionate on ruminal epithelial cells in vitro. *Small Ruminant Research,* **32**, 261-268.

Baldwin, R.L. (2000) Sheep gastrointestinal development in response to different dietary treatments. *Small Ruminant Research,* **35**, 39-47.

Baumrucker, C.R., Hadsell, D.L. and Blum, J.W. (1994) Effects of dietary rhIGF-I in neonatal calves: Intestinal growth and IGF receptors. *Journal of*

Animal Science, **72**, 428-433.

Beharka, A. A., Nagaraja, T. G. and Morrill, J. L. (1991) Performance and ruminal function development of young calves fed diets with Asperigillus oryzae fermentation extract. Journal of Dairy Science, **74**, 4326-4336.

Beharka, A. A., Nagaraja, T. G., Morrill, J. L., Kennedy, G. A. and Klemm R. D. (1998) Effects of form of the diet on anatomical, microbial, and fermentative development of the rumen of neonatal calves. *Journal of Dairy Science,* **81**, 1946-1955.

Blowey, R.W. (2005) Management of diseases in calves and heifers. In *Calf and Heifer Rearing* (Ed. P.C. Garnsworthy), pp 191-195. Nottingham University Press, Nottingham.

Blum, J.W. (2005) Bovine gut development. In *Calf and Heifer Rearing* (Ed P.C. Garnsworthy) pp 31-52. Nottingham University Press, Nottingham.

Carson, A. F., Dawson, L. E. R. and Gordon, F. J. (2002b). Research points the way for heifer rearing. *Proceedings of a seminar held at the Agricultural Research Institute of Northern Ireland*, March 2002, pp. 1-16.

Carson, A. F., Dawson, L. E. R., McCoy, M. A., Kilpatrick, D. J. and Gordon, F. J. (2000a). Effects of rearing regime on body size, reproductive performance and milk production during the first lactation in high genetic merit dairy herd replacements. *Animal Science*, **74**, 553 – 565.

Carson, A. F., Dawson, L. E. R., Wylie, A. R. G. and Gordon, F. J. (2004). The effect of rearing regime on the development of the mammary gland and claw abnormalities in high genetic merit Holstein-Friesian dairy herd replacement. *Animal Science*, **78**, 181-185.

Caugant, I., Toullec R., Formal M., Guilloteau P., Savoie, L (1993) Digestibility and amino acid composition of digesta at the end of the ileum in preruminant calves fed soyabean protein. *Reproduction Nutrition and Development*, **33**, 335-347.

Chelikani, P. K., Ambrose, J. D. & Kennelly, J. J. (2003) Effect of dietary energy and protein density on body composition, attainment of puberty, and ovarian follicular dynamics in dairy heifers. *Theriogenology*, **60,** 707-725.

Chua, B., E. Coenen, J. van Delen, and D. M. Weary. (2002) Effects of pair versus individual housing on the behavior and performance of dairy calves. *Journal of Dairy Science,* **85**, 360-364.

Church, D.C. (1971) The Ruminant Animal. In *Digestive Physiology and Nutrition*. Prentice Hall, Englewood Cliffs, NJ.

Davis, C. L. and J. K. Drackley. (1998) *The Development, Nutrition, and Management of the Young Calf.* Iowa State University Press, Ames, Iowa.

Dawson, L.E.R. and Carson, A.F. (2005) Grazing systems for dairy herd replacements. In *Calf and Heifer Rearing* (Ed. P.C. Garnsworthy), pp 253-276. Nottingham University Press, Nottingham.

Diaz, M. C., Van Amburgh, M. E., Smith, J. M., Kelsey, J. M. and Hutten, E. L.

(2001) Composition of growth of Holstein calves fed milk replacer from birth to 105 kilogram body weight. *Journal of Dairy Science,* **84**, 830-842.

Diskin, M. G., Mackey, D. R., Roche, J. F. & Sreenan, J. M. (2003) Effects of nutrition and metabolic status on circulating hormones and ovarian follicle development in cattle. *Animal Reproduction Science*, **78,** 345-370.

Ekblad, E. and Sundler, F (2002) Innervations of the small intestine. In *Biology of the Intestine in Growing Animals*, pp *235-270.* Edited by R. Zabielski, P.C. Gregory and B. Weström. Elsevier Science, Amsterdam.

Elrod, C. C. & Butler, W. R. (1993) Reduction of fertility and alteration of uterine pH in heifers fed excess ruminally degradable protein. *Journal of Animal Science*, **71,** 694-701.

Emmons, D.B., Lister E.E. (1976) quality of protein in milk replacers for young calves. I. Factors affecting in vitro curd formation by rennet (chymosin, rennin) from reconstituted skim milk powder. *Canadian Journal of Animal Science*, **56**, 317-325.

Esslemont, R. J. and Kossaibati, M. A. (1996) Incidence of production diseases and other health problems in a group of dairy herds in England. *Veterinary Record*, **139**, 486-490.

Franklin, S. T., D. M. Amaral-Phillips, J. A. Jackson, and A. A. Campell. (2003) Health and performance of Holstein calves that suckled or were hand-fed colostrums and were fed one of three physical forms of starter. *Journal of Dairy Science,* **86**, 2145-2153.

Funaba, M, K. Kagiyama, T. Iriki, and M. Abe. (1994) Changes in nitrogen balance with age in calves weaned at 5 or 6 weeks of age. *Journal of Animal Science,* **72**, 732-738.

Galfi, P., Neogrady, S. and Sakata, T. (1991) Effects of volatile fatty acids on the epithelial cell proliferation of the digestive tract and its hormonal mediation. In: *Physiological Aspects of Digestion and Metabolism in Ruminants,* pp 49-59. Edited by T. Tsuda, Y. Sasaki and R. Kawasaki. Academic Press, San Diego, CA.

Graulet, B., Gruffat-Mouty, D., Durand, D., Bauchart, D. (2000) Effects of milk diets containing beef tallow or coconut oil on the fatty acid metabolism of liver slices from preruminant calves. *British Journal of Nutrition*, **84**, 309-318.

Greenwood, R. H., J. L. Morrill, E. C. Titgemeyer, and G. A. Kennedy. (1997) A new method of measuring diet abrasion and its effect on the development of the forestomach. *Journal of Dairy Science,* **80**, 2534-2541.

Grongnet, J-F., Patureau-Mirand, P., Toullec R., Prugnaud J (1981) Utilisation des protéines du lait et du lactosérum par le jeune veau préruminant. Influence de l'âge et de la dénaturation des protéines de lactosérum. *Annales de Zootechnie*, **30**, 443-464.

Guilloteau, P., Le Huërou-Luron, I., Toullec, R., Chayvialle, J.-A., Zabielski,

R. and Blum J.W. (1997) Gastrointestinal regulatory peptides and growth factors in young cattle and sheep. *Journal of Veterinary Medicine A*, **44**, 1-23.

Guilloteau, P., Biernat, M., Wolinski, J. and Zabielski, R. (2002). Gut regulatory peptides and hormones of the small gut. In *Biology of the Intestine in Growing Animals*, pp 325-362. Edited by R. Zabielski, P.C. Gregory and B. Weström). Elsevier Science, Amsterdam.

Hill, T. M., Aldrich, J. M. and Schlotterbeck, R. L. (2005) Nutrient sources for solid feeds and factors affecting their intake by calves. In *Calf and Heifer Rearing* (Ed. P.C. Garnsworthy), pp 113-133. Nottingham University Press, Nottingham.

Hodgson, J. (1968). The relationship between the digestibility of a sward and the herbage consumption of grazing calves. *Journal of Agricultural Science, Cambridge*, **70**, 47 - 51

Jagannatha, S., Keown, J.F. and Van Vleck, L.D. (1998) Estimation of relative economic value for herd life of dairy cattle from profile equations. *Journal of Dairy Science,* **81**, 1702-1708.

Jasper, J. and D. M. Weary. (2002) Effects of ad libitum milk intake on dairy calves. *Journal of Dairy Science,* **85**, 3054-3058.

Jenkins, K.J., Kramer, J.K.G. (1986) Influence of low linoleic and linolenic acids in milk replacers on calf performances and lipids in blood plasma, heart and liver. *Journal of Dairy Science*, **69**, 1374-1386.

Kampheus, J., Stolte M., Rust, P. (2002) High sulfate content in whey products and milk replacers – a potential reason for changed faeces composition/ watery faeces in calves. Abstract In *Proceedings of the XXII World Buiatrics Congress, August 19-23rd 2002*, Hanover, Germany. World Association of Buiatrics.

Kenny, D. A., Humpherson, P. G., Leese, H. J., Morris, D. G., Tomos, A. D., Diskin, M. G. & Sreenan, J. M. (2002) Effect of elevated systemic concentrations of ammonia and urea on the metabolite and ionic composition of oviductal fluid in cattle. *Biology of Reproduction*, **66,** 1797-1804.

Lallès, J-P. (1993) Nutritional and antinutritional aspects of soyabean and field pea proteins used in veal calf production : A review. *Livestock Production Science*, **34**, 181-202.

Lammers, B. P., Heinrichs, A. J., Aydin, A. (1998) The effect of whey protein concentrate or dried skim milk in milk replacer on calf performance and blood metabolites. *Journal of Dairy Science*, **81**, 1940-1945.

Lane, M.A. and Jesse, B.W. (1997) Effect of volatile fatty acid infusion on development of the rumen epithelium in neonatal sheep. *Journal of Dairy Science*, **80**, 740-746.

Lane, M.A., Baldwin, R.L. and Jesse, B.W. (2000) Sheep rumen metabolic development in response to age and dietary treatments. *Journal of Animal Science*, **78**, 1990-1996.

Lane, M.A., Baldwin, R.L. and Jesse, B.W. (2002) Developmental changes in ketogenic enzyme gene expression during sheep rumen development. *Journal of Animal Science,* **80**, 1538-1544.

Le Heurou-Luron, I., Guilloteau, P., Wicker-Planquart, C., Chayvialle, J-A., Burton, J., Mouats, A., Toullec, R. and Puigserver, A. (1992) Gastric and pancreatic enzyme activities and their relationship with some gut regulatory peptides during postnatal development and weaning in calves. *Journal of Nutrition*, **122**, 1434-1445.

Leplaix-Charlat, L., Durand, D., Bauchart, D (1996) Effects of diets containing tallow and soyabean oil with and without cholesterol on hepatic metabolism of lipids and lipoproteins in the preruminant calf. *Journal of Dairy Science*, **79**, 1826-1835.

Longenbach, J. I., Heinrichs, A. J. (1998) A review of the importance and physiological role of curd formation in the abomasum of young calves. *Animal Feed Science and Technology*, 73, 85-97.

Losinger, W. C. and Heinrichs, A. J. (1997) Management practices associated with high mortality among preweaned dairy heifers. *Journal of Dairy Research*, **64**, 1-11.

Mackey, D. R., Sreenan, J. M., Roche, J. F. & Diskin, M. G. (1999) Effect of acute nutritional restriction on incidence of anovulation and periovulatory estradiol and gonadotropin concentrations in beef heifers. *Biology of Reproduction*, **61,** 1601-1607.

Mahaut, M., Jeantet, R., Schuck, P., Brulé, G. (2000b). Produits déshydratés In *Les produits industriels laitiers*, pp 49-90. Tec & Doc, Paris.

Mann, G.E., Lamming, G.E., Robinson, R.S. & Wathes D.C. (1999) The regulation of interferon-ô production and uterine hormone receptors during early pregnancy. *Journal of Reproduction and Fertility Supplement*, **54**, 317-328.

McLeod, K.R. and Baldwin, R.L. (2000) Effects of diet forage: concentrate ratio and metabolizable energy intake on visceral organ growth and in vivo oxidative capacity of gut tissues in sheep. *Journal of Animal Science*, **78**, 760-770.

Montagne, L., Toullec, R. Lallès, J-P (2001) Intestinal digestion and endogenous proteins along the small intestine of calves fed soyabean or potato. *Journal of Animal Science*, **79**, 2719-2730.

Morin, D. E., G. C. McCoy and W. L. Hurley. (1997) Effects of quality, quantity, and timing of colostrum feeding and addition of a dried colostrum supplement on immunoglobulin G1 absorption in Holstein bull calves. *Journal of Dairy Science*, **80,** 747-753.

National Animal Health Monitoring System. (1996) *Dairy Herd Management Practices Focusing on Preweaned Heifers.* USDA, Animal and Plant Health Inspection Service, Veterinary Services, Fort Collins, CO, USA.

National Research Council (2001) *Nutrient Requirements of Dairy Cattle.* Seventh rev. Ed., National Academy of Science, Washington, DC, USA.

Pryce, J. E., Royal, M. D., Garnsworthy, P. C. and Mao, I. L. (2004). Fertility in the high yielding dairy cow. *Livestock Production Science*, **86,** 125-135.

Pryce, J. E., Simm, G. and Robinson, J. J. (2002). Effects of selection for production and maternal diet on maiden dairy heifer fertility. *Animal Science*, **74,** 415-421.

Quigley, J. D., Hammer, C. J., Russell, L. E. and Polo, J. (2005) Passive immunity in newborn calves. In *Calf and Heifer Rearing* (Ed. P.C. Garnsworthy), pp 135-157. Nottingham University Press, Nottingham.

Quigley, J. D., and J. J. Drewry. (1998) Nutrient and immunity transfer from cow to calf pre- and post-calving. *Journal of Dairy Science*, **81**, 2779-2790.

Quigley, J. D., Steen, T. M. and Boehms, S. I. (1992) Postprandial changes of selected blood and ruminal metabolites in ruminating calves fed diets with and without hay. *Journal of Dairy Science*, **75**, 228-235.

Rauprich, A.B.E., Hammon, H. M., and Blum, J.W. (2000a) Influence of feeding different amounts of first colostrum on metabolic, endocrine, and health status and on growth performance in neonatal calves. *Journal of Dairy Science,* **78**, 896-908.

Rauprich, A.B.E., Hammon, H. M., and Blum, J.W. (2000b) Effects of feeding colostrum and a formula with nutrient contents as colostrum on metabolic and endocrine traits in neonatal calves. *Biology of the Neonate*, **78**, 53-64.

Robinson, J. J., Sinclair, K. D., Randel, R. D. & Sykes, A. R. (1999). Nutritional management of the female ruminant: mechanistic approaches and predictive models. In: H. -J. G. Jung and G. C. Fahey (Ed). *Vth International Symposium on the Nutrition of Herbivores.* pp 550-608 American Society of Animal Science, Savoy, IL, USA.

Robinson, R.S., Mann G.E., Lamming G.E. & Wathes, D.C. (2001) Embryonic-endometrial interactions during early pregnancy in cows. *British Society of Animal Science Occasional Publication*, **26,** 289-295.

Schillo, K. K., Hall, J. B. & Hileman, S. M. (1992) Effects of nutrition and season on the onset of puberty in the beef heifer. *Journal of Animal Science*, **70,** 3994-4005.

Scott, R. R., Hall, G. A., Jones, P. W. and Morgan, J. H. (2004). *Calf Diarrhoea in Bovine Medicine*, 2nd Ed, pp 185, Blackwell Scientific, Oxford, UK.

Sejrsen, K. (1978) Mammary gland development and milk yield in relation to growth rate in the rearing period in dairy and dual purpose heifers. *Acta Agriculturae Scandinavica*, **28**, 41.

Sejrsen, K. (2005) Mammary development and milk yield potential. In *Calf and Heifer Rearing* (Ed. P.C. Garnsworthy), pp 237-251. Nottingham University Press, Nottingham.

Sejrsen, K. and Purup, S. (1997) Influence of prepubertal feeding level on milk yield potential of dairy heifers: a review. *Journal of Animal Science*,

75, 828-835.

Sejrsen, K., Purup, S., Vestergaard, M. and Foldager, J. (2000) High body weight gain and reduced bovine mammary growth: physiological basis and implications for milk yield potential. *Domestic Animal Endocrinology*, **19**, 93-104.

Sinclair, K.D. and Webb, R. (2005) Fertility in the modern dairy heifer. In *Calf and Heifer Rearing* (Ed. P.C. Garnsworthy), pp 277-306. Nottingham University Press, Nottingham.

Sinha, Y.N. and Tucker, H.A. (1969) Mammary development and pituitary prolactin level of heifers from birth through puberty and during oestrus cycle. *Journal of Dairy Science*, **52**, 507-512.

Steen, R. W. J. (1994). A comparison of pasture grazing and storage feeding, and the effect of sward surface height and concentrate supplementation from 5 to 10 months of age on the lifetime performance and carcass composition of bulls. *Animal Production*, **58**, 209 – 219.

Tanan, K.G. (2005) Nutrient sources for liquid feeding of calves. In *Calf and Heifer Rearing* (Ed. P.C. Garnsworthy), pp 83-112. Nottingham University Press, Nottingham.

Tarlton, J. (2004). *Proceedings of the 13th International Symposium on Disorders of the Ruminant Digit*, Maribor, Slovenia, p. 88.

Tikofsky, J. N., van Amburgh, M. E. and Ross, D. A. (2001) Effect of varying carbohydrate and fat levels on body composition of milk replacer-fed calves. *Journal of Animal Science*, **79**, 2260-2267.

Toullec, R. (1989) Veal Calves. In *Ruminant nutrition. Recommended allowances and feed tables*, pp. 109-120. Edited by Jarrige R. John Libbey Eurotext, London- Paris.

Toullec, R., Grongnet, J-F. (1990) Remplacement partiel des protéines du lait par celles du blé ou du maïs dans les aliments d'allaitement : influence sur l'utilisation digestive chez le veau de boucherie. *INRA Production Animales*, **3**(3), 201-206.

Toullec, R., Lallès, J.P., Bouchez, P. (1994) Replacement of skim milk with soya bean protein concentrates and whey in milk replacers for veal calves. *Animal Feed Science and Technology*, **50**, 101-112.

Troccon, J. L. (1993). Dairy heifer rearing with or without grazing. *Annales de Zootechnie*, **42**, 271 – 288.

Tyler, J. W., Hancock, D. D., Thorne, J. G., Gay, C. C., Gay, J. M. (1999) Partitioning the mortality risk associated with inadequate passive transfer of colostral immunoglobulins in dairy calves. *Journal of Veterinary Internal Medicine*, **13**, 335-337.

Ullerich, A., Kampheus, J. (1998) Effects of high concentrations of sodium and potassium in milk replacers on intestinal processes and faeces composition in young calves. *Journal of Animal Physiology and Animal Nutrition*, **80**, 194-200.

Van Amburgh, M. and Drackley, J. (2005) Current perspectives on the energy and protein requirements of the pre-weaned calf. In *Calf and Heifer Rearing* (Ed. P.C. Garnsworthy), pp 67-82. Nottingham University Press, Nottingham.

Warner, R. G. (1991) Nutritional factors affecting the development of a functional rumen – a historical perspective. In *Proceedings of the Cornell Nutrition Conference for Feed Manufacturers*. Rochester, NY. pp 1-12.

Webb, R., Nicholas, B., Gong, J. G., Campbell, B. K., Gutierrez, C. G., Garverick, H. A. & Armstrong, D. G. (2003) Mechanisms regulating follicular development and selection of the dominant follicle. *Reproduction, Supplement*, **61,** 71-90.

Webster, A.J.F. (2001) Effects of housing and two forage diets on the development of claw horn lesions in dairy cows at first calving and in first lactation. *Veterinary Journal*, **162**, 56-65.

Wrenzycki, C., De Sousa, P., Overstrom, E. W., Duby, R. T., Herrmann, D., Watson, A. J., Niemann, H., O'Callaghan, D. & Boland, M. P. (2000) Effects of superovulated heifer diet type and quantity on relative mRNA abundances and pyruvate metabolism in recovered embryos. *Journal of Reproduction and Fertility*, **118,** 69-78.

Yaakub, H., O'Callaghan, D. & Boland, M. P. (1999) Effect of roughage type and concentrate supplementation on follicle numbers and in vitro fertilisation and development of oocytes recovered from beef heifers. *Animal Reproduction Science*, **55,** 1-12.

Zanming, S., Seyfert, H.-M., Löhrke, B., Schneider, F., Zitnan, R., Chudy, A., Kuhla, S., Hammon, H.M., Blum, J.W., Martens, H., Hagemeister, H., and Voigt, J. (2004) Effect of nutritional level on rumen papillae development and IGF-1 and IGF type 1 receptor in young goats. *Journal of Nutrition*, **134**, 11-17.

LIST OF DELEGATES

The University of Nottingham is grateful to Provimi Ltd for supporting the conference; to Karine Tanan and Norman Downey for help in formulating the programme; to Jill Hoyland and Karen Wright for publicity and marketing; to Susan Golds for secretariat work; to Peter Buttery, John Newbold and Norman Downey for chairing sessions; to all the speakers for their presentations and written papers; and to the following delegates who registered for the Easter School.

Agnew, Dr K	United Feeds, 8 Northern Road, Belfast BT3 9AL, UK
Aguilar Perez, Mr C	University of Nottingham, Sutton Bonington Campus, Loughborough, Leics LE12 5RD, UK
Andersson, Mr T	Svenska Lantmannen, Hinnesta Vastergard, 61032 Vikbolandet, Sweden
Bakke, Mr M J	Custom Dairy Performance Inc, P O Box 99, Clovis CA93613-099, UK
Barnev, Mr V	Provimi Moscow, 72 Varshavskoye Shosse, Block 2 Ru Moscow, 113556, Russia
Bartram, Dr C	Pye-Bibby, ABN House, PO Bx 250, Oundle Rd, Woodston, Peterborough PE2 9QF, UK
Bastard, Mr W	Lillico Attlee, Beddow Way, Aylesford, Maidstone ME20 7BT, UK
Batley, Mrs M	Rumenco Ltd, Stretton House, Derby Rd, Burton-upon-Trent, Staffs DE13 ODW, UK
Benoist, Mr J	Celtic, Parc D'Activités de Ferchaud, F35320 Crevin, France
Blowey, Mr R	Wood Veterinary Group, St Oswalds Rd, Gloucester GL1 2SJ, UK
Blum, Prof J	University of Berne, Div of Nutrition and Physiology, Bremgartenstrasse 109a, Berne CH-3012, Switzerland
Bolvari, Mr A	Agrokomplex, PO Box 1, H 8112 Zichyujfalu, Hungary
Boning, Dr I	Cremer Futtermuhlen, Getreidestr. 9, D-28217 Bremen, Germany

Boothman, Mr S	Cogent Ltd, Woodhouse, Aldford, Chester CH3 6JD, UK
Bouet, Mr M	Lillico Attlee Ltd, Beddow Way, Aylesford, Kent ME20 7BT, UK
Brady, Miss S	Lakeland Dairies, Bailieboro, Co Caven, Ireland
Brameld, Dr J	University of Nottingham, Sutton Bonington Campus, Loughborough, Leics LE12 5RD, UK
Bramley, Ms P	Provimi Ltd, SCA Mill, Dalton Airfield Ind Est, Dalton, Thirsk, N Yorks YO7 3HE UK
Brickell, Miss J	Royal Veterinary College, Hawkshead Lane, North Mymms, Hatfield, Herts AL9 7TA, UK
Brown, Mr S	Wynnstay Group plc, Eagle House, Llansantffraid SY22 6AQ, UK
Bryan, Ms C	Provimi Ltd, SCA Mill, Dalton Airfield Ind Est, Dalton, Thirsk, N Yorks YO7 3HE, UK
Bull, Mr R	R J Bull, Westside Dairy, North Waltham Basingstoke RG25 2DD, UK
Buttery, Prof P J	University of Nottingham, Sutton Bonington Campus, Loughborough, Leics LE12 5RD, UK
Byers, Mr D J	HST Feeds, 4th Avenue, Weston Rd, Crewe CW1 6BN, UK
Carrick, Mr I	Trouw Nutrition, 36 Ship Street, Belfast BT15 1JL, UK
Carson, Dr A	ARINI, Large Park, Hillsborough, Co Down BT26 6JF, UK
Cloy, Mr F C	Harbro Ltd, Markethill, Turriff AB53 4PA, UK
Cole, Mr J	BFI Innovations Ltd, 1 Telford Court, Chester Gates, Dunkirk Lea, Chester CH1 6LT, UK
Cole, Dr M	Provimi Ltd, SCA Mill, Dalton Airfield Ind Est, Dalton, Thirsk, N Yorks YO7 3HE, UK
Colquhoun, Mr G	Harbro Ltd, Markethill Road, Turriff, Aberdeenshire AB53 4PA, UK
Conway, Mr C	Mixrite (I) Ltd, Benettsbridge, Co. Kilkenny, Ireland
Corless, Mr J	Farrington Agri, Rathcoffey, Donadea Co. Kildare, Ireland
Dale, Mr N	Scotmin Nutrition Ltd, 13 Whitfield Drive, Heathfield Industrial Estate, Ayr KA8 9RX, UK

Davies, Mr B	Dugdale Nutrition, Bellman Mill, Salthill, Clitheroe BB7 1QW, UK
Davies, Dr Z	DEFRA, Room 701, Cromwell House, Dean Stanley Street, London SW1P 3JH, UK
Dinsdale, Miss N	I'Anson Bros, Thorpe Raod, Masham, Ripon, N Yorks HG4 4JB, UK
Dooren, Mr P	Joosten, Industriekade 34, NL 6001 Se Weert, The Netherlands
Downey, Mr N	Provimi Ltd, SCA Mill, Dalton Airfield Industrial Estate, Dalton, Thirsk, N Yorks YO7 3HE, UK
Drackley, Prof J	University of Illinois, 260 Animal Sciences Laboratory, 1207 W. Gregory Drive, Urbana IL 61801, USA
Edge, Mr J	Provimi Ltd, SCA Mill, Dalton Airfield Ind Est, Dalton, Thirsk, N Yorks YO7 3HE, UK
Elek, P	Agrokomplex, PO Box 1, H 8112 Zichyujfalu, Hungary
Eouzan, Mr J	Centralys, PO Box 108 F78191 Trappes, France
Every, Mr M	I'Anson Bros, Thorpe Road, Masham, Ripon, N Yorks HG4 4JB, UK
Fallon, Dr R J	TEAGASC, Grange Research Centre, Dunsany, Co Meath, Ireland
Featherstone, Mrs C A	VLA Thirsk, West House, Station Rd, Thirsk, N Yorkshire YO7 1PZ, UK
Fishbourne, Miss E	Stock 1st, 8 The Courtyard, Holmbush Farm, Faygate, West Susex RH12 48E, UK
Fletcher, Miss V	Meadow Quality, Bearley Hill, Bearley, Stratford on Avon CV3 70SA, UK
Forbes, Mr A	Merial Animal Health, P O Box 327, Harlow CM19 5TG, UK
Fouladi, Dr A	University of Nottingham, Sutton Bonington Campus, Loughborough, Leics LE12 5RD, UK
Fowers, Miss R	Frank Wright Ltd, Blenheim House, Blenheim Rd, Ashbourne DE6 1HA, UK
Frentrup, Ms M	SCA Deutschland (Provimi), Muhienstrasse 8, 48282 Emsdetten, Germany

Fry, Dr C	LHV Spelle, Hafenstr. 1-3, D-48480 Spelle, Germany
Furmidge, Miss V	Volac International Ltd, Volac House 50 Fisher's Lane, Orwell, Royston, Herts SG8 5QX, UK
Gardner, Dr N	Volac International Ltd, Volac House, 50 Fisher's Lane, Orwell, Royston, Herts SG8 5QX, UK
Garnsworthy, Dr P C	University of Nottingham, Sutton Bonington Campus, Loughborough, Leics LE12 5RD, UK
Gerke, Mr G	Rudolf Peters Landhandel, Luhdorfer Str. 115, D-21423 Winsen / Luhe, Germany
Gilbert, Mr H	Volac International Limited, Volac House, 50 Fisher's Lane, Orwell, Royston, Herts SG8 5DX, UK
Golds, Mrs S	University of Nottingham, Sutton Bonington Campus, Loughborough, Leics LE12 5RD, UK
Gonzales, Mr A	Nutral, Polig. Industrial Sur, C/. Cobal P 261-265, Apartado De Correos 60 E 28770 Colmenar Viejo, Madrid, Spain
Gould, Mrs M	Volac International Ltd, Volac House, 50 Fisher's Lane, Orwell, Royston, Herts SG8 5QX, UK
Guilloteau, Dr J	INRA Rennes, 65 rue de Saint Brieuc, CS 84215, RENNES-Cedex F-35042, France
Harrington, Mr C	Meadow Feeds, P O Box 262, Westhoven St, Paarl 7620, South Africa
Harris, Mr C	Consultant, 18 Priest Hill, Caversham, Reading, Berks RG4 7RZ, UK
Heinrichs, Prof J	Penn State University, 324 Henning Building, University Park, PA 16802, USA
Hellberg, Mr	Svenska Lantmannen, Hinnesta Vastergard, 61032 Vikbolandet Sweden, Sweden
Helliwell, Mr D	W E Jameson & Son Ltd, Leyburn Rd, Masham, Ripon HG4 4ER, UK
Henderson, Mr I	Scotmin Nutrition Ltd, 13 Whitfield Drive, Heathrow Ind Estate, Ayr KA8 9RX, UK
Hennessy, Mr P	J.H. Roche & Co Ltd, Dock Road, Limerick, Ireland
Heron, Mr A	Pye-Bibby Agriculture, Lansil Way, Lancaster LA1 3QY, UK

Hill, Dr M	Akey, P O Box 5002, Lewisburg, Ohio 45338, USA
Hilliard, Mr M	W L Duffield & Sons, Saxlingham Thorpe Mills, Norwich, Norfolk NR15 1TY, UK
Hogan, Miss C	Pfizer Animal Health, Walton Oaks, Dorking Rd, Tadworth, Surrey KT20 7NS, UK
Hollick, Mr N	Honesberie Farms, The Grange, Priors Marston, Southam, Warks CV47 7SG, UK
Hooley, Mrs E	University of Nottingham, Sutton Bonington Campus, Loughborough, Leics LE12 5RD, UK
Hough, Mr T	NWF Agriculture Ltd, Wardle, Nantwich, Cheshire CW5 6AG, UK
Hoyland, Mrs J	Provimi Ltd, SCA Mill, Dalton Airfield Industrial Estate, Dalton, Thirsk, N Yorks Y07 3HE, UK
Huelin, Miss V	Trinity Manor Farm Ltd, Trinity, Jersey JE3 5JP, UK
Hyam, Mr J M	Qualivet, Avda de los Reyes Catolicos 6, Oficina 16 A, Majadahonda 28220, Spain
Jackson, Mr J	Agricultural Consultant, Planning & Env Dept, P O Box 327, Houard Davis Farm, Trinity JE4 8UF, UK
Johns, Mr N	HST Feeds, 4th Avenue, Weston Road, Crewe CW1 6BN, UK
Kafi, Dr M	University of Nottingham, Sutton Bonington Campus, Loughborough, Leics LE12 5RD, UK
Keane, Mr N	Countrywest Trading Ltd, Underlane, Holsworthy, Devon EX22 6EE, UK
Lawrence, Prof A	SAC Edinburgh, Animal Biology Division, Bush Estate, Penicuik, Midlothian EH26 0QE, UK
Lawrence, Ms C	Farmers Weekly, Quadrant House, The Quadrant, Sutton, Surrey, UK
Lawson, Dr D	Davidson Animal Feeds, Gray Street, Shotts, Lanarkshire ML7 5EZ, UK
Lee, Miss E	University of Nottingham, Sutton Bonington Campus, Loughborough, Leics LE12 5RD, UK
Leversha, Mr S	Provimi Ltd, SCA Mill, Dalton Airfield Ind Est, Dalton, Thirsk, N Yorks YO7 3HE, UK

Long, Mr J	Pye-Bibby Agriculture, Lansil Way, Lancaster LA1 3QY, UK
Lovell, Mr S	NuTec Ireland (Provimi), Monread Road, Naas, Co Kidare, Ireland
Macer, Mr P	Kite Consulting, The Mill, 6 Church St, Melbourne, Derbys DE73 1ET, UK
MacLeod, Mr D	Provimi Ltd, SCA Mill, Dalton Airfield Ind Est, Dalton, Thirsk, N Yorks YO7 3HE, UK
Margerison, Dr J	University of Plymouth, Seale-Hayne Faculty, Newton Abbot TQ12 6NQ, UK
Marsh, Mr S	Harper Adams University College, Newport, Shropshire TF10 8NB, UK
Marsman, Dr G	Borculo Domo Ingredients, Hanzeplein 25, P O Box 449, Zwolle 8000AK, The Netherlands
Martinho, Mr C	Provimi Portuguesa, Estada do Adarse, Apartado 26P 2616 - 953 Alverca, Portugal
Martyn, Mr S	BFI Innovations Ltd, 1 Telford Court, Chester Gates, Dunkirk Lea, Chester CH1 6LT, UK
Masson, Dr L	Provimi Ltd, Norvite Mill, Wardhouse, Insch, Aberdeenshire AB52 6YD, UK
Mawhinney, Mr D	Dept of Agriculture & Rural Dev. (DARD), Greenmount Campus, Cafre, 22 Greenmount Rd, Antrim, NI BT41 4PU, UK
May, Mr A	Mole Valley Farmers Ltd, Huntworth Mill, Bridgwater, Somerset TA6 6LQ, UK
McConnell, Mr M	Vistavet (Ireland) Ltd, 211 Castle Road, Ranbalstown, Co Antrim BT41 2EB, UK
McDonald, Mr I	Provimi Ltd, Norvite Mill, Wardhouse, Insch, Aberdeenshire AB52 6YD, UK
McFarlane, Mr R	Davidson Animal Feeds, Gray Street, Shotts, Lanarkshire ML7 5EZ, UK
McIlmoyle, Dr D	Countrywide Farmers, Bradford Road, Melksham SN12 8LQ, UK
Meijer, Mr R	Nutreco, Veerstraat 38, Boxmeer 5830 MA, The Netherlands

Mellor, Miss S	Dairy & Beef Magazine, Reed Business Information, Hanzestraat 1, Doetinchem 7006 RH, The Netherlands
Metais, Mr D	Néolait (Provimi), BP 1, F 22121 Yffiniac Cedex, France
Moore, Mr R W A	John Thompson & Sons Ltd, 35-39 York Road, Belfast BT15 3GW, UK
Morris, Mr V	Lillico Attlee, Parsonage Mill, Dorking, Surrey RH4 1EL, UK
Morris, Mr W	BOCM Pauls Ltd, 1st Ave, Royal Portbury Dock, Portbury, Bristol BS20 9XS, UK
Mulder, Mr R	Lillico Attlee, Beddow Way, Aylesford, Kent ME20 7BT, UK
Mulligan, Dr F	University College Dublin, Dept of Animal Husbandry & Production, Belfield, Dublin D4, Ireland
Newbold, Dr J	Provimi R & T, Lenneke Marelaan, 2 B 1932 Sint-Stevens-Woluwe, Belgium
O'Dwyer, Mr A	Irish Cement Ltd., Cooperhill Farm, Clarina, Co Limerick, Ireland
O'Keefe, Mr F	A.W. Ennis Ltd, Ballyconnell, Co Cavan, Ireland
Oldenziel, Mr H	Sloten BV, Postbus 474, 7400 AL, Deventer, The Netherlands
Onions, Miss V	University of Nottingham, Sutton Bonington Campus, Loughborough, Leics LE12 5RD, UK
Packington, Mr A	DSM Nutritional Products, Heanor Gate, Heanor, Derbyshire DE75 7SG, UK
Pearce, Dr G P	Dept of Veterinary Medicine, University of Cambridge, Maddingley Rd, Cambridge CB3 OES, UK
Peppard, Ms H	Greenvale Animal Feeds, Thurles, Co Tipperary, Ireland
Peters, Mrs E	Schils GB Ltd, Heidestraat ug, Susteren 6114AA, The Netherlands
Phelps, Mr S	Meadow Quality, Bearley Mill, Bearley, Stratford on Avon CV3 70SA, UK
Pine, Dr A P	Premier Nutrition, The Levels, Rugeley WS15 1RD, UK
Pittet, Mr O	Protector (Provimi), Zone Industrielle, CH 1522 Lucens, Switzerland

Piva, Prof G	Universita Cattolica Del Sacvro Cuore, Largo Agostino Gemelli, 20123 Milano Italy,
Pocza, Mr S	Agrokomplex (Provimi), PO Box 1, H 8112 Zichyujfalu, Hungary
Porter, Ms R	Cow Management & Farm Business, 52 Brook Lane, Felixstowe, Suffolk IP11 7LG, UK
Quigley, Dr J	APC Inc, 227 Trailridge Rd, Ames IA 50014, USA
Retter, Dr W	Heygate & Sons, Bugbrooke Mill, Bugbrooke, Northampton NN7 3QH, UK
Reynolds, Mr D	ProviCo Pty Ltd, RMB 5 Brunsskill Road, NSW 2650 Wagga Wagga, Australia
Rhodes, Ms A	Dairy Farmer, Sovereign House, Sovereign Way, Tunbridge, Kent, UK
Richards, Dr S	Provimi Ltd, SCA Mill, Dalton Airfield Ind Est, Dalton, Thirsk, N Yorks YO7 3HE, UK
Robinson, Mr S	Nottingham University Press, Manor Farm, Church Lane, Thrumpton, Nottingham NG11 OAX, UK
Rowland, Mr T	Pen Mill Feeds Ltd, Babylon View, Pen Mill Trading Estate, Yeovil, Somerset BA21 5HR, UK
Russell, Mrs H	University of Nottingham, Sutton Bonington Campus, Loughborough, Leics LE12 5RD, UK
Saito, Dum A	Zen Raku Ren Cooperative, Chikusan Kaikan Bldg, 9-2, 4 Chonme, Giuza Chua-Ku, Tokyo 104-0061, Japan
Sandiford, Mr M	The George Veterinary Group, 20 High St, Malmsbury, Wilts SN16 9AU, UK
Sejrsen, Dr K	Danish Inst. of Agricultural Sciences, Department of Animal Nutrition and Physiology, Foulum DK-8830, Denmark
Sinclair, Dr K	University of Nottingham, Sutton Bonington Campus, Loughborough, Leics LE12 5RD, UK
Sinclair, Mr W	British Denkavit Ltd, Denkavit Hiouse, 6 The Aplha Centre, Upton Rd, Poole, Dorset BH17 7AG, UK
Slippers, Mr S C	Meadow Feeds RSA, 56 Ohrtmann Road, Willowton, Pietermaritzburg 3201, South Africa

Sprent, Ms M	Provimi Ltd, SCA Mill, Dalton Airfield Ind Est, Dalton, Thirsk, N Yorks YO7 3HE, UK
Sridhar, Dr V	Vetcare, IS-40, KHB Industrial Area, Yelahanka New Town Bangalore 560 064, India
Stockhill, Mr P	Provimi Ltd, SCA Mill, Dalton Airfield Industrial Estate, Thirsk, N Yorks YO7 3HE, UK
Swali, Miss A	Royal Veterinary College, Hawkshead Lane, North Mymms, Hatfield, Herts AL9 7TA, UK
Tanan, Dr K	Provimi R & T, Lenneke Marelaan, 2, Sint-Stevens-Woluwe B1932, Belgium
Thornton, Mr D	Rumenco, Stretton House, Derby Rd, Burton-on-Trent DE13 ODW, UK
Trainer, Mr J	Grain Harvesters Ltd, The Old Colliery, Wingham, Canterbury, Kent CT3 1LS, UK
Trebble, Mr J	Mole Valley Farmers Ltd, Huntworth Mill, Bridgwater, Somerset TA6 6LQ, UK
Tucker, Mr N	Wresource, The Old Wain House, Canon Frome, Ledbury, Hereford HR8 2TE, UK
Twigge, Dr J	Nutreco Ruminant Research Centre, c/o Trouw Nutrition, Wincham, Northwich CW9 6DF, UK
Uprichard, Mr J	Trouw Nutrition, 36 Ship Street, Belfast BT15 1JL, UK
Van Amburgh, Prof M	Cornell University, 272 Morrison Hall, Ithaca NY 14853, USA
Van Den Bighelaar, Mr H	BFI Innovations Ltd, 1 Telford Court, Chester Gates, Dunkirk Lea, Chester CH1 6LT, UK
Van der Valk, Mr L	Denkavit International Bv, c/o British Denkavit Ltd, 6 Alpha Centre, Upton Rd, Poole BH17 7AG, UK
Van der Vliet, Dr H	ABCTA, Postbus 91, Hochem 7240 AB, The Netherlands
Vecqueray, Mr R	Dugdale Nutrition, Bellman Mill, Salthill, Clitheroe BB7 1QW, UK
Veth, Mr P	Provimi BV, Veerlaan 17-23, NL 3072 An Rotterdam, The Netherlands
Waldmann, Dr A	Estonian Agricultural University, Kreutzwakdi 64, Tartu 51014, Estonia

Walkland, Mr C	British Dairying, 148 Windy Arbour, Kenilworth CV8 2BH, UK
Walsh, Mr H	Farmers Guardian, Fullwood, Preston, Lancs, UK
Wathes, Prof D C	Royal Veterinary College, Hawkshead Lane, North Mymms, Hatfield, Herts SG8 8RG, UK
Webb, Prof R	University of Nottingham, Sutton Bonington Campus, Loughborough, Leics LE12 5RD, UK
Whalley, Mrs L	Tangerine Group, Docklands, Dock Road, Lytham, Lancs FY8 5AQ, UK
Whitaker, Mr A	Newline (Farm Partnerships) Ltd, Unit 1A, Highfield Business Park, Tewkesbury Rd, Deerhurst, Gloucs GL19 4BP, UK
Whitaker, Miss C	Countrywest Trading, Underlane, Holsworthy, Devon EX22 GLP, UK
Wicks, Dr H C F	ARINI, Hillsborough, Large Park, Hillsborough, Co Down BT26 6JF, UK
Wilde, Mr D	Alltech (UK) Ltd, Alltech House, Ryhall Road, Stamford, Lincs PE9 1TZ, UK
Wisden, Mr P	British Denkavit Ltd, Denkavit House, 6 The Alpha Centre, Upton Rd, Poole, Dorset BH17 7AG, UK
Wright, Ms K	Karen Wright PR, Manor Farm House, Sudborough, Northants NN14 3BX, UK
Wynn, Dr R	ABNA Ltd, Oundle Road, Woodston, Peterborough PE2 9QF, UK
Yeates, Dr M	Kingshay Farming Trust, Bridge Farm, West Bradley, Glastonbury BA6 8LU, UK
Young, Mr C	Provimi Ltd, Norvite Mill, Wardhouse, Insch, Aberdeenshire AB52 6YD, UK

INDEX

F

G

H

I

J

L

U

V

W

Y